Melt-Quenched Nanocrystals

Melt-Quenched Nanocrystals

A. M. Glezer and I. E. Permyakova

CISP

CRC Press
Taylor & Francis Group
Boca Raton London New York

CRC Press is an imprint of the
Taylor & Francis Group, an **informa** business

CRC Press
Taylor & Francis Group
6000 Broken Sound Parkway NW, Suite 300
Boca Raton, FL 33487-2742

First issued in paperback 2019

© 2013 by Taylor & Francis Group, LLC
CRC Press is an imprint of Taylor & Francis Group, an Informa business

No claim to original U.S. Government works

ISBN-13: 978-1-4665-9414-2 (hbk)
ISBN-13: 978-0-367-37992-6 (pbk)

Visit the Taylor & Francis Web site at
http://www.taylorandfrancis.com

and the CRC Press Web site at
http://www.crcpress.com

Contents

Preface

The method of quenching from the liquid state (melt quenching) provides fundamentally new opportunities for producing advanced materials with a unique combination of properties. Quenching from the solid state, when the material is heated to temperatures below the solidus prior to rapid cooling, is a classic method of heat treatment of steels and alloys that undergo phase transitions in the solid state. Its history goes back many centuries. The concept of melt quenching refers to the operation associated with rapid cooling of the molten material. The primary method of cooling in this method is fast heat transfer through the solid substrate, which allows for the effective rate of decrease in temperature up to 10^{10} deg/s. As a rule, in melt quenching it is required to obtain a certain critical cooling rate to produce a specific structural state of the material. The rate depends on the type of alloy and the nature of the metastability of the chosen state

Systematic research and development of th melt quenching method started in 1959 by Soviet scientists led by I.S. Miroshnichenko. An apparatus for melt quenching by bilateral cooling of the melt at a rate of 10^6 deg/s was constructed. P. Duvez later created a 'gun' which ejected a liquid drop on a cooled substrate (splatting). Crystalline materials whose structure and phase composition differed significantly from equilibrium were initially produced. In addition, it was possible to expand considerably the temperature and concentration ranges of solid solutions and produce new intermediate phases and unusual morphological features. Finally, it became possible to obtain the amorphous (non-crystalline) state of metallic alloys where the processes of crystallisation are completely suppressed in melt quenching.

Despite the obvious importance of these scientific results, they were met with little enthusiasm, because 'piglets' unsuitable for serious physical studies – the first samples of rapidly quenched materials – seemed to be exotic materials rather than the subject of in-depth studies of the structure and physical and mechanical properties of crystalline and amorphous alloys. The 'amorphous boom' started only in the late 60's, when a group of Japanese scientists, led by T. Masumoto, applied the spinning method for producing amorphous alloys. Melt quenching on a rapidly spinning disc, or other similar methods, made it possible to produce reproducible amorphous and crystalline materials suitable for large-scale studies, with, as it turned out, the unique combination of physical, chemical and mechanical properties.

This monograph is the fruit of thirty years of research conducted at the G.V. Kurdyumov Institute of Metals Science and Metals Physics and the Institute of Precision Alloys of the I.P. Bardin Central Scientific Research Institute of Ferrous Metallurgy. Studies of rapidly quenched alloys began with more exotic amorphous alloys with a number of unusual and very promising physico-chemical properties. The spectrum of research was then significantly expanded to include a systematic analysis of the structure and properties of ultrafine crystalline materials, obtained by melt quenching (the term 'nanomaterial' did not exist in the early 80's). Today we would like to pay tribute to the outstanding scientists of the I.P. Bardin Central Scientific Research Institute of Ferrous Metallurgy who stood at the cradle of successful development of the melt quenching method: B.V. Molotilov, V.T. Borisov, V.P. Ovcharov, A.I. Zusman, A.F. Prokoshin, V.P. Makarov, V.V. Sadchikov, and many others. We would like to thank our colleagues and students who have taken an active part in the research, included in this book: V.I. Goman'kov, AV. Shelyakov, V.A. Pozdnyakov, H. Rösner, V.A. Fedorov, L.S. Metlov, A.V. Shalimov, N.S. Perov, S.G. Zaichenko, V.M. Kachalov, O.L. Utevskaya, M.I. Yaskevich, I.V. Maleev, V.B. Sosnin, Yu.E. Chicherin, O.M. Zhigalin, E.N. Blinov, M.R. Plotnikov, N.A. Shurygin, and R.V. Sundeev. We would also like to express deep gratitude to the leading Russian and foreign scientists for the fruitful discussion of the results which certainly contributed to a better understanding of these results: R.A. Andrievskii, V.I. Betekhtin, B.S. Bockstein, G. Wilde, Y.I. Golovin, L. Greer, S.V. Dobatkin, A.E. Ermakov, M.I. Karpov, A. Kozlov, Yu.R. Kolobov, V.P. Manov, M.M. Myshlyaev, L.E. Popov, S.D. Prokoshkin, V.G. Pushin, Yu.A. Osip'yan, V.V. Fish, M.A. Shtremel' and E.I. Estrin. A substantial part of the research was supported by the Russian Foundation for Basic Research, to whom we are grateful. We will appreciate all the comments and suggestions of readers reading the monograph.

We dedicate this book to the memory of our teacher, an outstanding Russian scientist, Academician Georgy Vyacheslavovich Kurdyumov, whose bright personality and unique talent will show us, his disciples, the way to new scientific challenges for many years to come.

Introduction

1. General classification of nanostructured states

In 1989, H. Gleiter introduced a new term into scientific practice –'nanocrystal' – and the corresponding group of the materials was referred to as nanocrystaline (nanostructured) [1–3]. In the initial stage, the nanostructured materials were the materials consisting of nanoregions (structural components with the dimensions in the nanometre range) separated by the boundaries (two-dimensional regions with a different structure) [1]. In many cases, the conventional upper limit of the nanosized range of the dimensions of the structure was 100 nm. The nanostructured materials include single-phase and multi-phase nanocrystalline (nanophased) materials (NM), represented by the polycrystalline solids with the grain size of the order of tens of nanometres (no more than 100 nm).

The structural characteristics of the main types of nanocrystalline materials are shown in the scheme (Fig. I.1), proposed by H. Gleiter. It may be seen that there are four varieties as regards the chemical composition and distribution (single-phased, statistical multiphase composition with identical and non-identical interfaces, and the matrix compositions), and three groups of the form of the structure (laminated, columnar and the structures containing equiaxed inclusions). In fact, the difference in the structural types can be even greater as a result of mixed variants, the presence of porosity, polymer matrixes, etc. The most widely used structures are the single-phase and multi-phase matrix and statistical objects, columnar and multilayer structures. The latter (the so-called superlattices) are characteristic of films. In the nanocrystalline materials, the manifestation of the special features of the boundary states is as distinctive as that of the volume effects because the number of the

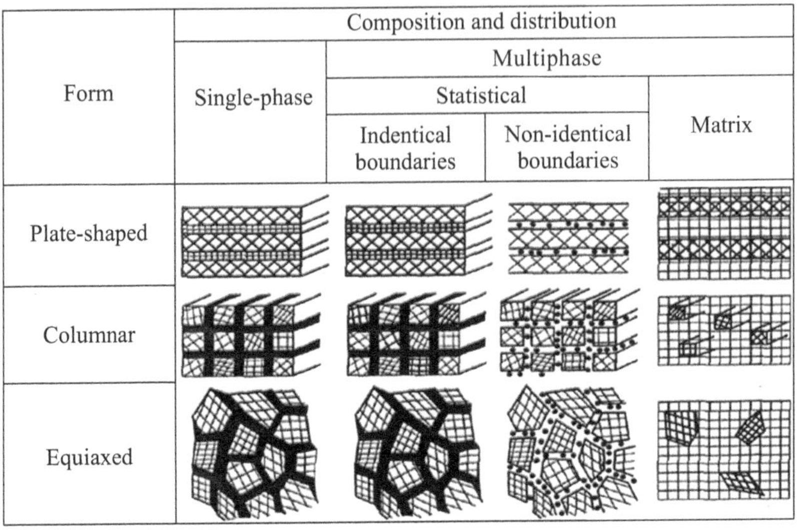

Form	Composition and distribution			
	Single-phase	Multiphase		
		Statistical		Matrix
		Indentical boundaries	Non-identical boundaries	
Plate-shaped				
Columnar				
Equiaxed				

Fig. I. 1.   The structural classification of nanomaterials.

grain boundary atoms in such systems is comparable with the number of the volume atoms [47]. The fraction of the volume, occupied by intergranular and/or interphase boundaries, in the nanocrystalline materials may reach 50%.

We examine in greater detail the relationship between the relative fraction of the crystalline matrix and the relative fraction of the grain boundaries which is evidently high in the nanocrystals. The volume fraction of the total intergranular space (IGS) is equal to [5]

$$V_{IGS} = 1 - \left(\frac{L-s}{L}\right)^3,$$
(I.1)

where L and s are the grain size and the thickness of the grain boundary, respectively. If we conditionally separate the intergranular space into two components, belonging to the grain boundaries (GB) and triple joints (TJ), the corresponding volume fractions will be:

$$V_{GB} = \frac{3s(L-s)^2}{L^3}$$
(I.1a)

and

$$V_{TJ} = \left(V_{IGS} - V_{GB}\right)$$
(I.1b)

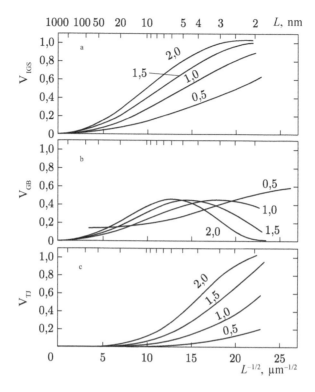

Fig. I.2. Calculated values of the volume fractions of the individual structural components V_{IGS} (a), V_{GB} (b) and V_{TJ} (c) in relation to the value $L^{1/2}$, where L is the size of the nanocrystals at different values of the thickness of the intecrystalline boundaries s (0.5; 1.0; 1.5 and 2.0 nm) [5].

Figure I.2 shows the dependence of the values of V_{IGS}, V_{GB} and V_{TJ}, calculated using the equations (I.1), (I.1a) and (I.1b) in relation to the grain size L for different fixed values of the thickness of the grain boundaries s. It may be seen that the volume fraction of the intergranular space rapidly increases with the reduction of the grain size (Fig. I.2.a). For example, at $s = 1$ nm and $L < 15$ nm, the volume fraction is more than 0.5. In other words, the fraction of the intergranular space increases with the increase of the proportion of the grains. If we analyse the graphs, shown in Fig. I.2b and Fig. I.2c, it may be seen that the reduction of the grain size is accompanied by a smooth increase of the proportion of the grain boundaries to the maximum value of 0.45 and the subsequently reduction of L is accompanied by a smooth reduction of this value. Thus, the widely held view according to which the reduction of the grain size of the nanocrystals results in an increase of the proportion of the 'grain boundary' phase is incorrect. At low values of L the fraction of the volume occupied by the grain boundaries decreases (does not increase), but this is accompanied by the start of a rapid increase of the fraction of the volume occupied by other defects – triple joins (Fig. I.2c). Consequently, in the nanometre range of the grain size (less than 15–20 nm), the controlling role in the individual processes (including the processes of deformation and failure) should be played not as much by the grain boundaries as by the triple joints (and, possibly, joints of higher orders) – the dominant elements of the structure in this size range of the nanocrystals.

The concept of the nanomaterial can also be determined on the basis of the physical features. This is a material, whose grain size (the structural components) is comparable with the characteristic correlation scale of the specific physical process and/or contains structural elements whose size is smaller than the characteristic value, at which the mechanism of the investigated physical phenomenon changes. One of the representatives of the nanostructured materials are nanostructured pseudoalloys [2].

Recently, the interest in the nanomaterials has increased considerably as a result of their unusual physical–chemical properties. However, the extremely high popularity of the nanomaterials often results in terminological confusion. In the structural aspect, they are often regarded as synonymous with nanocrystalline, nanostructured, nanophased, amorphous and other materials, which in fact is not accurate. Evidently, this misunderstanding originated from the term 'nanocrystal', proposed by H. Gleiter, who was the first to introduce the term 'nanocrystal' and regarded it immediately as identical with the term 'nanostructured' material [1]. However, is the nanostructured material always nanocrystalline? We will try to remove these contradictions and provide a unified structural classification of all types of nanomaterials.

The most accurate definition of the concept 'nanomaterial' has been provided, in our view, by the RUSNANO Corp. experts [8]: 'the nanomaterial is a variety of products of the nanoindustry in the form of materials, containing structural elements with the nanometre dimensions whose presence results in the significant improvement or appearance of qualitatively new mechanical, chemical, physical, biological and other properties, determined by the manifestation of the nanoscale factors'. As indicated by this definition, the nanomaterial is not always nanocrystalline, i.e. consisting mainly of the crystals with the nanoscale dimension in at least one measurement. The nanostructured structural elements, which dramatically change the

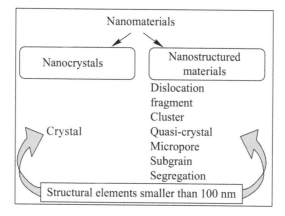

Fig. I.3. General classification of nanomaterials.

properties, can be not only the crystals but also fragments, pores, clusters, atomic segregations, dislocation ensembles, products of phase transformations, dendrites, quasicrystals, etc.

Figure I.3 shows the classification of nanomaterials resulting from this definition. It can be seen that the *nanocrystals* are only one of many types of nanomaterials. Other nanomaterials are conventionally divided into the group of *nanostructured materials*. However, strictly speaking, the nanocrystals are also nanostructured materials. We will pay special attention to the main types of these materials.

Nanofragmented materials (Fig. I.4) are the materials containing dislocation fragments or subgrains whose size does not exceed 100 nm. In particular, they include metals and alloys subjected to megaplastic (severe) deformation [9].

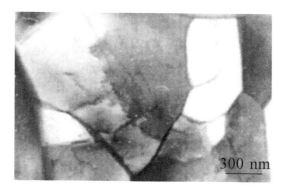

Fig. I.4. Structure of the nanofragmented material. Fe–Si alloy, produced by quenching from the liquid state; transmission electron microscopy (TEM).

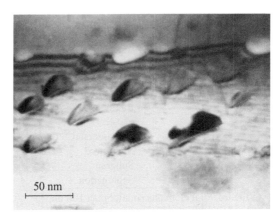

50 nm

Fig. I.5. The structure of a nanoporous material. The nanopores are distributed at the grain boundaries of the polycrystalline Fe–Al alloy, produced by melt quenching; TEM.

Nanoporous materials (Fig. I.5) are the materials with a high volume density of the nanopores with the size not exceeding 100 nm, distributed in the body of the conventional grains or, as in the present case (Fe–Al alloy, quenched from the melt [10]), at their boundaries.

Nanodendritic materials (Fig. I.6) are the materials containing the products of dendritic solidification in the form of nanosized dendrites or, as in the present case, in the form of degenerated dendrites (dendrite cells), formed in rapid solidification of Fe–Si [11].

Nanodislocation materials (Fig. I.7) are the materials characterized by the high volume density of nanoscale dislocation ensembles or configurations of a specific type. In the present case, the crystals are characterized by a very high (10^{11} mm^{-3}) volume density of the dislocation loops of the vacancy origin [11].

50 nm

Fig. I.6. The structure of a nanodendritic material. Nanosized dendrite cells, distributed inside the grains of the Fe–Si alloy, produced by quenching from the liquid state (melt quenching), are clearly visible; scanning electron microscopy (SEM).

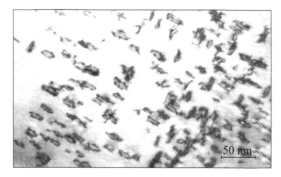

Fig. I. 7. The structure of a nanodislocation material. The Fe-Cr-Al alloy, quenched from the liquid state, shows the high volume density of prismatic dislocation loops of the vacancy origin; TEM.

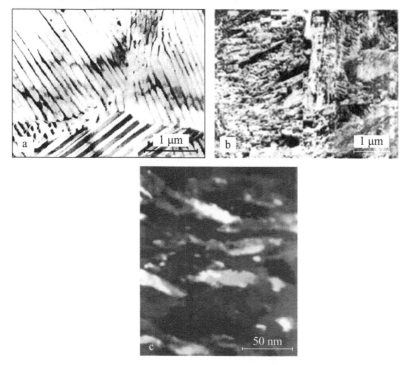

Fig. I.8. The structure of the nanophase materials. The nanophases are represented by the products of pearlitic (a), bainitic (b) and martensitic (c) transformations in steels, TEM [12].

Nanophase materials (Fig. I.8) are the materials containing nanosized products of phase transformations. The photographs show the nanoscale structures, formed as a result of pearlitic (a), bainitic (b) and martensitic (c) transformations in steels. In this case, the not quite correct term 'nanosteel' is sometimes used [13].

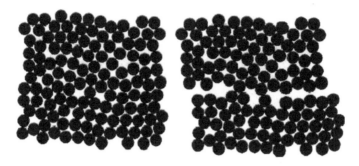

Fig. I. 9. Schematic representation of the Nanocluster structure in the amorphous alloys after melt quenching (left) and after additional plastic deformation (right) [15].

Nanosegregation materials are the materials containing grain boundary or other segregations of specific components, with the nanoscale length in at least one measurement. The scheme shown here shows the examples of the formation of grain boundary segregations, greatly reducing or increasing the cohesion strength of the grain boundaries.

Nanocluster (amorphous) materials (Fig. I.9). In accordance with the current views, the multicomponent amorphous metallic glasses have the nanocluster structure [14]. The clusterisation of the amorphous alloys is even more distinctive after the processes of local plastic yielding [15]. In this connection, the amorphous state of the alloys, produced by melt quenching, should be regarded as the nanostructured state.

Nanocrystalline materials (Fig. I.10). These materials consist mainly of nanosized (in at least one direction) crystals. In the case shown in Fig. I.10, the bulk nanocrystalline alloy was produced by controlled annealing of the amorphous state produced by melt quenching [16]. A partial case of this type of nanomaterials are *nano-*

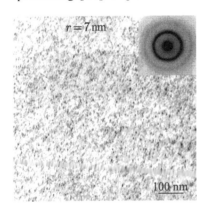

Fig. 1.10. The structure of the nanocrystalline material after controlled annealing of Ni–Fe–Co–Si alloy; TEM.

quasicrystalline materials consisting of poly-quasicrystals with the nanosized grains [17].

In conclusion, we will discuss the nanocrystalline materials as a variety of nanomaterials.

Three aspects should be mentioned:

- the above list of the possible types of nanostructured materials is obviously incomplete and may be regarded only as a basis for subsequent correction and supplementation.
- An important condition of the classification of any material as the nanostructured material is not only the existence in its structure of the nanoscale structural elements but also the strong effect of these elements on the properties of the material.
- The specific features of the influence of the dimensional effect (the nature of the dependence of any physical–chemical characteristic of the effective size of the structural element) depends almost completely on the nature of structural element (crystal, dislocation fragment, pore, segregation, etc). For the nanomaterials of different types the dimensional effect of a specific physical–chemical or mechanical characteristic is in most cases completely different.

The materials with the grain sizes (phase sizes), exceeding the nanometre range, but with the considerably smaller grain size in the conventional coarse-grained (~100 μm) materials, are also placed in a separate group which includes microcrystalline (1–10 μm) and sub-microcrystalline (ultrafine grained) (100 nm–1 μm) materials [18], which sometimes of course also belong to the nanostructured materials [13]. They represent the mesoscopic (intermediate) scale level of the structural state. By analogy with the nanomaterials, this group of the materials can be referred to as 'micromaterials' [19]. The micromaterials are classified not only on the basis of the scale but also physical features. In the microcrystalline materials, the parameters of the metastable structural state depend strongly on the method of preparation of the specimen and on its prior history [7]. The reduction of the size of the structural components of the materials to the level of 1–10 μm usually results in qualitative changes of a number of the properties, sensitive to the condition of the grain boundaries. They are characterized by the anomalies of the dependences of the properties, for example, the anomaly of the Hall–Petch ratio which links the level of the deformation stress with the grain size of the polycrystalline ensemble [20]. In addition to this, structural superplasticity is detected only at the grain sizes smaller than 10 μm [21]. The grain boundaries of the microstructural ceramic materials and of melt-quenched metallic materials, as also of the appropriate nanostructured materials, may have an amorphous structure or represent other phases [4, 22]. Analysis of the form of the Mössbauer spectra of submicrocrystalline (0.22 μm) iron shows in particular the presence of the so-called grain boundary phase in the structure [23].

The macroscopic scale level of the structural state is represented by the materials with the grain size (the fragments of the structure) greatly exceeding the micrometre range. Thus, the three scale levels of the structural state of solids are clearly evident. The materials in which at least one of the dimensions of the crystals or structural components does not exceed 100 nm belong in the group of the nanostructured materials [1–3]. The classification of the materials with greatly differing dimensions of the structural components in a separate group – 'nanomicrostructured' materials is more consistent. They may include low-dimension structures, such as nanowires, 'stressed superlattices', and others. The concept of nanostructured glasses was introduced in [2] by analogy with the nanocrystalline materials. The nanoamorphous metallic materials (metallic nanoglasses) represent one of the groups of the nanostructured materials. The nanoamorphous solids are produced, for example, by compacting of amorphous nanoparticles [24]. The structure and composition of the surface layers of the particles differ from the appropriate volume parameters. Consequently, a bulk amorphous material, consisting of nanoregions with a single amorphous structure, forms, and the boundaries between these regions are characterized by a different amorphous structure and/or atomic density. The nanoamorphous materials can also be produced by spinodal breakdown of the amorphous structure into two amorphous structural components with the nanometre dimensions or by plastic deformation of amorphous alloys [9].

Recently, special attention has been paid to the production and investigation of the properties of amorphous and crystalline metallic nanowires with the transverse size of the order of tens of nanometres [24]. The length of the wires may be in the range of a micron or greater. By pressing the nanowires of the amorphous alloys, it is possible to produce 'nanomicroamorphous' materials – metallic glasses with two characteristic scales of structural heterogeneity – micrometre and nanometre [24]. Thus, the materials with different scale levels of the structures can be placed in a separate group: nanomicroscopic, nanomacroscopic and micro-macroscopic.

Quasicrystals – solids with the quasiperiodic translational order and long-range orientation order – were discovered in 1984 [26]. Recently, the structure and properties of the quasicrystalline materials have been studied extensively. In particular, experiments were carried out with the synthesis of nano-quasicrystalline materials – poly-quasicrystalline materials with the grain size (crystal size) of the order of 10 nm [27]. It should be mentioned that the quasicrystalline state appears to be the intermediate state between the crystalline and amorphous states.

Recently, special interest has been given to the grain boundaries with the quasicrystalline structure and to the possibilities of existence of these boundaries [28]. The structure of the quasiperiodic boundaries of the grains is characterized by the irrational ratios of the numbers of the different structural elements forming the boundary [29]. The parameters of the disoriented boundaries can be both rational and irrational. If all the disorientation parameters, defining the relationship between the crystallographic basis of the adjacent grains are rational, i.e. the basal vectors of the two bases are linked by the linear relationships with the irrational coefficients, these boundaries are periodic and the ratio of the numbers of different structural elements is rational. In the opposite case, the boundary is quasiperiodic. It has been proposed to place the nanocrystalline materials with the quasicrystalline boundaries in a separate group of the materials [30]. The possibilities of existence of the boundaries with the amorphous structure must also be taken into account.

In the materials with the ultrafine structure in which the volume proportions of intragranular material and the material associated with the boundaries (surface) are comparable, it is necessary to take into account the boundary and surface structures and investigate the generalised structural-scale states [31]. In conventional materials, the macroscopic, crystalline and grain boundary structures are almost completely independent of each other and can be characterized and investigated separately. For the nanostructured materials, this classification loses its meaning: the structural state is characterized by the complex of the volume, boundary and surface structures. With the reduction of the grain size of the nanomaterials, the volume, boundary and surface structures should become more and more interrelated and depend on the grain size. For example, examination of nanocrystalline materials showed that the reduction of the grain size results in changes of the parameters and degree of tetragonality of the crystal lattice [16].

It is important to take into account the surface structural states for the low-dimensional systems in which the volume fraction of the atoms, adjacent to the free surface, is comparable with the volume fraction of the atoms inside the volume of the specimen – nanoparticles, nanowires and nanofilms. The crystalline, quasi-crystalline or amorphous materials, containing nanopores, where the mean distance between the nanopores is of the order of tens of nanometres, can also be regarded as nanostructured (nanoporous) materials [3], and it is also important to take into account the internal surface state of these materials.

On the basis of the combination of the volume and boundary structural states we can discuss the possibility of existence of various nanostructured materials, such as nanocrystalline with amorphous, quasicrystalline or crystalline boundaries; nano-quasicrystalline with quasicrystalline or amorphous boundaries and, finally, nanoamorphous materials.

The boundary structural state should have the atomic structure disordered at least to the extent of disordering of the structure of the bulk material. For example, the quasicrystalline grains can have boundaries with the quasicrystalline or amorphous structure. Otherwise, such a material should be regarded as a composite material. For example, if compacting of the amorphous powder is accompanied by surface crystallisation of the nanoparticles, we obtain a composite nanoamorphous material with nanocrystalline boundaries.

In accordance with the definition of the nanostructured materials, described previously, these materials do not include the materials whose structural components are characterized by the nanoscopic dimensions, but the distance between the components is greater than 100 nm. For example, the traditional dispersion-hardened alloys, containing nanocrystalline inclusions of other phases, with the smaller volume fraction of these phases (no more than 10%), do not belong evidently in the group of the nanostructured materials. If the amorphous alloys contain nanocrystalline inclusions, but the volume fraction of these alloys is far smaller than the unity, these materials should be referred to as amorphous crystalline materials or amorphous alloys with nanocrystalline inclusions. However, if the volume fraction of the nanoinclusions is large, and the mean distance between them is approximately 10 nm, they should be included in the group of the nanoamorphous crystalline materials [32], and they should also be regarded as representatives of the nanostructured materials (for example, Finemet type alloys [5]).

The heterophase nature of the generalised structural state is based on the presence in the material of different volume structural states, different boundary (surface) or simultaneously volume and boundary (surface) states. Important representatives of the heterophase nanostructured materials are nanoamorphous–crystalline and nano-amorphous–quasicrystalline alloys based on zirconium (aluminium) [33], produced by melt quenching.

An important stage of the classification is the consideration of different distribution and morphologies of the structural or phase components [2], in particular, plate-shaped, column and equiaxed forms of the elements of the bulk multiphase structures with different types of boundary structures. The identical investigation of different morphologies and distributions can also be carried out for structural components of the boundary and surface heterophase structures – banded, linear and equiaxed two-dimensional structural components.

I.2. Melt quenching

At present, several methods of melt quenching are used. They make it possible to produce using different modifications of nanocrystalline alloys in the form of ribbons up to 100 μm thick and up to 300 mm wide [34]:

1. Quenching of a cylindrical stream of the melt on a rapidly spinning disc;
2. Quenching of a flat stream on a rapidly spinning disc;
3. Quenching of a stream in rotating rollers;
4. Extraction of the melt.

The spinning method is used most widely [35]. In this method, the melt is supplied under pressure on a rapidly spinning disc-cooler (Fig. I.11). Consequently, the method produces ribbons with the thickness of 20–100 μm with the structure which depends on the composition of the alloy and cooling rate. Regardless of the apparently simple procedure, the spinning process is quite complicated and the effective quenching rate is determined by a large number of technological parameters (Fig. I.11).

Prior to the start of crystallisation, the melt should be supercooled below the equilibrium crystallisation temperature because there is an energy barrier for the formation of a nucleus of the crystal phase. The degree of supercooling of the metal depends on several

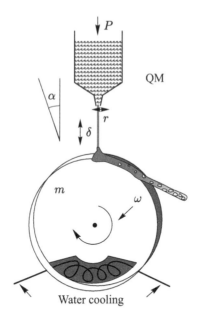

Fig. I. 11. Diagram of equipment for melt spinning with the technological parameters of the effect on the structure and properties of melt-quenched nanocrystalline alloy: P – excess pressure, r – the width of the orifice in the nozzle, δ – the size of the air gap, ω – the spinning speed of the quenching disc, m – the material of the quenching disc, α – the angle of the nozzle, T – the temperature of the cooling liquid, QM – quenching medium (inert gas, vacuum, air).

factors, including the initial viscosity of the melt, the rate of increase of viscosity with the reduction of temperature, the temperature dependence of the difference of the free energies of the supercooled melt and the crystal phase, the energy of the melt–crystal interface boundary, the volume density of the centres of heterogeneous nucleation of crystals and, finally, the actual cooling rate of the melt. The rate of growth of the crystals in the metallic melts is very high and, consequently, the crystallisation processes can be suppressed only in the case of highly efficient removal of heat into the surrounding medium. In addition to this, the increase of the cooling rate reduces the width of the crystallisation temperature range [36].

With the increase of the cooling rate the resultant crystal structure greatly changes. Initially, the polycrystalline structure is extensively refined and, subsequently, depending on the composition of the alloy, the solubility of the components in the solid solution increases and metastable crystalline phases can form. Finally, if the cooling rate is very high, crystallisation is completely suppressed as a result of the insufficient time available and the shear viscosity of the system continuously and smoothly increases during cooling. In the final form, the atomic structure, typical of the liquid state, is removed away from the thermodynamic equilibrium and at the so-called glass transition temperature T_g it is homogeneously frozen [37]. The amorphous state, formed in this procedure, is one of the key factors in the quenching experiments.

In accordance with the kinetic approach to the process of amorphisation, any metallic alloy, containing at least several percent of the second component, can be amorphised when a specific high cooling rate is reached [38]. Thus, the susceptibility of the alloy of the given chemical composition to amorphisation is characterised by the critical cooling rate v_{cr} which in the case of the spinning method corresponds with the critical thickness of the quenched ribbon t_{cr} [35]. For the limiting cooling rate of 10^6 °C/s, which can be obtained in spinning, the value of t_{cr} for the majority of the amorphised systems is usually up to 70–80 μm [39].

It should be mentioned that a number of easily amorphised multicomponent metallic systems (mostly based on Pd–Cu, Ti–Zr, Zr–Cu, Mg–Cu) for which the values of v_{cr} are so low that they can be produced in the form of thick bars and ingots [40] have recently been developed. However, the large majority of the amorphous and nanocrystalline alloys used in practice can be produced only in the form of ribbons with the thickness of several tens of microns.

The aim of all the methods of producing amorphous and nanocrystalline alloys by melt quenching is in fact the attempt to transfer, with the maximum possible speed, a certain amount of the melt to reliable contact with the cold surface which efficiently removes the heat (the disc-cooler in Fig. I.11) in order to produce the effective and uniform distribution of the melt on the surface and ensure reliable thermal contact. An increase of the value of v_{cr} for the given alloy reduces the thickness of the quenched cross-section of the solid and also the length of the section in which heat is removed. The cooling rate obtained in practice is evidently the function of the heat transfer coefficient. For the spinning method, the heat transfer coefficient equals $\approx 10^5$ W/m² K [35].

In addition to this, an important factor in quenching is the duration of contact τ with the heat-conducting surface. The value of τ for producing the amorphous or nanocrystalline state should be such that the material can be quenched to the temperature close to T_g and also to even lower temperature, without disrupting the contact. A reduction of τ results either in the material being solidified directly from the melt or, transferring into the amorphous state, the material will solidify subsequently during relatively slow cooling due to inefficient contact with the disc-cooler [41].

Melt quenching on the spinning disc-cooler is a high-rate process which depends on many physical and technological factors, with the majority of the factors shown in Fig. I.11. By changing or correcting these parameters, it is possible to ensure the optimum conditions of melt quenching and exert a direct effect on the structural state of the quenched materials of the given chemical composition. At the same time, the development of optimum technology for every alloy in the given equipment is an important research task. The available empirical relationships and experimental results obtained in other systems can indicate only the direction of search for the optimum quenching conditions in a specific research system as a result of the complicated nature and existence of a large number of factors in the experiment and also due to the fact that it is not possible to fully reproduce all the parameters, influencing the melt quenching process.

I.3. Classification of melt-quenched nanocrystals

The process of transition from the liquid (amorphous) to nanocrystalline state may be regarded as the transition of the order–disorder type [42]. In principle, the process can be realised either during cooling from the melt at a rate close to critical or in the thermal or deformation effect on the solid phase amorphous state produced in turn by melt quenching. In this case, the crystallisation process takes place in the conditions of constant heat supply (at a constant or continuously increasing temperature) taking into account the additional thermal energy generated during crystallisation. Consequently, the structure consists of two distinctive structural components: amorphous and nanocrystalline [43], which form in the system in the majority of cases in a specific stage of thermal or deformation treatment. In this case, the character of the structure depends to a certain extent on the quenching rate from the melt and subsequent heating rate and also on the temperature and annealing atmosphere or on the parameters of the deformation effect.

A completely different morphological type of the structure can be produced in the early stages of crystallisation in the conditions of rapid cooling of the melt characterized by effective removal of heat from the crystallising system. Similar amorphous–crystalline formations have been studied only in a very small number of cases, but the mechanical properties obtained in this case can be regarded as unique [44].

In principle, in melt quenching, depending on the value of v_{cr} and the heat removal parameters, four different scenarios can form and, consequently, four different types of nanocrystals with different structures and properties can be produced [11] (Fig. I.12):

1. The crystallisation of the liquid phase takes place completely in the process of melt quenching and we are concerned here with the single- or multiphase nano- or sub-microcrystalline structure (type I nanocrystals).

2. Melt quenching is accompanied by the formation of the amorphous state which in the process of subsequent cooling below the point T_g manages to crystallise partially or completely. Another variant is also possible: crystallisation takes place directly from the melt simultaneously with the transition of other regions of the melt to the amorphous state. Consequently, the amorphous–nanocrystalline structure (type II nanocrystals) is produced.

3. Melt quenching results in the formation of the amorphous state. In this case, the nanocrystals can be produced as a result of the subsequent thermal effect in the appropriate conditions (type III nanocrystals).

Types of nanocrystals	Description	Method of production	Structure
Type I	Complete crystallization in MQ	MQ	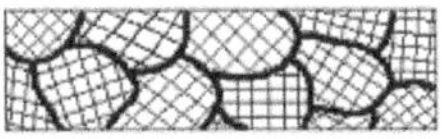
Type II	Partial crystallization in MQ	MQ	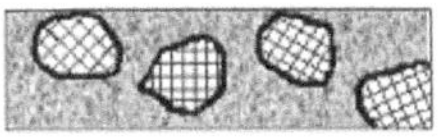
Type III	Crystallization in annealing amorphous state produced by MQ	MQ + HT	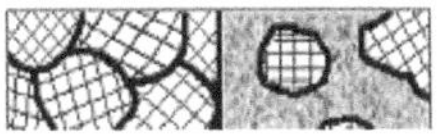
Type IV	Nanocrystallization in deformation of amorphous state produced by MQ	MQ + MPD	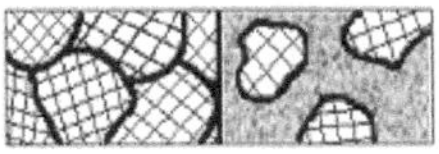

Fig. I.12. General classification of nanocrystals (NC) in quenching from the liquid state (QLS): HT – heat treatment, MPD – megaplastic (severe) deformation, MQ – melt quenching.

4. Melt quenching leads to the formation of the amorphous state. In this case, the nanocrystals can be produced as a result of the subsequent deformation effect in the appropriate conditions (type IV nanocrystals).

It can be seen that the type I and II nanocrystals form in a single stage (melt quenching), and the type III and IV nanocrystals form in two stages (melt quenching + heat treatment or plastic deformation). The classification of the melt-quenched nanocrystals, presented in this section, makes it possible to use a physically more justified approach to describing the structural special features typical of these materials. It should be stressed that the nanocrystalline state, produced by melt quenching, always forms by the 'from bottom to top' principle (the nanocrystals grow from the melt or from the amorphous matrix) which unites melt quenching with the most advanced nanotechnologies built on the same principle [45].

References

1.   Gleiter H., Progress in Material Science, 1989, V. 33, P. 223–315.
2.   Gleiter H., Nanostruct. Mater., 1992, V. 1, 1–19.
3.   Gleiter H., Nanostruct. Mater., 1995, V. 6, 3–14.
4.   Andrievsky R.A., Glezer A.M., Fiz. Met. Metalloved., 1999, V. 88, No. 1, 50–73.
5.   Andrievsky R.A., Glezer A.M., Fiz. Met. Metalloved., 2000, V. 89, No. 1. 91–112.
6.   Siegel R.W., J. Phys. Chem. Solids, 1994, V. 55, No. 10, 1097–1106.
7.   Wen S., Yan D., Ceramics Intern., 1995, V. 21, 109–112.
8.   Abstracts of the participants of the Second International Forum Nanotechnology (6–8 October 2009) – BM, BI, 2009, 728 c.
9.   Glezer, A.M., Izv. RAN. Ser. fiz., 2007, V. 71, No. 12, 1764–1772.
10.   Glezer A.M., Pozdnyakov V.A., et al., Mater. Sci. Forum, 1996, V. 225–227, 781–786.
11.   Glezer, A.M., Materialovedenie, 1999, No. 3, 10–19.
12.   Glezer, A.M., Functional nanocrystalline materials for technical and medical applications, Proc. Scientific. works of the Russian school-conference of young scientists and teachers, Biocompatible Nanostructured materials and Coatings for Medical Applications, Belgorod, Belgorod State University Publishing House, 2006, 23–33.
13.   Valiev R.Z., Aleksandrov I.V., Nanostructured materials obtained by severe plastic deformation, M, Lotos, 2000.
14.   Glezer A.M., Permyakova I.E., Gromov V.V., Kovalenko V.V., Mechanical asymptotic behaviuor of amorphous alloys, Novokuznetsk, Izd SibGIU, 2006.
15.   Gleiter H., Nanoglasses: A Way to Solid Materials with Tunable Atomic Structures and Properties, Mater. Sci. Forum. Nanomaterials by Severe Plastic Deformation IV, Eds. Yu. Estrin and H.J. Maier, 2008. V. 584–586, Part 1, 41–48.
16.   Lu K., Mater. Sci. Eng. Reports, 1996, V. 16, 161–221.
17.   Milman V., Goncharov I.V., Advanced Materials, Moscow, MISIs, 2009, V. 3, 5–54.

18. Konev N.A., Zhdanov A.N., Kozlov E.V., Izv. RAN, Ser. fiz., 2006, V. 70, No. 4, 577–580.
19. Pozdnyakov V.A., Crystallography Reports, 2003, V. 48, No. 4, 701–704.
20. Shtremel' M.A., Strength of alloys, Moscow, MISiS, 1997, V. 2.
21. Kaibyshev O.A., Valiev R.Z., Grain boundaries and properties of metals, Moscow, Metallurgiya, 1987.
22. Valiev R.Z., Islamgaliev R.K., Fiz. Met. Metalloved., 1998, v. 85, No. 3, 161–177.
23. Clarke D.R., J. Amer. Ceram. Soc., 1987, V. 70, Issue 1, 15–22.
24. Shabashov V.A., Ovchinnikov V.V., Mulyukov R., Valiev P.Z., Fiz. Met. Metalloved., 1998, V. 85, No. 3, 100–112.
25. Doudin B., Ansermet J.Ph., Nanostruct. Mater., 1995, V. 6, Issues 1–4, 521–524.
26. Shechtman D., Blech I., Gratias D., Cahn J.W., Phys. Rev. Lett., 1984. V. 53, 1951–1953.
27. Inoue A., Nanostruct. Maters., 1995, V. 6, 53–64.
28. Sutton A.P.. Acta Metall., 1988, V. 36, Issue 5, 1291–1299.
29. Gratias D., Phil. Mag. Lett., 1988, V. 57, Issue 2, 63–68.
30. Ovid'ko I.A., Fiz. Tverd. Tela, 1997, V. 39, No. 2, 306–312.
31. Brekhovskikh S., Scientific basis of materials science, Moscow, Nauka, 1981, 63–75.
32. Pashchenko N.V., Talanov V.M., Crystallography, 1995, V. 40, No. 6, 973–988.
33. Fan C., Li C., Inoue A., Haas V., Phys. Rev. B, 2000, V. 61, R3761–R3763.
34. Amorphous metallic alloys, Collection, ed. F. E.Lyuborskii, Moscow, Metallurgiya, 1987..
35. Filonov M.R., Anikin J.A., Levine J.B., Theoretical fundamentals of production of amorphous and nanocrystalline alloys by ultrafast quenching, Moscow, MISiS, 2006..
36. Borisov V.T., Theory of the two-phase zone of metallic ingots, Moscow, Metallurgiya, 1987.
37. Zolotukhin I., Kalinin Yu.E., Usp. Fiz. Nauk, 1990, V. 160, No. 9, 75–110.
38. Suzuki, K., Fujimori, H., Hashimoto K., Amorphous metals, Ed. C. Masumoto, Moscow, Metallurgiya, 1987.
39. Rapidly quenched metals, Collection, Ed. B.Kantor, Moscow, Metallurgiya, 1983.
40. Inoue A., Bulk amorphous alloys, Amorphous and nanocrystalline materials: Preparation, properties and applications (Advances in Materials Research), eds.: A. Inoue, K. Hashimoto, Springer-Verlag, Berlin; Heidelberg; N.Y., 2001, 1–51.
41. Wang G.X., Matthus E.F., J. Heat Trans., 1996, V. 118, No. 1, 157–163.
42. Frenkel' Ya.I., Introduction to the theory of metals, Leningrad, Nauka, 1972.
43. Glezer A.M., Ross. Khim. Zh., 2002, V. 66, No. 5, 57–63.
44. Glezer A.M., Chicherin Yu.E., Ovcharov V.P., Fiz. Met. Metalloved., 1987, V. 64, No. 6, 1106–1110.
45. Nanomaterials: Synthesis, Properties and Applications, eds. A.S. Edelstein, R.C.Kamarata, Bristol, Institute of Physics, 1996.

Melt-quenched nanocrystals (type I nanocrystals)

1.1. Main special features of the structure and properties

Structure

The main structural special features of the rapidly quenched crystalline materials formed by melt quenching are as follows [1–3]: reduction of the number of concentration heterogeneities; formation of very fine grains (more disperse structure); formation of a highly developed subgrain structure; a large increase of the solubility limit; formation of supersaturated solid solutions and new metastable phases; the non-equilibrium state of the interphase and intergranular boundaries; the possibility of formation of a high concentration of vacancies and defects formed during their coalescence.

The grain size of the materials, produced by superfast cooling of the melt, changes from tens of nanometres to tens of microns, and the size of the subgrain structure is correspondingly characterised by a fine scale. The products of melt quenching contain four structural zones. The freezing zone – the microdisperse structure, formed as a result of the multiple nucleation of crystals. The zone is adjacent to the quenched surface. The zone of the columnar structure – a system of greatly elongated grains occupying usually the main part of the cross-section of the specimen. In the immediate vicinity of the quenched surface this zone is directed along the normal to the quenched surface and then deviates in the direction of movement of the ribbon or fibre. In most cases, the central part of the cross-section of the specimen is characterized by the formation of a dendritic or dendritic–cellular structure. The fourth zone, which may form during crystallisation is the region of equiaxed randomly oriented nanograins.

Other important special features of the structure of the grains include:

1. The existence of a highly developed subgrain structure with different degrees of perfection;
2. Formation of the non-equilibrium structure of the grain boundaries, as indicated by the highly developed surface of the grain boundaries characterised by a high dislocation density, and the frequently changing orientation of their plane;
3. High thermal stability of the grain boundaries;
4. Selective high-temperature growth of the grains accompanied by the intensification of the crystallographic texture which is not very distinctive in the quenched state (quasisecondary recrystallisation).

Melt quenching includes the stage of rapid crystallisation. The effective cooling rate consists of the product of the temperature gradient at the crystallisation front and the linear speed of travel of the crystallisation front (the melt–crystal phase boundary). Depending on the ratio of the temperature gradient and the speed of movement of the front, the increase of the effective quenching rate may result in the formation of either dendritic, dendritic–cellular, or flat solidification front [4] (Fig. 1.1). As shown by the experimental results [5],

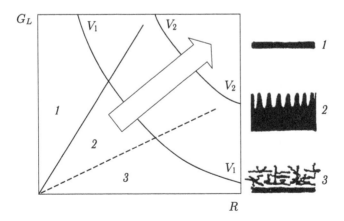

Fig. 1.1. Diagram of possible crystallisation mechanisms in relation to the temperature gradient at the crystallisation front G_L and the linear speed of movement of the crystallisation front R: 1) the flat crystallisation front; 2) the dendritic–cellular crystallisation front; 3) the dendritic crystallisation front. The diagram of the appropriate structures is shown on the right.

solidification in melt spinning is almost always accompanied by the movement of the dendritic–cellular solidification front due to the existence of the zone of concentration supercooling of the melt. As a result, dendritic cells form inside the grains, and the size of the cells is approximately an order magnitude smaller than the grain size (Fig. I.6). The grain boundaries and the cell boundaries form two subsystems with different scales which do not always coincide with each other. The reason for this mismatch is evidently the fact that the grain boundaries are capable of conservative rearrangement under the effect of quenching stresses, whereas the cell boundaries are not. Thermal effects are accompanied by the diffusion dissolution of the dendritic cells and complete equalisation of the chemical composition.

The rapidly quenched materials often contain (for example, [6]) two types of dendritic cells (Fig. I.6). In the first case, the form of the cells is close to regular hexagonal and they are associated with the movement of the crystallisation front away from the contact surface. When the thickness of the ribbon is sufficiently large (more than 20–25 µm), the crystallisation front also forms in the individual local regions on the contact surface and the dendritic cells have the characteristic fan-shaped form with the tips on the free surface.

The defects, formed in the process of rapid quenching, can be conventionally divided into two groups [7]: the defects formed by excess vacancies, and the defects associated with the relaxation of stresses generated in quenching. The most characteristic feature of the structure of many rapidly quenched alloys is the high density of dislocation prismatic small loops formed as a result of vacancy coalescence (Fig. I.7). The mean size of the loops in relation to the alloy and quenching conditions varies in the range 20–40 nm. The bulk density of the loops is $\sim 10^{11}–10^{12}$ mm^{-3}. Calculation of the vacancy concentration in the lattice at the melting point on the basis of the data on the mean size and the density of the loops gives the value $(1–2) \cdot 10^{-4}$ which is in good agreement with the theoretical calculations for BCC metals.

The special features of the dislocation loops formed during quenching are very important. Firstly, the prismatic loops form with a high probability at the grain boundaries and the boundaries of the dendritic cells. In addition to this, the concentration of the loops during annealing initially slightly increases. This indicates evidently that a large part of the non-equilibrium vacancies is retained in the lattice after melt quenching. In addition to the dislocation loops, the excess quenching vacancies form nano- and submicroscopic pores with the size of up to 0.1 μm, and in the case of materials with a low stacking fault energy – tetrahedrons of the stacking faults. Discs of antiphase boundaries (APB) form inside the loops in the ordered alloys. It can also be mentioned that, in a number of alloys, the dislocation prismatic loops of the quenched origin generally do not form at all. These are usually alloys undergoing phase transitions (ordering, breakdown), or alloys in which the vacancy–atom complexes of the solid solution can easily form. In these cases the vacancies are retained in the solid solution or segregate at the interphase boundaries.

The relaxation of the stresses formed in quenching leads to the formation of a high dislocation density. The wide temperature range in which the nucleation and propagation of the dislocations take place determines the wide spectrum of the resultant dislocation structures: from dislocation clusters to sub-boundaries (Fig. I.4). The boundaries of the subgrains and of the dendritic cells often coincide and, consequently, defects form in quenching at the cell boundaries.

Experiments with the annealing of melt-quenched alloys revealed the effect of the anomalous reduction of the grain size [8], manifested in the formation of a network of low-angle boundaries metallographically identical with the primary grain boundaries.

Investigation of rapidly quenched Fe–Cr–Al alloys showed the formation of four characteristic types of structure [8]. The increase of the cooling rate resulted in the formation of: 1. branched dendrites, 2. cells with low-angle boundaries; 3. grains containing parallel bands of vacancy dislocation loops along the $\langle 100 \rangle$ direction; 4) homogeneous grains without any characteristic features.

The broadening of the region of existence of the FCC γ-phase in the binary $Fe_{100-x}C_x$ ($x = 0–17$) and pseudo-binary $(Fe_{1-x}M_x)_{92}C_8$ alloys (M = Co, Cr, Cu, Mn, Ni, Pt; $x < 0.4$) was detected in the specimens produced by spinning in the argon atmosphere. The experimental results show that the lattice parameter of austenite changes in a non-linear manner with the increase of the carbon content of the alloy. The increase of the lattice parameter is attributed to the increase of the amount of the carbon dissolved in austenite, and the reduction – to the precipitation of cementite [9].

Phase transitions
Melt quenching may have a strong effect on the conditions of realisation and the nature of the structural phase transformations of different types.

It is interesting to investigate the course of the process of ordering and the possibility of suppressing this process in high-speed melt quenching. In [10] investigations were carried out into the rapidly quenched Fe–Al alloys in which atomic ordering can take place as a phase transition of the first or second kind. Depending on the composition, the superstructures B2 and DO_3 form in the investigated alloys. The experimental results show that melt quenching cannot be used to suppress the atomic ordering taking place as the phase transition of the second kind. Only the degree of the long-range order can be slightly reduced. The process of ordering by the mechanism of the phase transition of the first kind can be either completely suppressed or its course can be changed. For example, in Fe–22 at% Al alloy, melt quenching suppresses the equilibrium transition of the first kind $A2 \rightarrow DO_3 + A2$ and causes the $A2 \rightarrow B2 + A2$ phase transition which is characteristic of the alloy with a higher aluminium content [10, introduction Ref. 11].

There is a number of interesting special features of the structure of the antiphase boundaries in the melt-quenched superstructures.

1. The formation or disappearance of the effects of local deformation of the crystal lattice in the regions in the vicinity of the boundaries. It is justified to claim that the observed effects are directly connected with the application of the very high cooling rates and with the segregation of excess vacancies at thermal antiphase boundaries.

2. The 'two-phase' nature of coalescence of the antiphase boundaries in annealing. In the process of coalescence of domains two 'phases' appear to coexist within the limits of the same grain: the sections of the initial structure with small domains and the sections of the structure with large domains. The 'phase' with the large domains grows at the expense of the other 'phase' with the increase of the annealing time from the boundaries.

3. The elongated form of the antiphase boundaries. In the alloys with a not too high Kurnakov point, the structure of the antiphase boundaries is often elongated and not equiaxed. This very strange morphology of the domains is removed quite rapidly in annealing as soon as the coalescence effects become strong.

In some cases (Fe–Co, Fe–Cr–Al alloys) the effects of short-range ordering are detected fast quenching. These effects are not detected in the conventional state [Introduction Ref.11]. In this case, the structure is usually completely free from dislocation prismatic loops of the quenched origin. It may be assumed that a certain role in the formation of this short-range order is played by excess vacancies, playing the role of the additional component.

In most cases, melt quenching suppresses the precipitation of excess phases and supports stabilisation of the supersaturated solid solution. However, the authors of [8] reported the reversed effect – stimulation by quenching of the precipitation of a new phase in the Fe–23%Cr–10%Al alloy. Immediately after melt quenching, the alloy contained a large amount of the Cr_2Al phase forming a frame from plate-shaped precipitates at the boundaries of the dendritic cells [8]. In chromium and aluminium, the coefficient of distribution in iron is greater than the unity and, consequently, the cell boundaries are enriched with these elements. In addition to this, the process of precipitation of the phase at the cell boundaries is also supported by the higher density of the prismatic loops at these boundaries. In the melt-quenched Al–Fe alloys, the nanocrystals of the excess phase also precipitate at the boundaries of the dendritic cells [10, I11].

Spinning of the Fe–17 at.% Mo and Fe–17 at.% W alloys was performed to produce modulated structures, typical for the spinodal region of the equilibrium diagram [11]. In these alloys, produced by the traditional methods, the spinodal breakdown was not detected because the alloys with the limiting concentration of the solid solutions are situated outside the spinodal breakdown range. The large increase of the degree of supersaturation of solid solutions in melt quenching made it possible to 'penetrate' into the spinodal region. Ageing of the alloys with the modulated structure results in the precipitation of the stable phases: the η-phase in the Fe–Mo alloy and the λ-phase in the Fe–W alloy.

The special features of the kinetics and morphology of diffusion-less (martensitic) transformations for the type I nanocrystals have been studied most systematically and in detail in the alloys of the classic binary system Fe–Ni [12–14]. As in the rapidly quenched iron, the martensite in the Fe–(0–20)% Ni alloy had the form of packets consisting of laths with low-angle boundaries. However, in contrast to rapidly quenched iron or the Fe–Ni alloys, quenched in the solid state, many laths contained groups of thin parallel internal twins 1 nm wide, parallel to {112} and similar to the twins detected usually in the lath martensite of the alloy steels. The small size of the austenite grains, formed in crystallisation in the conditions of fast cooling, resulted in a large reduction of the temperature M_s of the start of martensitic transformation, reaching 250 K for the Fe–16% Ni alloy. In the alloys, containing (16–20)% Ni, the reduction of M_s was considerably smaller. After melt quenching, the alloys with (20–40)%Ni had at room temperature the structure of retained austenite with the grain size of 1–5 μm. The investigations carried out on the melt-quenched Fe–31%Ni alloy, showed that the formation of the nanocrystalline structure of the high-temperature γ-phase results in the suppression of the formation of isothermal and surface martensite in subsequent cooling, and the temperature of the start of athermal transformation rapidly decreases [15]. The high heterogeneity of the nickel concentration in the thickness of the ribbon leads to different conditions for the transformation in the cross section of the ribbon. The high heterogeneity of the nickel concentration in the thickness of the ribbon results in different conditions of the transformation in the cross-section of the ribbon.

The Fe–Mo–C alloys with a low carbon content contain both lath and twinned martensite [1]. The fraction of retained austenite increases with the increase of the carbon and molybdenum content. Thus, both the value of M_s and the morphology of martensite depend on the cooling rate and composition. The experimental results show that the structure of martensite in the thin films of iron-based alloys and in the thick material differs [1]. In [15, 16] investigations were carried out into the structural special features and martensitic transformation in the alloys of the systems of TiNi–TiCu and TiNi–TiFe quasi-boundary sections with shape memory, produced by the spinning method. The most radical changes in the microstructure both in austenite and in the subsequent martensitic transformation took place in the alloys with the content Cu > 25%. At a quenching rate of 10^{5}°C/s, nanograins were found in the micron-size grains. The quenching of the alloys with 28 ⩽ Cu ⩽ 34 at.% resulted in the formation of mainly a nanograined structure with the austenite grain size of ~30 nm and, consequently, of a nanostructured martensitic phase. The constructed diagrams of the martensitic transformations of the rapidly quenched alloys were used to determine the sequences and critical temperatures of direct and reversed martensitic transformations in these alloys.

Mechanical properties

All the iron-based alloys investigated in this section are usually brittle after conventional heat treatment. Melt quenching greatly increases the ductility of these alloys. In some cases, even the ductile–brittle transition takes place [17]. It is characteristic that in addition to the increase of ductility, melt quenching is also accompanied by a large increase of the strength of rapidly quenched alloys.

The fact that melt quenching increases both the strength and ductility of metallic materials is not at all paradoxical, as erroneously assumed by a number of researchers [18, 19]. In accordance with the well-known Hall–Petch ratio [Introduction, Ref. 5], the increase of the deformation stresses is undoubtedly associated with the reduction of the grain size in melt quenching and with the appearance of a developed intragranular substructure free from the long-range stress fields. The increase of ductility is also caused by the same structural reasons which can be supplemented by a number of other reasons (appearance of atomic ordering, homogenisation of the composition on the microlevel, etc). If the criterion of susceptibility to plastic yielding is represented by the widely used parameter – the ductile–brittle transition temperature – the modified classic Stroh–Trefilov equation gives [20]:

$$T_{db} = A - Bd^{-1/2},$$

where T_{db} is the temperature of the ductile–brittle transition, d is the grain size, and A and B are the physical constants which do not depend on the grain size.

Consequently, the ductile–brittle transition temperature should decrease linearly with the reduction of the parameter $d^{-1/2}$.

Magnetic properties

Melt quenching greatly influences the magnetic properties of materials. If we digress the amorphous state which results in drastic changes in the magnetic properties the formation of the nanocrystalline structure with new metastable phases and the unusual crystallographic texture has a strong effect on the magnetically soft and magnetically hard properties. A suitable example are the magnetically hard Nd–Fe–B alloys. Melt quenching drastically increases the magnetic properties of these unique alloys important for practice [21, 22]. However, evident progress is also observed in the magnetically soft alloys based on iron, nickel and cobalt, especially for the properties measured at high frequencies [23]. Below, we examine in detail the effect of melt quenching on the structure and properties of the magnetic materials important for practice.

1.2. Sendast alloy (Fe–Si–Al)

The iron alloy with 10 at.% Al and 16–17 at.% Si, widely known under the name Sendast and developed by Japanese investigators, is situated on the concentration triangle in the vicinity of the intersection of the lines of zero values of the constants of magnetic anisotropy K and magnetostriction λ_s (Fig. 1.2). Consequently, the alloy is characterised by the uniquely high magnetic softness ($\mu_0 =$ 30 000, μ_{max} = 200 000, $H_c \leqslant$ 1 A/m) and is an ideal material for the cores of magnetic heads because it combines the high level of the magnetic properties with high hardness and wear resistance. Unfortunately, the alloy is excessively brittle in a wide temperature range and its application in practice is associated with a number of serious difficulties.

Atomic ordering processes
It is well-known that the Fe–Si–Al alloys in the Fe–Fe$_3$Si–Fe$_3$Al concentration triangle, starting at specific concentrations of silicon and aluminium, are the ordered solid solutions of silicon and aluminium in the α-Fe lattice, and the Si and Al alloys together play the role of the second component in the superstructures B2

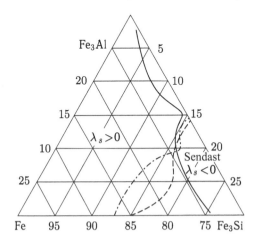

Fig. 1.2. Equilibrium diagram of the Fe–Fe$_3$Al–Fe$_3$Si system; λ_s is the magnetostriction constant; K is the magnetic anisotropy constant; μ_0 is the initial magnetic permittivity. The solid line K = 0, the dashed line μ_0 = max, the dot-dash line λ_s = 0. The concentration of Fe, Al, Si is in at.%.

and DO_3, formed in different temperature–concentration ranges. All the alloys of the Fe_3Si–Fe_3Al system (Fig. 1.2) show at room temperature the ordering effects in accordance with the type DO_3. Using the results of high-temperature x-ray diffraction investigations carried out in [24], it was shown that in cooling of the alloys from the region of the disordered BCC solid solution A2, the process of ordering of the DO_3 type is preceded by the formation of the B2 superstructure. The critical ordering temperatures T_x (A2 $\leftrightarrow$ B2) and T_y (B2 $\leftrightarrow$ DO_3) increased monotonically when the Al atoms were replaced by the Si atoms. The sequence of the phase transition A2 $\leftrightarrow$ B2 $\leftrightarrow$ DO_3 in cooling of the alloys situated in the Fe_3Si–Fe_3Al section, is retained for the alloys containing less than 12.5 at.% Si. According to [24], at higher Si concentrations the B2 superstructure forms directly from the melt, and in cooling the B2$\rightarrow DO_3$ transition takes place in the system.

The Sendast alloy is situated slightly away from the Fe_3Si–Fe_3Al section. Its classic composition can be expressed as $Fe_{75-1.5}Si_{15+1.5}Al_{10}$, i.e. it differs from the stoichiometric composition for the DO_3 superstructure with the composition $Fe_{75}(Si, Al)_{25}$ by the fact that the iron atoms in the amount of 1.5 at.% are substituted in the alloy by the silicon atoms. Therefore, from the viewpoint of occurrence of the atomic ordering processes, the Sendast alloy can be regarded with a small error as synonymous with the stoichiometric $Fe_{75}Si_{50}Al_{10}$ alloy. According to [24], this alloy should show the transition B2$\rightarrow DO_3$ ($t = 1000$–$1050°C$) in the process of cooling from the liquid state, and the B2 superstructure forms directly from the melt. However, there have been investigations in which the authors concluded that the nature of the phase transitions in the Sendast alloy is different. For example, in [25] it was concluded that the $Fe_{75}Si_{16}Al_9$ alloy is characterised by the occurrence of the A2$\rightarrow DO_3$ phase transition during cooling from high temperatures. In [26] no long-range order was found in the Sendast alloy after melt quenching. This result contradicts the data published in [24] because in the precipitation of the ordered phase directly from the melt its formation cannot evidently be suppressed by any (even high-speed) quenching (of course, when examining the crystalline and not amorphous state).

Figure 1.3 shows the Mössbauer spectra of the 'conventional' Sendast alloy, produced by vacuum melting followed by slow cooling (approximately 100°C/h) and by melt quenching. The same figure shows the spectrum for pure iron. It can be seen that the nature of the environment in the first coordination sphere in the Sendast alloy,

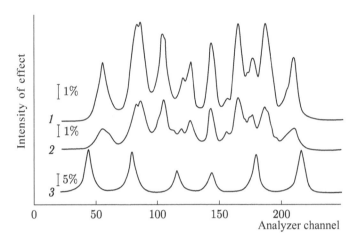

Fig. 1.3. Mössbauer spectra of Sendast alloy, obtained by vacuum melting (*1*), melt quenching (*2*) and pure iron (*3*).

produced by the two methods, is almost identical, but the width of the lines in the case of melt quenching is considerably greater. The thickness of the specimen of the quenched Sendast alloy is half the thickness of the other specimen and, consequently, the broadening of the lines is not associated with the possible effect of the thickness of the investigated specimens.

Figure 1.4 shows the microdiffraction pattern of the quenched alloy with the strong superstructural reflections, indicating the ordering of the matrix by the DO_3 type. The intensity of the superstructural reflections with even and odd indexes is a relatively

Fig. 1.4. The electron diffraction micropattern corresponding to the alloy produced by melt quenching. The reflections 200 and 111 are superstructural. The orientation of the foil (110): 000 is the zero reflection.

high and approximately the same which, together with the results of Mössbauer studies (Fig. 1.3) indicates the similar and high values of the long-range order of parameters, characterising ordering by the type DO_3 in the first and second coordination spheres [27].

The dark field images of the same section of the foil of the quenched alloy, produced under the effect of the superstructural reflections of different types, are shown in Fig. 1.5. Figure 1.5a (the acting reflection 111) shows the system of thermal antiphase boundaries (APB), forming the 'Swiss cheese' type structure characteristic of the redistribution of the atoms in two ordering sublattices. In Fig. 1.5b (reflection 222) the conventional diffraction contrast from APB does not form. The thermal APBs do not show any tendency to the distribution in specific crystallographic planes and do not form the anomalous diffraction contrast under the effect of the main reflections. The special features of the domain structures also remain after annealing of the quenched alloy in the entire temperature range.

Examination of the electron microscopic contrast from the APB showed an important relationship – the effect of the reflection 222 (and, generally, of any superstructural reflection with even indexes) results in the formation of a slight contrast from the APBs which form the 'normal' diffraction contrast under the effect of superstructural reflections with the odd indexes (in the reflection 111 in Fig. 1.5). This weak contrast is clearly visible in, for example, Fig. 1.5b (region A). It may be seen that the detected contrast is banded and is not the consequence of the effects of selective etching of the APB in the preparation of the thin foils because it is of the diffraction nature. Later, this contrast in Fig. 1.5 will be referred to as anomalous. The anomalous contrast completely disappears after annealing at 300°C and higher temperatures.

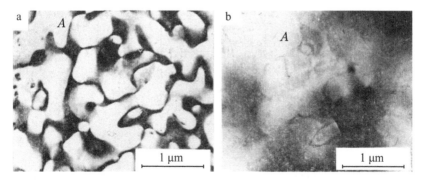

Fig. 1.5. Dark field images of the section of the crystal of the Sendast alloy under the effect of superstructural reflections 111 (a) and 222 (b).

The mean effective size of the domains D_{eff}, separated by the boundaries, visible in Fig. 1.5a, equals 50–100 nm, depending on the specific melt quenching conditions. The value D_{eff} does not change greatly up to the annealing temperature of 500°C and this is followed by rapid increase of D_{eff} (Fig. 1.6), indicating the large increase of the mobility of the domain boundaries at this temperature.

The formation of the DO_3 superstructure (Fig. 1.7) from the disordered BCC solid solution A2 can take place by two different mechanisms. In the first case, in cooling from the region of the phase A2 the formation of the superstructure is preceded by the formation of the B2 superstructure. In this case, the process of the formation of the four-domain structure of the DO_3 type appears to

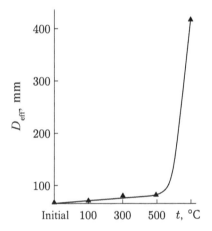

Fig. 1.6. Dependence of the effective size of the antiphase domains on the temperature of isothermal annealing for 0.5 h.

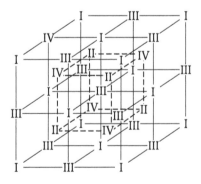

Fig. 1.7. Ordering sublattices in the formation of the B2 and DO_3 superstructures. The Al and Si atoms are situated in the sublattices II and IV (superstructure B2) or only in the sublattice IV (superstructure DO_3).

be divided into two stages and, consequently, the internal regions of the two-domain 'Swiss cheese' structure, formed by the APB with the antiphase shear vector $a/2\langle 111 \rangle$ which forms in the A2→B2 phase transition [28], shows the formation of a fine domain structure of the 'Swiss cheese' type which in this case is formed by the APB with the antiphase vector $a\langle 100 \rangle$ in the B2→DO$_3$ phase transition [28]. Thus, the subsequent phase transition A2→B2→DO$_3$ results in the formation in the crystal of a domain structure whose morphology is identical with that of the structure formed in the B2 superstructures but on two different scale levels in the study [29].

In the second case, the DO$_3$ superstructure may form also directly from the disordered state. In the A2→DO$_3$ transition four types of domains should form simultaneously in the initial matrix and, correspondingly, the so-called foam structure, predicted by Bragg as early as in the 30s, also forms [28]. It is important to note that in this case the effective size of the domains, restricted by the APB of the type $a/2\langle 111 \rangle$, should be comparable with that for the regions limited by the APB of the type $a\langle 100 \rangle$ [28].

The photographs shown in Fig. 1.5 show that the B2↔DO$_3$ phase transition takes place in the investigated alloy and the critical temperature of this transition, according to direct measurements [24], is 1000–1050°C. The boundaries of the type $a\langle 100 \rangle$, shown in Fig. 1.5, form in particular during this phase transformation. The conclusion of the author of [25] on the occurrence in the Sendast alloy of the A2↔DO$_3$ phase transition is not accurate because in this case after cooling the alloy from high temperatures the APB of the type $a/2\langle 111 \rangle$ should form with approximately the same density as the APB of the type $a\langle 100 \rangle$. Taking into account the fact that even very high quenching speeds, used in the given investigation, did not result in the formation of the thermal APB of the type $a/2\langle 111 \rangle$ in the structure, it may be assumed that these APBs are not likely to form (in this case, we do not consider naturally the extremely unlikely case in which the nuclei of the crystal phase have the same orientation). Since the ordered phase B2 forms directly from the melt, this conclusion is in complete agreement with the data obtained in high-temperature measurements [24] and in fact indicates that the thermal APBs of the type $a/2\langle 111 \rangle$ coincide with the boundaries of the grains formed during crystallisation. The difference between the crystallisation temperature of the Sendast alloy (1260°C) and the point of the B2↔DO$_3$ phase transition (1000–1050°C) is not large so that it cannot be used to assume the occurrence of the complete annihilation of the APBs of the type $a/2\langle 111 \rangle$ at very high temperatures of the A2→B2 phase transition, because the bulk density of the APBs of the type $a\langle 100 \rangle$, formed at temperatures higher than 1000°C, is very high.

Thus, the ordered phase B2 in the Sendast alloy forms during crystallisation and, consequently, the processes of atomic ordering in this alloy cannot be suppressed, of course if we do not consider the possibility of transition from the crystalline to amorphous state. Therefore, the conclusion in study [26] regarding the absence of atomic ordering in the rapidly quenched Sendast alloy should be regarded as erroneous.

We now transfer to discussing the nature of the detected anomalous contrast on the thermal APBs of the type $a\langle 100 \rangle$. On the whole, the morphology of the domain structure ('crystallographic nature' of the APBs and the presence of effects of near-boundary deformation) in the Sendast alloy is the same as that found in the Fe–Al alloys [29] and differs from the domain structure found in the thermal APBs in the Fe–Si alloys [27]. The boundaries have isotropic orientation and there are no near-boundary deformation effects causing the formation of the banded contrast from the APBs under the effect of main reflections [27]. Thus, the anomalous contrast, detected in Fig. 1.5, cannot be associated with these effects.

One of the possible reasons for the formation of the anomalous contrast is assumed to be the appearance of multibeam dynamic effects of scattering of electrodes and in which the acting reflection (in the present case, the superstructural reflection with the odd indexes) can be characterised by some of the characteristics of other acting reflections (for example, superstructural with even indexes). Evidently, in this case we can expect the formation of a weak residual contrast in the formation of the dark field image in the reflections which should not form the contrast from the APBs in the two-beam scattering conditions. If the two-beam scattering conditions are fulfilled, the following equation can be written for the contrast from a stacking fault or an APB [30]

$$B_{hkl} = \frac{\pi V_C}{\lambda F_{hkl}} \text{ctg } \varphi \sqrt{1 - \beta N}$$

where B_{hkl} is the distance between the two nearest dark or bright bands on the image of the defect under the effect of the reflection hkl; V_C is the volume of the elementary cell; λ is the wavelength of the electrons; F_{hhl} is the structural factor for the given reflection; φ is the angle between the plane of the foil and the plane in which the defect is situated; $N = h^2 + k^2 + l^2$; $\beta = \lambda^2/4a^2$; a is the lattice parameter.

Considering the ratio B_{hkl} for the acting reflections 222 and 111 for the same section of the APB using the above equation at $\beta = 1.113 \times 10^{-4}$, we obtain

$$B_{222}/B_{111} = 1.15.$$

The ratio B_{222}/B_{111}, measured in the experiments and shown in Fig. 1.5, is also equal to 1.15 which shows that the two-beam scattering conditions are satisfied in the case of the formation of the anomalous contrast. Consequently, the observed effects cannot be linked with the dynamic multi-beam scattering. This conclusion is confirmed by the results obtained in [31] where it is shown that in the effect of the reflection 111 or 222 in the alloys with similar compositions and identical crystallographic structure, with the accurate Bragg scattering conditions satisfied, the contribution of the systematic reflections to the extinction length can be ignored.

Finally, we examine another possible reason for the formation of the anomalous contrast – a change in the nature of ordered distribution of the atoms in the nodes of the crystal lattice. The fact that the effect of the 222 superstructural reflection results in the formation of a weak contrast from the APB of the type $a\langle 100\rangle$ means that these boundaries also contain the antiphase shift $a/2\langle 111\rangle$, and the concentration of one of the components, which determines this 'shift', is not high. Therefore, high-speed melt quenching results in the considerable supersaturation of the solid solution with quenching vacancies and it may also assumed that these vacancies will have the tendency for the ordered distribution in the sublattices of the DO_3 superstructure. It is characteristic that the vacancies in the investigated alloy do not segregate on the APB and remain in the solid solution. As shown in [32], the segregation effects in the Cu_2MnAl alloy should lead to the formation of a banded contrast from the APB under the effect of the main reflections.

Consequently, it is justified to propose the following structural model of the observed phenomenon. At the B2→DO_3 phase transition temperature (possibly, also at higher temperatures) a surplus of Si and Al atoms forms in the corresponding ordering sublattice (for example, IV in Fig. 1.7) as a result of a small deviation of the composition of the Sendast alloy from the A_3B stoichiometry and, correspondingly, a shortage of Fe atoms appears in other sublattices. The shortage of the Fe atoms can not be compensated by the surplus component

(Si + Al), which normally takes place in the 'normal' conditions, and it can be compensated by the non-equilibrium vacancies whose concentration at such high quenching rates is sufficiently high. The vacancies, playing in fact the role of the third component, are distributed in only two sublattices (not three sublattices), occupied by the Fe atoms. This results in the situation in which the contact of the domains, ordered by the DO_3 type, results, in addition to the 'normal' antiphase shift $a\langle 100 \rangle$, in the 'anomalous' antiphase shift of the type $a/2\langle 111 \rangle$ between the sublattices occupied by the vacancies and the sublattices free from vacancies. In the three-component solid solution where the excess vacancies are one of the components, the resultant structural state can be regarded as a completely new crystallographic type of the DO_3 superstructure whose formation in the three-component solid solutions was not previously detected.

In subsequent annealing the surplus vacancies travel away from the solid solution to the sinks of dislocation origin, present in the structure, or to the free surface and the metastable superstructure DO_3 transforms to the DO_3 equilibrium superstructure.

Quenching defects
As already mentioned in the section 1.1, the crystal structure defects, formed in the course of high-speed quenching, can be conventionally divided into two groups – the defects determined by the excess density of the dislocations, and the defects associated with the thermal stresses, formed in quenching.

The defects belonging to the first group in the BCC metals and in the alloys based on these metals have been studied to a considerably lesser extent than those in the metals and alloys with the BCC and HCP lattices. Doubts were even cast on the possibility of existence of these defects as a result of the high energy of formation of vacancies in the BCC metals which should determine in turn the low concentration of the vacancies in the vicinity of the melting point. In addition to this, the high mobility of the dislocations may cause that only a small part of these vacancies are quenched [33]. Some of these doubts were partially removed after experimental confirmation of the presence of the traces of quenched vacancies in molybdenum [34] and tungsten [35]. For example, in quenched molybdenum, ageing was accompanied by the formation of dislocation loops with the Burgers vector $a\langle 100 \rangle$ [34]. The formation of quenching defects in superfast quenching from the melt is even more likely [36]. The data were obtained by the method of direct resolution electron microscopy indicating the formation of vacancy clusters in the melt-quenched material [36].

The most typical feature of the fine structure of the investigated alloy is the high density of small dislocation loops (Fig. 1.8) distributed non-uniformly both in the volume of the each crystal and in the thickness of the ribbon specimens. Electron microscopic analysis shows [30] that the dislocation loops are, as expected, of the vacancy origin. The mean size of these loops, determined to increase the accuracy of measurements on the images obtained in the weak beam mode, is 20–40 nm in the initial (quenched) conditions.

The bulk density of the dislocation loops was calculated by the method of standard statistical analysis, based on the measurement of the thickness of the foil using the extinction contours and counting the number of loops in the regions inside the grains with the size of approximately 1 μm^2 [30].

The bulk density of the loops in the central regions of the samples is approximately 10^{11} mm^{-3} and examination of both the quenched condition and of the individual stages of subsequent annealing did not show any tendency to the preferential distribution of the loops in the crystal planes of a specific type. In some cases, only characteristic 'lines' were found – straightening of the loops along the specific crystallographic directions. It is also important to note that the density of the loops at the grain boundaries and in the volumes in the immediate vicinity of the boundary is considerably higher than in the body of the grain, and the mean size of the loops is higher than the size of the loops found in the matrix (Fig. 1.8b). The calculated concentration of the vacancies in the lattice and the melting point determined on the basis of the mean size and density

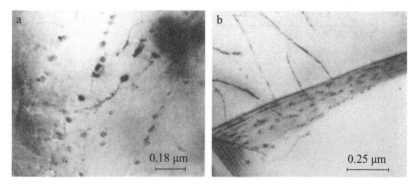

Fig. 1.8. Dislocation prismatic loops of the vacancy origin in the rapidly quenched Sendast alloy, light field images.

of the dislocation loops is $1.2 \cdot 10^{-4}$, which is in good agreement with the results published in [33] where the density of the vacancies in the BCC metal at the melting point was theoretically estimated.

As shown by the results of electron microscopic studies (Fig. 1.9), the antiphase boundaries with the antiphase shift vector $a/2\langle 111 \rangle$ form irregular neighbourhoods in the first coordination sphere inside the dislocation loops and in the crystal lattice of the alloy, ordered in accordance with the DO_3 type, [37]. Figure 1.9 shows a series of dark field images of the same 'line' of dislocation loops under the effect of the matrix reflection 440 (a) and the 222 superstructural reflection (b). In the latter case, a contrast forms from APB 'discs', situated inside the loops, inside the dislocation loops, shown in Fig. 1.9. The fact that the diffraction contrast from the APB forms under the effect of superstructural reflections with the even indexes indicates unambiguously that the vector of the antiphase shift at these boundaries is $a/2\langle 111 \rangle$. At the same time, it may be concluded, even without carrying out conventional gb-analysis, that crystallographic shift $a/2\langle 111 \rangle$ forms in the plane of the loop and the dislocation loops themselves are characterised by the same Burgers vector.

An important special feature of the dislocation loops in the investigated alloy is their high thermal stability. During annealing, the bulk density of the dislocation loops initially slightly increases and then decreases (Fig. 1.10), but even after annealing at 800°C the density of the dislocation loops at the grain boundaries does not change with the size of the quenched dislocation loops slightly increasing (50–100 nm), Fig. 1.9.

In addition to the dislocation defects, the alloy is also characterised by the presence of a certain number of submicropores with the size not exceeding 0.1 µm (Fig. 1.11a). The photograph shows the

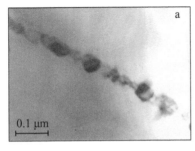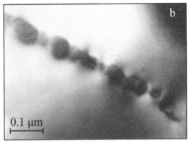

Fig. 1.9. Dislocation prismatic loops under the effect of the matrix reflection 440 (a) and the 222 superstructural reflection (b). The dark field images of the structure of the Sendast alloy, annealed at 800°C, $\tau = 0.5$ h.

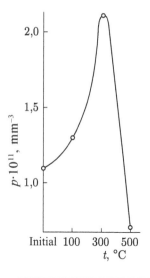

Fig. 1.10. Dependence of the bulk density of the dislocation loops on the isothermal annealing temperature in the melt-quenched Sendast alloy.

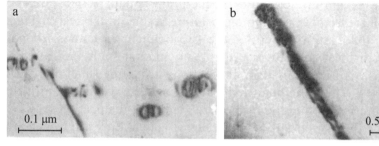

Fig. 1.11. Pores in quenched Sendast alloy: a) dark field image in the 400 matrix reflection; b) bright field image.

characteristic contrasts from the spherical submicropores of a small size which corresponds to the calculated profile of the intensity of the submicropores in the birefringent dynamic approximation [30]. The density of the submicropores rapidly increases in transition from the central to surface regions of the ribbon specimens. The thin foils of the surface regions of the ribbon specimens were produced by the method of one-sided polishing. In some cases, the subsurface regions are characterised by the formation of a continuous 'frame' from the submicropores along the grain boundaries form in the process of rapid quenching of the crystals (Fig. 11.1b).

The stresses, formed in quenching, lead to the processes of local plastic yielding resulting in the presence of a high dislocation density in the investigated alloy (Fig. 1.0). The wide temperature range in which the dislocations nucleate and undergo interaction determines the wide range of the dislocation structures differing in different degrees of the intensity of the relaxation processes. For example, the grains show a branched system of sub-boundaries of different degrees of perfection (Fig. 1.12a). On the other hand, individual regions of the crystal contain disordered dislocation clusters characterised by relatively extensive local rotations (several tens of degrees) and determining the high elastic stresses in the head of the clusters, as indicated by the formation of the bending extinction contours in Fig. 1.12b. The most distinguishing feature of the dislocation structure is the formation of a large number of isolated helicoids or helicoids grouped at the sub-boundaries (Fig. 1.12c) which evidently form as a result of the combined effect of the quenching stresses and the excess concentration of the vacancies in the solid solution. A characteristic element of the structure is also the presence of the often detected 'sticking up' dislocations (Fig. 1.12d), i.e. elongated dislocation segments distributed in the direction normal to the plane of the grain boundaries and in fact ending at the grain boundary.

An increase of annealing temperature results in the rearrangement of the dislocation structure: the sub-boundaries consist of more perfect dislocation networks, the disorientation angle increases, and the density of the individual dislocations inside the subgrains greatly decreases. The helicoidal dislocations acquire a more equilibrium concentration in the gliding process, emitting dislocation loops.

Since the crystal lattice of the investigated alloy after quenching from the melt is ordered in accordance with the DO_3 type [37], the single dislocations with the Burgers vector $a/2\langle 111 \rangle$ do not ensure the complete shift by the translation vector of the superstructure and each of them should generate a semi-infinite band of the APBs. In most cases, the dark field images under the effect of the superstructural reflections of different types shows that the dislocations end at the thermal APBs of the type $a\langle 100 \rangle$, which form in the alloys during the $B2 \rightarrow DO_3$ transition in the temperature range at approximately 1000°C [37], and the dislocation themselves are either present in pairs with the Burgers vector of $a/2\langle 111 \rangle$ each, or as single dislocations with the Burgers vector $a\langle 100 \rangle$ (inside the dislocation networks). Nevertheless, superdislocations, consisting of four single dislocations with the Burgers vector $a/2\langle 111 \rangle$, can sometimes be detected (Fig. 1.12e). Similar dislocation complexes are unstable in plastic yielding [38] and may form in the structure, evidently, as a result of dynamic recovery processes.

The process of melt quenching can be divided into two consecutive stages: rapid solidification (crystallisation) and quenching in the

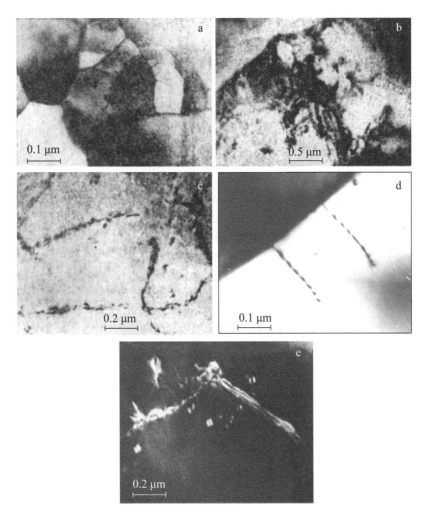

Fig. 1.12. The dislocation structure in the Sendast alloy, determined by thermal quenched vacancies: a, b) bright field images; c, d) dark field images in the reflection 400 and 440, respectively; e) the dark field image in the 400 reflection in the low-intensity beam mode.

solid state. In the processes of movement of the 'melt–solid phase' boundary and after completion of the crystallisation process, the vacancies diffuse to the grain boundaries as a result of the general non-equilibrium concentration of the vacancies and the non-equilibrium concentration in the volume of the grain, formed during rapid movement of the 'melt–solid phase' boundary [39]. This increases the excess concentration of vacancies at the grain boundaries where quenching dislocations additionally arrive from the previously solidified regions. An efficient method of the relaxation of quenched dislocations is 'closure' of the dislocations in prismatic dislocation loops. It is also possible that the 'sticking up' of dislocations, shown in Fig. 1.12d, also act as effective vacancy sinks from the moving crystallisation front and grow as a result of settling of excess vacancies on them.

Similar processes result in a considerably higher concentration of dislocation loops both at the grain boundaries and in the volumes in the immediate vicinity of these boundaries. The size of these defects is greater than the size characteristic of the loops away from the boundaries.

Attention should be given to the fact that the bulk density of the dislocation loops slightly initially increases in the process of low-temperature (300°C) annealing and only then slightly and gradually decreases with increase of the intensity of the thermal effects. This indicates that a large part of the non-equilibrium quenching vacancies are situated in the solid solution after quenching and, according to [37], occupy specific positions in the ordering sublattices.

Unexpected and very important results include the high thermal stability of the dislocation loops, 'surviving' in the process of high-temperature (700–800°C) annealing. Evidently, this is supported by the fact that the internal parts of the dislocation loops contain the APB discs which reduce their mobility. Since the loops are dislocation formations, the loops with the APB discs are therefore a variety of the superstructural dislocations. Since the specific surface energy of the APBs in the investigated alloy is very high (150 mJ/m^2), the mobility of these superdislocation loops is low, as in other superdislocation complexes, since the climbing process and the process of subsequent annihilation of the loops are made more difficult and require considerably stronger thermal fluctuations and, consequently, higher temperatures in comparison with the mobility of the dislocation loops in the disordered alloys.

In the case of rapid quenching from the melt there are several different groups of the mechanisms of formation of prismatic dislocation loops [39, 40]:

1. agglomeration mechanisms;
2. vacancy mechanisms of the multiplication of dislocation loops, associated with dislocation climb in the conditions of supersaturation of the crystal lattice with vacancies;
3. mechanisms of formation of the dislocation loops in the field of external stresses together with supersaturation of the crystal lattice with vacancies;

4. The Jones–Mitchell mechanism of prismatic extrusion.

The examination of the prismatic loops in the sections of the matrix showing no signs of dislocation sliding, with no preferential distribution of the dislocation loops in a specific crystal plane, confirms that the agglomeration mechanisms dominate in the process of rapid quenching from the melt. Evidently, the effect of these mechanisms is caused by the high rate of formation of the vacancy clusters in the alloys with a large dimensional mismatch in the conditions of melt quenching.

The presence of helicoidal dislocations with different degrees of perfection and the stroke configuration of the rows of prismatic loops indicate the possible operation of the mechanisms 2 and 3. The probability of the mechanism 2 operating in this case is the highest in low-temperature annealing because this requires a considerably lower degree of supersaturation with the vacancies. The effect of the mechanism 4 is not very probable because this mechanism can operate in the rapidly quenched alloys if they contain micro-heterogeneities. Since the processes of crystallisation and ordering in the Sendast alloy take place simultaneously [37], the dislocation structure, associated with the formation of the quenching stresses, forms in the ordered crystal. Consequently, the majority of the detected dislocations form paired dislocation complexes. It is characteristic that the same types of superdislocations cause plastic yielding in the Fe–Si-based alloys under active loading [41]. The high vacancy density results in the formation of left thread helicoids of different degrees of perfection with the small size of the turn (approximately 20 nm). The continuation of non-conservative motion of the dislocations in low-temperature annealing increases the degree of perfection of the helicoidal dislocations and sub-boundaries based on them and, consequently, vacancy sources of multiplication of dislocations may appear.

Special features of the structure of crystals

The magnetic and mechanical properties of the Sendast alloy are determined to a large extent by the form and crystallographic orientation of the crystals, formed in the solidification and under subsequent thermal effects. Evidently, in the conditions of superfast melt quenching, the structure of crystals will have a number of special features.

The special features of the grain structures were studied in detail in [42]. The mean grain size d_m, determined on the sections taken from the surface of the ribbon, was 2–12 μm, depending on ribbon thickness. No large variations were detected along the length and width of the ribbon. A large part of the grains in this cross section contained an equiaxed structure, but approximately 20% of the grains were oriented along the transverse direction of the ribbon (Fig. 1.13a).

Examination of pre-etched end sections, produced in the direction parallel to the ribbon axis, shows that on the side which was in contact with the disc the grain size was slightly smaller (by a factor of ~1.2) than the grain size in the centre of the ribbon and on the side of the free surface [43]. In addition to this, the 'end' section was characterised by the formation of a developed columnar structure of the grains and by a tendency to the deflection of the axes of the columnar crystals by the angle of 20° in the direction of movement of the ribbon when leaving the disc (Fig. 1.13b).

The magnetic properties of the Sendast alloy are sensitive to the changes in chemical composition. This is associated with the fact that the deviation from the classic composition results in non-zero values of the magnetostriction and magnetic anisotropy constants and, consequently, in a large reduction of magnetic permittivity and an increase of coercive force [44]. At the same time, it is well-known that crystallisation of single-phase solid solutions at relatively high cooling rates results in concentrational delamination [45]. The crystallisation cells are etched on the contact side of the ribbon. They are elongated with the axis oriented in the direction normal to the direction of movement of the ribbon during melt quenching (Fig.

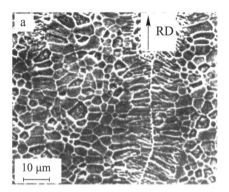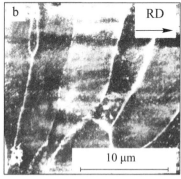

Fig. 1.13. The structure of grains in the melt-quenched Sendast alloy on the surface of the ribbon (a) and in the 'butt-end' longitudinal section (b). The arrow indicates the 'rolling' direction (RD) (movement of the ribbon during quenching on a cooling disc). Scanning electron microscopy in reflected electrons.

1.14). The presence of the cellular substructure in the quenched state regardless of the cooling conditions indicates that solidification takes place by the movement of the cellular crystallisation front. Since the cellular nature of the crystallisation front is determined by the existence of the zone of concentrational supercooling of the melt, there are obviously some differences in the chemical composition of the cell boundaries where enrichment with the atoms of the dissolved element (or elements) should take place, and in their central regions. The role of the impurities, accumulated at the crystallisation front and subsequently at the cell boundaries is played by the silicon atoms and, to a lesser degree, by aluminium alloys. The latter was confirmed in experiments by x-ray microanalysis.

It has been observed that the changes in the composition from the cell boundaries to its centre take place within the limits of the solid solution [46]. Local analysis makes it possible to determine a certain degree of enrichment of the grain boundaries with the atoms of the dissolved components (Si, Al), by 1–4 at.%. It should be mentioned that enrichment with the silicon atoms is greater than that with the

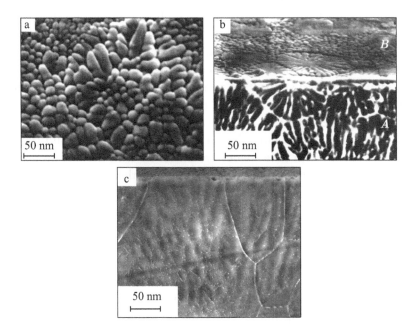

Fig. 1.14. The crystallisation substructure of the melt quenched Sendast alloy, on the side of the free surface (a), simultaneously with the free surface and the 'butt-end' cross-section (b) and on the 'butt-end' longitudinal section (c). Scanning electron microscopy in reflected electrodes. A and B in the figure correspond to the 'butt-end' section on the free surface.

aluminium atoms (approximately by a factor of 2–3). Figure 1.14b
shows the image in the conditions of scanning of reflected electrons
of the inclined 'end' section (the free surface is also visible). The
end sections show cells of two types: elongated to different degrees
along the axis, deflected from the normal to the ribbon surface by
the angle of approximately 20° in the direction of rotation of the
disc, and fan-like distribution of the cells, converging to a single
area situated on the free surface. The formation of cells of two
types (Fig. 1.14) is evidently caused by the fact that crystallisation
takes place both on the side which is in contact with the cooling
disc and partially on the free surface. The extent of development of
these competing processes is determined by the ribbon thickness.
The increase of the thickness increases the volume of the regions
in which solidification took place on the free surface. On the other
hand, the reduction of the ribbon thickness reduces the volume of
these regions, and when the ribbon thickness is less than 17 µm the
alloy crystallises only on the side of the contact surface [47].

Another interesting special feature of the crystallisation
substructure is the correlation in the distribution of the previously
mentioned defects of the crystal lattice (submicropores and dislocation
loops) with the cell boundaries (Fig. 1.14c). The correlation is
determined by the fact that the boundary volumes are characterised
by the formation of a high dislocation density associated with the
movement of the crystallisation front. At the same density of the flow
of the non-equilibrium vacancies to the boundary between the solid
and liquid phases, the regions (the regions not situated on the surface
of the ribbon or in this immediate vicinity), in which the solidification
is the last process to take place, are enriched with the vacancies and
this results in the preferential distribution of the submicropores and
the dislocation loops at the cell boundaries. The examined mechanism
is very similar to that which the authors of the book proposed in [6]
to explain the higher density of defects of the vacancy origin in the
regions in the immediate vicinity of the grain boundaries. It is also
highly likely that these reasons lead to the banded distribution of the
dislocation loops and the submicropores inside the individual crystals,
as reported previously by the authors of the present book in [6]. In
addition to this, taking into account the considered mechanism of fast
crystallisation, the process of formation of the bands of the dislocation
loops in the direction [100], reported in [36], can be explained. These
loops are decorated with fine carbide particles and this can be used to
explained the boundaries of the crystallisation cells, decorated with

the carbide particles, observed in [48] on the rapidly solidified FCC solid solutions.

The texture was examined both on the contact side of the ribbon and on the side of the free surface [49]. Analysis of the pole figures {110} and {200} of the quenched alloy showed the presence of a planar cubic texture with the most distinctive component (100) [001], deflected in the plane through the angle of approximately 20° around the direction situated in the ribbon plane in the direction normal to the 'rolling' direction, and this also explains the presence of a very weak textural component with the plane {110} in the ribbon plane (Fig. 1.15).

Figure 1.16 shows the variation of the mean grain size in relation to the annealing conditions of the rapidly quenched alloys [42]. The

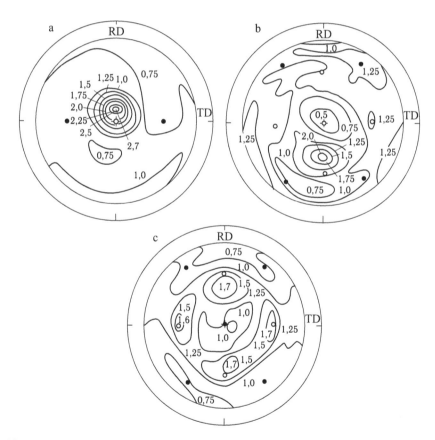

Fig. 1.15. Crystallographic texture of the microcrystalline melt-quenched Sendast alloy. The pole figure, obtained on the free surface, in the reflection {200} (a) and {110} (b) and the pole figure, produced from the contact surface in the reflection {110} (c): ● – (110)[001]; ○ – (100)[001].

high thermal stability of the grain boundaries may be associated, on the one hand, with the higher density of the submicropores and the dislocation loops in the boundary volumes and naturally in the grain boundaries and, on the other hand, with the decorating effect of the surface on the migration of columnar crystals. The reduction of the grain size (measured on the ribbon surface) in specific stages of the thermal effect is caused by the activation of the process of growth of columnar crystals throughout the entire thickness of the ribbon and by high-rate fragmentation, taking place at annealing temperatures of 400–500°C.

At the annealing temperatures higher than 900–1000°C the mobility of the grain boundaries rapidly increases. Selective growth takes place in the grains oriented with the {110} plane parallel to the ribbon surface. This is accompanied by rapid increase of the grain size from 10–15 to 100 µm. The kinetics and degree of development of the process depend on the mean grain size in the initial state. The texture of the alloy after annealing at 1100°C/0.5 h showed a large increase in the intensity of the orientations having the plane {110} in the plane of the ribbon, with the most distinctive component being (110)[001] (Fig. 1.17). In addition to this, the weak planar cubic texture with the component (100)[0kl] is retained.

The 'blurring' of the boundaries of the cells, formed in the process of rapid solidification, is an important process with the highest rate of the process recorded at the annealing temperatures starting from 600–700°C. Whilst after annealing at 800–900°C all the elements of

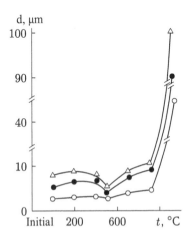

Fig. 1.16. Dependence of the mean grain size d_m on the preliminary annealing temperature t for 0.5 h for the ribbons with different thickness, µm: ∆ – 40, ● –27, ○ –17.

the cellular structure are retained, at temperatures higher than 900°C
cells disappear. The solidification substructure on the free surface
of the ribbon shows the highest thermal stability.

Annealing temperatures higher than 1000°C result in the processes
of selective growth of the crystals leading to an increase, by an
order of magnitude, of the mean grain size and to intensification
of the textural components with the {110} plane in the ribbon
plane, in particular (110)[001] (Fig. 1.17). Evidently, the nature
of this phenomenon is very similar to the processes of secondary
recrystallisation in the cold-deformed crystals and can be determined
as the phenomenon of quasi-secondary crystallisation in single-phase
microcrystalline metals and alloys, produced by melt quenching.
The role of the inhibitor phase, whose presence is essential for the
processes of secondary recrystallisation to take place, is played in
this case by the defects of vacancy origin, mostly submicropores,
whose density rapidly decreases in the temperature range in which
this phenomenon takes place. The process of quasi-secondary
recrystallisation in the melt-quenched Fe–6.5% Si alloy was observed
previously in [50] where the selective growth of the grains with
the orientation (100)[0kl] and (110)[$h\bar{h}l$] was detected. However, in
the present case, temperatures higher than 1000°C resulted in the
selective growth of grains only with the orientation (110)[$h\bar{h}l$]. The
increase in the intensity of the (100)[0kl] textural component with
no reflection on the plane is not accompanied in the present case by
any large increase of the mean grain size and evidently takes place
in the stage of normal (not selective) grain growth.

Mechanical properties
Figure 1.18 shows the variation of microhardness $H(a)$ and the
critical bending diameter in the three-point bend tests D_{cr} in relation
to the annealing temperature of the melt-quenched Sendast alloy
[51]. The same figure shows the dependence of the density of the
dislocation loops ρ_l and the effective grain size d_e. The correlation
between the dependence of the values of H and ρ_l on the annealing
temperature on the one side, and the values D_{cr} and d_e, on the other
side, is clearly visible. The maximum hardness is obtained at the
same annealing temperature as the temperature resulting in the
maximum bulk density of the dislocation loops (300°C), and the
minimum critical bending diameter (maximum plasticity) is obtained
at the same annealing temperature as the annealing temperature
resulting in the minimum effective grain size. It should be mentioned

Fig. 1.17. The crystallographic texture of the melt-quenched Sendast alloy after annealing at 1100°C, 0.5 h; the pole figure, produced from the free surface, in the reflection {200} (a) and {110} (b), and the pole figure, produced from the contact surface in the {200} reflection (c). The notation is the same as in Fig. 1.15.

that the value H provides a certain amount of information on the yield limit of crystalline materials and, therefore, it is evident that the resultant relationships also hold for the values of the yield limit of the investigated alloy.

The fact that the nature of the variation of hardness (yield limit) with the annealing temperature corresponds to the nature of variation of the number of dislocation loops (Fig. 1.18a) indicates that the dislocation prismatic loops of the quenched origin determine in particular the high strength properties of the Sendast alloy thus balancing the important role of the grain size. In fact, the prismatic loops represent an effective barrier to the dislocations causing plastic deformation of these alloys.

In this sense, the prismatic loops not capable of conservative gliding are the analogs of disc-shaped coherent particles of the second phase [52].

The form of the curves of the dependences of D_{cr} and d_e on annealing temperature (Fig. 1.18b) shows that the mean grain size in particular is the controlling factor of the plastic properties of the melt-quenched Sendast alloy. This is not surprising if it is taken into account that the temperature threshold of cold-shortness of the BCC metals and alloys is highly sensitive to the grain size [53].

Magnetic properties

The results of measurement of the main parameters of the magnetically soft properties at direct current (coercive force H_c, magnetic induction B_{10}, initial μ_0 and maximum μ_{max} magnetic permittivity) of the Sendast alloy, produced by melt quenching, are shown in Fig. 1.19.

In the initial (melt-quenched) conditions, the ribbon specimens of the Sendast alloy had the following properties [54]: $\mu_0 = 4000$–6000, $B_{10} = 0.8$–1.0 T, which are of considerable interest for applications. The dependence of the magnetic properties on annealing temperature, shown in Fig. 1.19, can be conventionally divided into three sections.

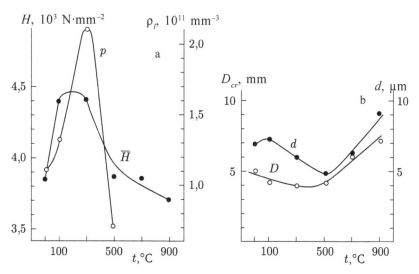

Fig. 1.18. Effect of annealing temperature (annealing time 30 min) on microhardness H and bulk density of dislocation loops ρ_l (a), the critical bending diameter D_{cr} and the mean grain size on the surface of the ribbon d_e (b) for the melt-quenched Sendast alloy.

1) at t <350°C the magnetic properties differ on this slightly from
the initial properties, 2) in the range t = 350–650°C the magnetic
permittivity slightly decreases and coercive force increases, 3) at
t > 650°C the magnetically soft characteristics greatly increase.
The highest level of the magnetic properties, obtained in [42],
corresponded to the following values: μ_0 = 22 000, μ_{max} = 178 000,
H_c = 2.4 A/m, B_{10} = 1.1 T.

In low-temperature annealing, there are several factors influencing
the magnetic properties. Annealing at temperatures below 350°C
increases the density of quenched dislocation loops and this evidently
results in a reduction of the permittivity and increase of the coercive
force. However, this is accompanied by a reduction of the density
of quenching vacancies in the solution and by the relaxation of the
quenching stresses resulting in the opposite effect. The effect of
processes such as the reduction of the effective grain size and the
growth of the columnar crystals throughout the entire thickness of the
ribbon specimen exerts different effects on the magnetic properties.
On the whole, the magnetic properties remain almost constant in the
process of low-temperature annealing. In the annealing temperature
range 500–700°C the processes reducing the initial and maximum
permittivity and increasing the coercive force are most distinctive.
In particular, these processes include intensive fragmentation at a
constant high density of the prismatic dislocation loops. With a
further increase of annealing temperature (above 700°C) the decrease

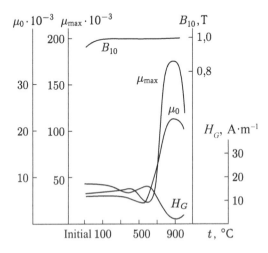

Fig. 1.19. Effect of annealing temperature (30 min) on the magnetic properties
(μ_0 and μ_{max} are the initial and maximum magnetic permittivity, respectively, B_{10}
is magnetic induction, H_c is coercive force) of the rapidly quenched Sendast alloy.

of the density of the loops decreases, the increase of the grain size and homogenising of the chemical composition (dissolution of the dendritic cells) result in a large increase of both permittivity values and also in a large reduction of coercive force.

Thus, melt quenching results in the optimum combination of the magnetic and mechanical properties of the Sendast alloy characterised by unique functional characteristics.

1.3. Fe–Si alloys

The Fe–Si alloys, containing 6–17% at.% Si, belong in the group of the magnetically soft materials. The combination of high permittivity, saturation magnetisation, specific electrical resistivity, low magnetostriction and low remagnetising losses results in extensive application of these materials in electrical engineering. The brittleness of the high-silicon materials greatly complicates the production technology and restricts the range of applications in industry. Previously, attempts were made to improve the ductility properties of this group of alloys by two methods: firstly, by the additional alloying elements [55] and, secondly, by the increase of temperature in deformation of the alloy [56].

The methods of melt quenching have been used to produce Fe–Si (Si >10 at.%) alloys in the form of thin ribbons with a thickness of 20–100 μm, bypassing here the relatively complicated technological process [50, 57].

Atomic ordering processes

In [58] the authors reported the suppression of A2→B2→DO$_3$ phase transitions by quenching, although in [59] it was reported that in the Fe–Si system with the silicon content of 11–22 at.%, the A2→B2 phase transition takes place by the mechanism of phase transitions of the second kind so that it is almost impossible to suppress the ordering processes of the atoms by the rapid quenching methods.

Transmission electron microscopy in the microdiffraction mode was used in [60] to produce electron diffraction patterns from different areas of thin foils, produced from ribbons of rapidly quenched alloys with 5.8–17.1 at.% Si. Studies of the diffraction patterns show that in the Fe–(16–17) at.% Si alloys the atoms are ordered by the DO$_3$ type and in the alloys with 11–13 at.% Si in accordance with the B2 type. The absence of superstructural reflections on the diffraction patterns of the Fe–5.8 at.% Si alloys indicates that the process of ordering of the atoms in this case does not take place [60]. The Fe–(11–13) at.% Si alloy is characterised by the relatively low-intensity blurred superstructural reflections corresponding to the ordered distribution of the atoms in accordance with the B2 type, and the blurring of the reflections is expressed differently even within the limits of the same ribbon. The minimum intensity of reflections with the even indexes was recorded for the Fe–11 at.% Si alloy [61].

Studies of the dark field images in the superstructural reflections with even indexes show that in the Fe–(11–13) at.% Si alloys the mean size of the antiphase domains (APD), separated by the APBs with the vector of the antiphase shift $R = a/2\langle 111\rangle$, is 15 nm, and in the alloys with 16–17 at.% Si it is 150 nm (Fig. 1.20). For the Fe–(16–17) at.% Si, the size of the domains separated by the APBs with $a\langle 100\rangle$ is ~30 nm. The anomalous contrast, formed in the matrix reflection from the APBs of the type $a/2\langle 111\rangle$ in the concentration range 11–17 at.% Si for the melt-quenched alloys of the same composition was insignificant.

The experimental results show that the anomalous contrast forms in fragments in the sections of APBs $a/2\langle 111\rangle$ in the Fe–(16–17) at.% Si alloys (Fig. 1.20c). In isothermal annealing at temperatures of 500–550°C (0.5–50 h) the Fe–(11–13) at.% Si showed an increase of the intensity of the superstructural reflections, corresponding to the ordered distribution of the atoms by the type B2, and the formation of new superstructural reflections with odd indexes, formed as a result of ordering of the atoms by the type DO_3. In the alloys with ≈17 at.% Si the increase of the annealing temperature above 500°C increased the grain size of the APBs, separated by the APBs of the type $a/2\langle 111\rangle$, to the size obtained at temperatures of 800–900°C, and the size of the separated APBs of the type $a\langle 100\rangle$, increased to 50–70 nm.

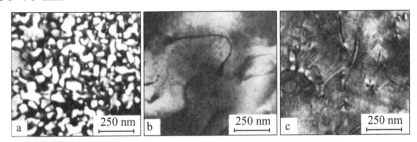

Fig. 1.20. Ordered domain structure of the melt-quenched Fe–17 at.% Si alloy. The dark field images under the effect of superstructural reflections 111 (a) and 222 (b) in the matrix reflection 220 (c). Foil orientation (112).

It is well-known that the iron alloys with the silicon content of 11–12 at.% Si undergo the following sequence of phase transitions in cooling: $A2 \rightarrow B2 \rightarrow DO_3$. It was shown in [59] that in this concentration range the $A2 \rightarrow B2$ phase transition takes place by the mechanism of the phase transitions of the second kind, and $B2 \rightarrow DO_3$ – by the mechanism of the phase transitions of the first kind. In the present study, it was shown that the $A2 \rightarrow B2$ phase transition is not suppressed by melt quenching and this has also been confirmed in general in [62]. Regions of the disordered phase A2 can form in the quenched Fe–11 at.% Si alloy only as a result of concentration heterogeneities, formed in quenching.

The absence of anomalous contrast in the matrix reflections, determined by the effect of local deformation of the AFB is caused evidently by the compensation of the resultant displacement as a result of the relaxation of the elastically deformed state in the conditions of higher concentration of thermal vacancies. Calculation of the additional displacement vector, formed at the APBs as a result of local deformation of the boundary layers for the alloys containing 11–17 at.% Si, was described in [63]. It is equal to 0.06–0.07 nm. It is therefore possible to estimate the excess concentration of the vacancies required for compensating this displacement of the atoms. Calculations give the value of the concentration of the thermal vacancies of $\sim 0.5 \cdot 10^{-3}$, which corresponds to the order of the value obtained previously in the determination of the concentration of the vacancies forming quenching defects of the type of submicropores and dislocation loops in the Fe–Si alloys, produced by the spinning method. All these considerations can be used to estimate the nature of local deformation of the APBs in the Fe–Si alloys produced by the conventional methods, because the segregation of the vacancies at the APBs can compensate only local boundary deformation of the type of dilation compression of the lattice.

Dislocation substructure

The entire spectrum of the investigated compositions is characterised by: 1) the presence of crystal lattice defects of the type of prismatic dislocation loops and the submicropores, distributed by different mechanisms depending on composition; 2) different degrees of fragmentation of the grains and also the presence of dislocations, determined by the effects of local plastic deformation in melt quenching (Fig. 1.21).

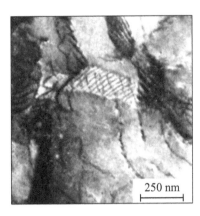

Fig. 1.21. Fragmented structure in the melt-quenched Fe–17 at.% Si alloy. The dark field image in the 400 matrix reflection. Foil orientation (001).

250 nm

If in the alloys with 11–17 at.% Si the submicropores and dislocation loops are distributed mostly in the volumes adjacent to the grain boundaries and directly in the boundaries, then in the Fe–6 at.% Si alloy, the distribution of these defects is random. In this alloy, grain fragmentation does not take place. The density of the prismatic dislocation loops in the investigated alloys also greatly differs. It should be mentioned that the ordered alloys are characterised by a considerably lower density of the dislocation loops in comparison with the Fe–6 at.% Si alloys [64].

The the size and volume fraction of the submicropores, distributed at the grain boundaries, increase on approach to the free surface. Calculations carried out using the mean size of the loops and the value of the Burgers vector of the loops show that in the Fe–Si alloy the excess vacancy density of 10^{-3}–10^{-4} is quenched, the accurate value depends on composition. In heat treatment, starting at a temperature of 400°C, the dislocation density inside the grains decreases up to complete 'cleaning' of the body of the grain to remove dislocation loops at t_{ann} = 700°C. However, the thermal stability of the defects, associated with the excess vacancy density at the grain boundaries, is relatively high (up to temperatures of 800–900°C). It should be mentioned that the size of the boundary defects increases with increasing annealing temperature.

The higher density of the defects of the type of submicropores and dislocation loops at the grain boundaries is determined by the 'trap' effect of the free volume (the melt→A2 transition and the vacancy sink in the quenching process), formed at joining of the boundaries of the adjacent cells in the processes of movement of the solidification front. The identical phenomenon, leading to the formation of pores in the interaxial space of the dendrites was observed in the development of dendritic crystallisation [65, 66].

The formation of the submicropores and the dislocation loops during melt quenching takes place mostly at the boundaries of the crystallisation cells enriched with the dissolved elements. The absence of defects of this type at the grain boundaries in the Fe–5.8 at.% Si alloy confirms the previously mentioned results: the flat crystallisation front assumes the statistically equal distribution of the vacancies, and the high-angle boundaries can work as an effective vacancy sink. Consequently, the density of the quenched defects, determined by the higher density of the thermal vacancies, greatly decreases in the regions in the immediate vicinity of the grain boundaries.

Calculations of the density of the dislocation loops ρ_l show that in the Fe–Si alloys of the investigated composition, the maximum density is obtained in the alloy with 5.8 at.% Si. Similar values of ρ_l in the Sendast alloy and the Fe–5.8 at.% Si alloy indicate that the APB size of ≈ 70 nm has no significant effect on the density of the thermal vacancies. The presence of a relatively high value of ρ_l in the Fe–(11–17) at.% Si melt-quenched alloys is caused probably by the formation of vacancy clusters at temperatures above the temperature of the A2→B2→DO$_3$ phase transition.

The considerably higher density ρ_l in the Fe–5.8 at.% Si alloy in comparison with the Fe–(11–17) at.% Si alloys is determined by the increase of the density of thermal vacancies resulting from the absence of the segregation of vacancies and the APBs in quenching in the solid state. The difference by almost an order of magnitude in the density of the dislocation loops and the submicropores in this alloys in the quenched state would lead, as it appears, to the unexpected result: a large reduction of the strain to fracture ε_p in the Fe–5.8 at.% Si alloy in comparison with the Fe–(11–12) at.% Si alloy [67]. Low-temperature annealing (500°C) increases the value ε_p in the Fe–5.8 at.% Si alloy and reduces this parameter in the Fe–(11–12) at.% Si alloy; secondly, the reduced density of the dislocation loops in the body of the grains which facilitates the process of plastic deformation of the material.

Electron microscopic studies of the thin foils taken from the ribbons, tested in uniaxial tension, show [63] that the plastic deformation in the Fe–Si ordered alloys takes place by the slip of paired superdislocations with the cubic Burgers vector. Figure 1.22 shows electron microscopic images of the same section of the foil of

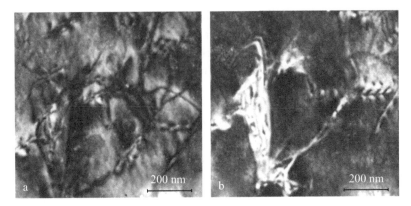

Fig. 1.22. Dislocation structure in the plastically deformed melt quenched Fe–12 at.% Si alloy. Dark field images were obtained under the effect of the reflections 040 (a) and 400 (b).

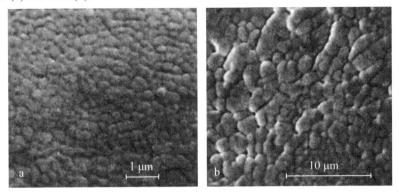

Fig. 1.23. Crystallisation cells in the melt quenched Fe–12 at.% Si (a) and Fe–17 at.% Si (b) alloys. Electron microscopic images on the side of the free surface in the regime of scanning of reflected electrons; d_c = 0.3 µm (a) and 1.5 µm (b).

Fe–12 at.% Si under the effect of the 400 and 040 matrix reflections. The photographs indicate that plastic deformation is takes place by gliding of the super dislocations with the Burgers vector of the single dislocations $a\langle100\rangle$, and two slip systems operate in this case.

Crystallisation processes and crystal structure

The Fe–(11–17) at.% Si alloys are characterised by the process of cellular crystallisation in the course of rapid solidification of the melt. Figure 1.23 shows the images of the free surface of the ribbons of the Fe–12 at.% Si and Fe–16.5 at.% Si alloys, produced in the conditions of scanning of reflected electrons. It may be seen that the reduction of the silicon content reduces the size of the crystallisation cell.

X-ray diffraction microanalysis and Auger spectroscopy of the etched free surface of the material showed a certain degree of enrichment of the cell boundaries with the atoms of the dissolved element (Si). The small size (0.3 μm) of the cells in the Fe–12 at.% Si alloy and the large difference in the cell size with the variation of the quenching rate indicate that in this alloy solidification can take place by the movement of the flat crystallisation front.

The crystallographic texture of the ribbon and the dynamics of the variation of the grain size of the melt-quenched Fe–Si alloys are very similar to those in the Sendast alloy (Fe–Si–Al) [68]. It should be mentioned that in the alloys with the silicon content of 11–12 and 6 at.% the intensification of the orientations $(100)[0kl]$ takes place at a slightly lower temperature in comparison with the Fe–17 at.% Si alloy (see Fig. 1.24), respectively, and the orientation $(110)[h\bar{h}l]$ develops at a lower temperature in the alloy with 11–12 at.% Si.

In [68] it was shown that the angle of deviation of the most distinctive component of the $(100)[001]$ crystallographic texture is determined by the angle of inclination of the crystallisation cells in the direction of movement of the ribbon in the process of spinning from the melt. The angle of inclination of the cells of the crystallisation substructure is determined by the specific geometry of movement of the solidification front: the crystallisation direction deviates from the normal to the surface of contact of the liquid metal with the cooling disc. The retention of the similar solidification geometry also determines the small differences in the textures of the Fe–Si and the Sendast alloy, rapidly quenched from the liquid state. The reduction of the cell size does not alter the crystallographic texture of the rapidly quenched Fe–Si alloys.

The sequence of the textural changes with the increase of the isothermal annealing temperature remains the same as in the Sendast alloy. Consequently, it may be assumed that annealing in the high-temperature range from 900 to 1100°C is accompanied by the operation of the mechanism of blocking of movement of the grain boundaries, mainly with the orientation $(100)[0kl]$ by the previously described defects formed at the cell boundaries in the process of rapid crystallisation and further quenching. The increase of the cell size which in this case may be used as a criterion for the development of cellular crystallisation approximately corresponds to the increase of the volume fraction of the ribbon solidified by this mechanism. Therefore, this is probably the reason why the alloys Fe–(16–17) at.% Si and the Sendast alloy in comparison with the Fe–

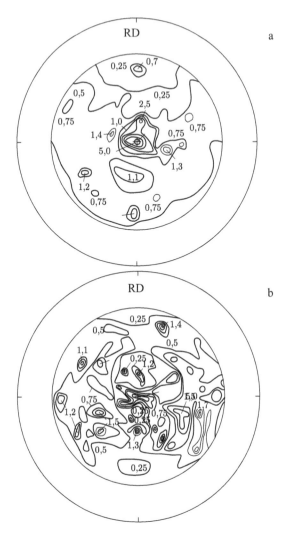

Fig. 1.24. a, b) crystallographic textures in the rapidly quenched Fe–(11–17) at.% Si alloys. The pole figures of the Fe – 12 at.% Si alloy, annealed at 1000°C (a) and 1100°C (b); a – reflection {200}; b – reflection {110}.

(11–12) at.% Si alloys are characterised by more efficient blocking of the movement of the high-angle grain boundaries, mostly with the orientation (100)[0*kl*]. It should be mentioned that annealing at high temperatures may result in the blocking of the grain boundaries by the quenching defects instead of the defects are distributed at the grain boundaries. Evidently, this process takes place in the Fe–5.8 at.% Si alloy and to a certain degree also in the Sendast alloy.

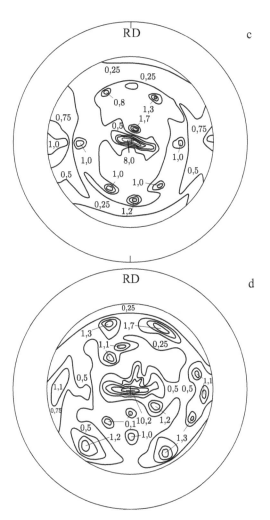

Fig. 1.24. c, d) crystallographic textures in the rapidly quenched Fe–(11–17) at.% Si alloys. The pole figures of the Fe–12 at.% Si alloy, annealed at 1100°C (a) and 1150°C (b); a – reflection {200}; b – reflection {110}. The pole figures of the Fe–12 at.% Si alloy, annealed at 1100°C (c) and 1150°C (d); c – reflection {200}; d – reflection {110}.

Table 1.1 summarises all the structural special features of the melt quenched Fe–Si alloys, determined in [61]. For comparison, the data for the Sendast alloy (Fe–Si–Al) are also given.

Fracture mechanism and mechanical properties
The specimens of the Fe alloys with 6–17 at.% Si, produced by melt quenching and by 'standard' technology (hot forging and warm rolling of ingots, melted in a vacuum induction furnace) were heat

Table 1.1. Structural characteristics of mechanical properties of the melt quenched Fe–Si alloys

Si content, at.%	Phase composition	APD size, nm	Loop density, mm^{-3}	Cell size, μm	Plasticity, %
5.2	A2	–	$5 \cdot 10^{11}$	–	0.5/2.5
11–12	B2/B2 + DO$_3$	15/25	$4 \cdot 10^{10}$	0.3	2.5/1.0
16–17	DO$_3$	25/70	$6 \cdot 10^{11}$	1.5	0.2/0
Sendast	DO$_3$	75	$3 \cdot 10^{11}$	2.5	0

Comment. 1. The numerator of the fraction shows the characteristic prior to annealing, the denominator after annealing at 500°C. 2. In the alloy with 5.8 at.% Si the distribution of the loops is uniform and there is a flat crystallisation front, in the other alloys the loops are distributed mainly at the cell boundaries of the grains and there is a cellular crystallisation front.

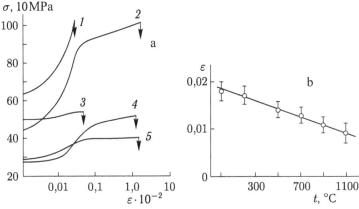

Fig. 1.25. The curves of uniaxial tensile loading of the alloys with 5–17 at.% Si, produced by different methods: a) dependence of true stress σ on plastic strain ε for the rapidly quenched alloys, containing 17 (1), 12 (2) and 5.8 (3) at.% Si, for the alloy with 11.2 at.% Si (4) and produced by 'normal technology', and also for the rapidly quenched alloy, containing 5.8 at.% Si (5) and annealed at 500°C; b) dependence of ε on the annealing temperature for the melt–quenched Fe–17 at.% Si alloy.

treated in [69] to ensure that they have approximately the same ratio of the mean grain size to the thickness of the investigated specimens and also similar ratios of the thickness of the specimens to their width. Consequently, it was possible to exclude the dimensional defects in the specimen–testing machine system in investigating the effect of melt quenching on the mechanical properties of the iron–silicon alloys. Figure 1.25 shows the dependences of true deformation stress σ on true strain ε for the alloys with 5.8, 11 and 16.5 at.% Si. Attention should be given to the following dependences.

1. The alloys produced by melt quenching have higher ductility but at the same time higher strength (by a factor of two) in comparison with the alloys produced by 'standard' technology.
2. The highest ductility is shown by the rapidly quenched alloy with 12 at.% Si and not with the minimum silicon content, as observed in the conventional production methods [63].
3. After melt quenching, the alloy containing 16.5 at.% Si has a low but noticeable plasticity. The conventional alloys with such a high silicon content are very brittle at room temperature.
4. The ductility of the rapidly quenched alloy with 12 at.% Si decreases after preliminary annealing at 500°C, and in the alloy with 5.8 at.% Si it rapidly increases.

The results of the bend tests of the quenched and annealed (at different temperatures) specimens of the Fe–16.5 at.% Si alloy are presented in Fig. 1.25b. After annealing at temperatures higher than 1000°C the alloy is greatly embrittled.

Studies of the fractographic features of the alloys with 12 at.% Si produced by melt quenching and by 'normal' technology show (Fig. 1.26) that in the first case, the process of plastic yielding is more intensive but it is also localised greatly along the length of the specimen subjected to uniaxial tensile loading. In the alloy with 16.5 at.% Si annealing at high temperatures results in the increase of the grain size of the fracture and this confirms the embrittlement detected in the bend tests.

Analysis of the fracture morphology and comparison of the $\sigma(\varepsilon)$ dependences of the Fe–11.2 at.% Si alloy, produced by melt quenching and by the 'normal' technology shows that fracture in the quenched state takes place with a higher preliminary plastic deformation. It should be mentioned that quenching also results in

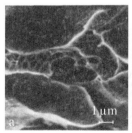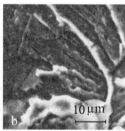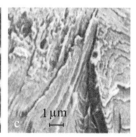

Fig. 1.26. The fracture morphology of the rapidly quenched Fe–12 at.% Si alloy in the initial condition (a) and after annealing at 700°C (b) and also of the alloy of the same composition but produced by 'normal' technology (c). The electron micrographs in the regime of scanning of reflected electrons.

extensive hardening of the material. The increase of the brittleness of the Fe–(16.17) at.% Si alloys and 'coarsening' of the cleavage fracture at fracture, detected at an annealing temperature of 1000°C, are determined mainly by the grain growth processes.

Thus, the Fe–Si alloys produced in the form of ribbons with a thickness of 20–40 μm are characterised by higher strength and ductility in comparison with the alloys produced by the 'conventional' method. These differences are not a consequence of the dimensional effect (small thickness of the investigated objects). The higher strength is determined by the developed fragmentation of the grains of the rapidly quenched alloys and by the presence of a high bulk density of the defects of the vacancy nature (prismatic loops and micropores), acting as obstacles in movement of the dislocations. The higher plasticity of the alloys is associated not only with the presence of the smaller grains but also with the formation of a developed polygonal structure and with the significant contribution brought into the process of plastic yielding by the more mobile dislocations with the Burgers vector $a\langle 100 \rangle$.

Magnetic properties
The effect of melt quenching on the magnetic characteristics was investigated mainly for two compositions which are important for practice: Fe–6 at.% Si (electrical engineering steel) and Fe–12 at.% Si (the alloy with the zero constant of saturation magnetostriction and high magnetic permittivity). The temperature dependence of the saturation magnetisation for the alloys with 11 and 13 at.% Si, produced by melt quenching, in the magnetisation field of 125 A/m is shown in Fig. 1.27, and Fig. 1.28 shows the effects of the annealing

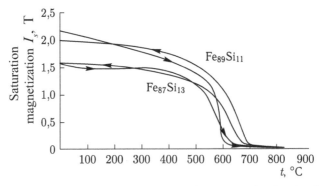

Fig. 1.27. Temperature dependence of saturation magnetisation in the field of 125 A/m of iron alloys with 11 and 13 at.% Si.

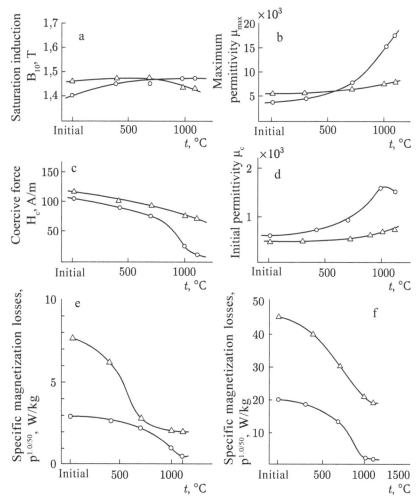

Fig. 1.28. Dependence of saturation induction (a), maximum permittivity (b), coercive force (c), initial permittivity (d) and specific magnetic losses at a frequency of 50 Hz (e) and 400 Hz (f) in the Fe–12 at.% Si alloy produced by melt quenching; spinning in air ($\circ$) and in argon (Δ).

temperature on the main magnetically soft characteristics of the alloy with 12 at.% Si. It may easily be seen that the nature of variation of the saturation induction (a), maximum (b) and initial (d) permittivity and coercive force (c) is the same as in the Sendast alloy (section 1.2). A similar behaviour is also shown by the specific losses in remagnetisation at a frequency of 50 Hz (e) and 400 Hz (f) important for practical application of the Fe–Si alloys. Measurements were taken for the specimens quenched in air and in an inert medium. The graphs in Fig. 1.27 show that the medium used for melt quenching

influences the individual magnetic characteristics, including the value
of specific losses which is most important for electrical engineering
steels (in this case, quenching in air is preferred).

Figure 1.29 shows the variation of coercive force H_c in high–
temperature annealing of the alloys with 6 and 12 at.% Si [70],
produced by quenching in rolls. After quenching the value H_c was
142 and 51 A/m, respectively, but after annealing at $t > 800°C$
the value H_c rapidly decreased because of the reasons which we
examine in greater detail when studying the magnetic properties of
the rapidly quenched Sendast alloy. It should only be mentioned
that the reduction of H_c in the alloy with 6 at.% Si is determined
mostly by the formation of the coarse-grained structure and by the
intensification of the favourable orientation (100)[0vw] of the crystals
in the ribbon plane.

On the whole, the lowest value of H_c for the alloy with
12 at.% Si was 1.5 times lower, and for the alloy with 6 at.%
silicon atoms higher than that obtained in the 'best' specimens of
the same composition produced by conventional technology [71]. The
unfavourable change of H_c in the alloy with 6 at.% Si, produced by

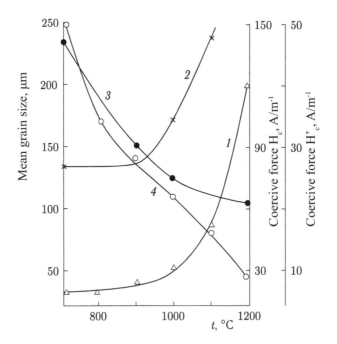

Fig. 1.29. Variation of the mean grain size (1, 2) and coercive force (3, 4) in relation
to the annealing temperature of the quenched Fe alloys with 6 (1, 3) and 12 (2, 4)
at.% Si; H_c – the scale for the curve 3; H_c' – the scale for the curve 4.

melt quenching and annealed subsequently at high temperatures, may be associated with the presence of pores and microcracks since the quenching of this composition is associated with difficulties because of the high melting point. In many cases, the magnetic properties of the rapidly quenched alloys are increased by additional cold rolling which stimulates secondary recrystallisation in subsequent annealing [62].

1.4. Fe–Al alloys

The Fe–Al alloys, rich in iron, have been studied in a large number of investigations which have been efficiently summarised in [63, 72]. Regardless of certain differences in the quantitative estimates, determined by the special features of the experimental procedures, the phase transitions in the Fe–Al system can be described quite reliably using the equilibrium diagrams shown in Fig. 1.30 [73]. Attention should be given to the two special features: the absence of the two-phase regions $A2_p$ (disordered paramagnetic solid solution with the BCC lattice) + B2 superstructures and B2 + DO_3 superstructure, and also a sharp break in the $A2_f$ region. It has been established that the ordering of the A2 ferromagnetic phase always takes place as a phase transition of the first kind, but the ordering of the paramagnetic phase is a phase transition of the second kind.

It is very interesting to consider the nature of atomic ordering and the possibility of suppressing this process in superfast melt quenching. The calculations, carried out in the framework of the

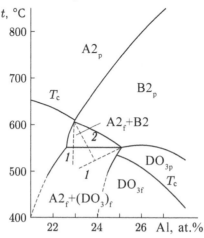

Fig. 1.30. Equilibrium diagram of the Fe–Al system; indexes 'f' and 'p' correspond to ferromagnetic and paramagnetic phases, respectively.

vacancy model of migration of the atoms, show that ordering in accordance with the type B2 takes place in these alloys at such high rates that it cannot be suppressed by cooling from the disordered state at rates lower than 10^7 deg·s^{-1}. The corresponding critical rate of suppression of ordering of the DO$_3$ type is considerably longer and equals 10^3 deg·s^{-1} [63]. At the same time, the experiments carried out by Ya.P. Selisski showed [74] that the disordered state in the Fe$_3$Al alloy (the stoichiometric composition for the DO$_3$ superstructure), produced by heating above 800°C, can be 'fixed' at room temperature by quenching at a cooling rate of approximately 10^4 deg·s^{-1}.

The investigations of the melt-quenched alloys of the Fe–Si system, described in section 1.3, which are quite similar to the Fe–Al alloys from the viewpoint of the relationships governing phase transformations, show that the ordering processes also take place at cooling with ultrahigh rates of 10^6–10^7 deg·s^{-1} [75]. It was also established that the tendency for the suppression of ordering is observed only in the phase transitions of the first kind.

The initial investigations of the Fe–Al alloys, produced by melt quenching or spraying, show that the high cooling rate has a strong effect on the structure and physical properties of the alloys [76–78]. The authors of [76] found a new phase with the L1$_0$ structure after fast quenching in the alloy corresponding to the region B2 on the equilibrium diagram. The phase formed in the grains of the ordered B2 phase in the form of needles or plates. In the sprayed films of the alloys with 6–15 at.% Al, corresponding to the region of the disordered solid solutions, x-ray diffraction examination revealed the DO$_3$ superstructure. At the same time, it should be mentioned that the regions of the composition from 18 to 40 at.% Al have not as yet been investigated from the viewpoint of the effect of melt quenching on the atomic ordering processes. It is important to stress that this concentration range contains the compositions of the alloys which are used widely as magnetic materials: alfer (23.6 at.% Al), alfenol (28.2 at.% Al) and alperm (29.8 at.% Al). Evidently, it is also important to investigate all the special features of the structure in the rapidly quenched Fe–Al alloys which are associated melt quenching (structural defects, special features of the structure of the crystals, the dendritic crystallisation structure, the crystallographic texture).

Figure 1.31 shows the electron diffraction micropatterns corresponding to the alloys in the quenched state indicating the extent of the atomic ordering process. The electron diffraction micropattern of the alloy containing 18 at.% Al shows the intensification of the

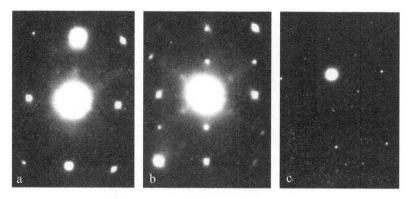

Fig. 1.31. Electron diffraction micropatterns of the rapidly quenched Fe–Al alloys; the axis of the zone $\langle 110 \rangle$: a – Fe–18 at.% Al, b — Fe–22 at.% Al, c – Fe–25 at.% Al.

diffusion background in the region of the inverse space, corresponding to the nodes of the type 111 and 222 of the DO_3 superstructure. The alloy with 22 at.% Al is characterised by the formation of relatively sharp superstructural reflections of the type 222 belonging to both the B2 superstructure and the DO_3 superstructure; in the regions, corresponding to the superstructural reflections of the 111 type and related only to the DO_3 superstructure there are distinctive maxima of diffusion scattering but sharp reflections are also visible. The investigation of the quenched alloys, containing 18–22 at.% Al, by the x-ray diffraction method shows that they are in the disordered state because no superstructural lines are detected on the appropriate x-ray diffraction patterns.

The dark field electron microscopic images, produced in the alloy with 22 at.% Al under the effect of the 222 superstructural reflection can nevertheless be used to evaluate the size of the resultant ordered regions (Fig. 1.32). The size of these regions is 2–4 nm. The stoichiometric alloy Fe_3Al (25 at.% Al) in the quenched state contains both types of superstructural reflections (both even and odd indexes) (Fig. 1.31c), and the size of the regions found in the second type reflection is 2–4 nm, i.e. approximately corresponds to the size of the ordered regions in the alloy with 22 at.% Al. The size of the ordered regions (domains), formed under the effect of the 222 reflection, is 70 nm (Fig. 1.33).

The increase of the aluminium content of the quenched alloys results in a large increase of the size of the domains of this type which are separated by the antiphase boundaries (APB) with the antiphase shift vector $\mathbf{R} = a/2\langle 111 \rangle$. For example, in the alloy with 28 at.% Al this dimension is already 350 nm. At the same time, the

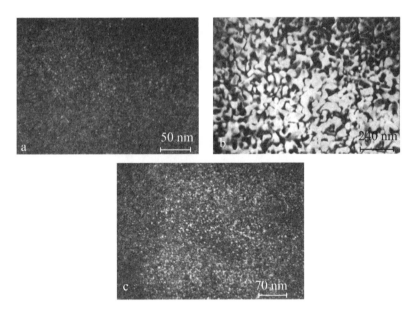

Fig. 1.32. Dark field electron microscopic images of the rapidly quenched alloys Fe–22 at.% Al (a), Fe–25 at.% Al (b, c): a, b – **g** = 222, c – **g** = 111.

size of the domains detected in the 111 reflection (they are separated by the APB with **R** = $a\langle100\rangle$) does not increase at all. In the alloy with 28 at.% Al this size is approximately 4 nm. In transition to the alloy with 31 at.% Al the size of the domains of both types does not change but, at the same time, the intensity of the 111 reflections becomes slightly lower in comparison with the alloy containing 25 or 28 at.% Al. In the alloy with 37 at.% Al there are only thermal APBs with **R** = $a/2\langle111\rangle$, and the appropriate size of the domains is the same as in the alloys with 31 at.% Al.

The systems of the thermal APBs of both types form the structure characteristic of two-domain superstructures indicating the two-stage process of ordering by the DO_3 type. As in the Fe–Si alloys, the sequence of the phase transitions is: A2 (disordered BCC solid solution) → B2 → DO_3.

The thermal APBs show no tendency to form in the specific crystallographic planes. At the same time, the thermal APBs with **R** = $a/2\langle111\rangle$ show the anomalous electron microscopic contrast formed under the effect of main (not superstructural) reflections. An example of such an anomalous banded contrast in the Fe–28 at.% Al alloy is shown in Fig. 1.33. By changing the diffraction conditions it was shown that the contrast is not the consequence of selective etching of the APBs and is of the diffraction origin.

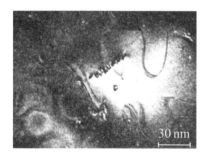

30 nm

Fig. 1.33. Anomalous contrast from the antiphase boundaries in the Fe–28 at.% Al alloy; $\mathbf{g} = 400$.

Detailed analysis of the ordered regions, found in the 111 reflection in the alloys with 25–31 at.% Al, shows that in this case we are concerned with the isolated particles of the DO_3 superstructure, distributed in the matrix ordered in accordance with the type B2. As in [79], the matrix deformation contrast from these regions is observed (Fig. 1.33).

The formation of the anomalous banded contrast from the APBs under the effect of the main reflections (Fig. 1.33) indicates the presence of additional shift $\mathbf{R}$; on the APBs with $\mathbf{R} = a/2\langle 111 \rangle$. Since this effect is not detected in the equilibrium Fe–Al alloys [68], it may be claimed that it is directly associated with the relaxation of very high cooling rates. In all likelihood, the excess vacancies segregate at the thermal APBs and this results in the formation of local displacements in the boundary regions. In principle, the identical effect was detected by the authors of [32] after quenching and annealing of $CuMn_2Al$ alloy, ordered according to the type $L2_1$. Attention should be given to the following special feature: in the equilibrium Fe–Si alloys, local deformation is found almost always on the thermal APBs and disappears only after melt quenching under the effect of excess quenching vacancies, segregated on the APBs.

It can be seen that in the Fe–Al alloys the situation is completely reversed: in the equilibrium alloys that there is no local deformation on the APB but deformation takes place under the effect of excess vacancies. There are two other circumstances confirming the presence of segregation effects on the thermal APBs: 1) transition from the disordered solid solutions to ordered solutions with the variation of aluminium concentration in the rapidly quenched alloys results in a large decrease of the bulk density of the quenched prismatic loops which are a consequence of the existence of vacancy supersaturation in the crystal lattice; 2) heating of the ordered rapidly quenched alloys above the Kurnakov point followed by slow cooling to room temperature leads to the complete disappearance of the local deformation on the thermal APBs.

Heat treatment at temperatures of 300 and 500°C below the temperature of the phase transitions A2 $\leftrightarrow$ B2, A2 $\leftrightarrow$ A2+ DO_3 leads to extensive changes in both the phase composition of the alloys and the morphology of the thermal APBs. In the alloy with 22 at.% Al high-intensity superstructural reflections with both even and odd indexes already appear on the electron diffraction micropatterns after annealing at 300°C. The appropriate dark field images show contacting domains of approximately the same size, ~5 nm (Fig. 1.34).

In the alloys containing 25 and 31 at.% Al, annealing at 300 and 500°C results in a small increase of the effective size of the domains separated by APBs with $\mathbf{R} = a/2\langle 111 \rangle$ (up to 400 nm after annealing for 1 h at 500°C). Annealing at 500°C changes the morphology of the thermal APBs with $\mathbf{R} = a\langle 100 \rangle$. Figure 1.34 shows an example of the process of coalescence of the domains, separated by the APBs of this type, in the Fe–28 at.% Al alloy. It may be seen that two 'phases' seem to coexist within the limits of the same grain: the sections of the initial structure with small domains and sections of the structure with larger domains. It should be stressed than the sections of the 'phase' with the larger domains form initially in the zones situated in the immediate vicinity of the high-angle grain boundaries (see Fig. 1.34). The 'two-phase' domain structure (Fig. 1.34) can form only at different rates of coalescence of the domains in different regions of the grain which in turn may be associated with concentrational heterogeneity or with the non-uniform distribution of the particles of the second phase [80]. It may be assumed that the preferential development of the B2 $\rightarrow$ DO_3 transformation in the boundary regions is determined by the non-uniform distribution of one of the components of the three-component solid solution where one of the components (with the boundary volumes depleted in this component) is represented by the excess vacancies. If the bulk density of the quenching vacancies in the sublattices of the B2 and

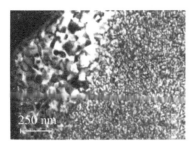

250 nm

Fig. 1.34. Effect of heat treatment on the morphology of antiphase boundaries in the Fe–28 at.% Al alloy (500°C, 1 h), $\mathbf{g} = 111$.

DO_3 superstructures differs, and is higher for B2, the process of the B2 + DO_3→DO_3 phase transition will take place at a higher rate in the regions adjacent to the high-angle boundaries.

In the study [81] it was shown on the example of the A2→B2 phase transition that the largest increase of the vacancy density is obtained at the temperatures close to the Kurnakov point. The vacancy density may increase by several orders. In the alloys containing 25–31 at.% Al, the temperature of the A2 + B2 phase transition is situated in the equilibrium diagram several hundreds of degrees higher than the temperature of the B2→DO_3 phase transition. Therefore, in melt quenching the B2 superstructure will show a considerably higher density of quenched vacancies.

Heat treatment below the appropriate Kurnakov point does not result in the lowering of the intensity of the anomalous banded contrast from the APBs under the effect of main reflections, but annealing above the Kurnakov point, followed by slow cooling to room temperature, greatly weakens or even completely removes the contrast.

Thus, the Fe–Al alloys, containing less than 20 at.% Al, show only the effects of short-range ordering. The pattern of diffusion scattering of the x-rays and the electrons, associated with the short-range order effects in these alloys, have been studied most extensively in [82, 83]. In addition to diffusion scattering, determined by the maxima at the nodes of the reciprocal lattice of the DO_3 superstructure, these alloys are also characterised by the diffusion scattering associated with the displacement of the atoms from the nodes of the ideal BCC lattice. On the basis of the observed pattern of diffusion scattering, the authors of [83] assume that the displacements of the atoms in the Fe–Al alloys, containing approximately 18 at.% Al, are associated with the formation of local atomic configurations in accordance with the anti-omega phase. The detailed analysis of the patterns of diffusion scattering, found in the rapidly quenched alloys with 18–20 at.% Al, shows that it can be interpreted within the framework of the assumptions of the short-range substitution order and the short-range displacement order with the formation of the atomic configurations in the form of the anti-omega phase. At the same time, in the individual cases, the electron microdiffraction patterns show the effects of diffusion scattering not fitting these considerations. Evidently, this is associated with the effect of the high concentration of the non-equilibrium quenching vacancies, capable of playing the role of the third component in the solid solution [37]. For example, the nature of diffusion scattering in the vicinity of the matrix reflections of the type 222, 400 is described more efficiently by the model of displacement of the atoms whose formation is determined, according to the authors of [84], by the interaction of the sublattices of the ordered solid solution with the vacancies.

In the Fe–22 at.% Al alloy, the equilibrium diagram in cooling at lower rates shows the occurrence of the $A2 \rightarrow DO_3 + A2$ phase transition (Fig. 1.30). After melt quenching, the disordered matrix shows the precipitates of the B2 phase with the retention of the effects of short-range substitution in accordance with the type DO_3 with the short-range order of the displacements. Thus, the superfast quenching results in the suppression of the equilibrium phase transition of the first kind $A2 \rightarrow DO_3 + A2$ and stimulates the $A2 \rightarrow A2 \rightarrow B2 + A2$ phase transition which is characteristic of the higher aluminium content.

The stoichiometric Fe–25 at.% Al alloy after melt quenching contains the nanoscale precipitates of the DO_3 phase in the matrix. In other words, the two-phase $B2+DO_2$-phase state forms in this case which is usually not detected on the equilibrium diagram. The morphology of the APBs, formed in ordering, indicates [37] that melt quenching is accompanied by the consecutive phase transformations $A2 \rightarrow B2 \rightarrow B2 + DO_3$. In the alloys, containing 28–31% Al, the phase composition remains the same as in the alloy with 25 at.% Al, and the sequence of the phase transitions is also the same. The volume fraction of the DO_3 phase slightly increases with the increase of the aluminium content, but the lever rule is not fulfilled in this case. In contrast to [37], the study of the alloy with 37 at.% Al showed only the ordered B2 phase.

Heat treatment at 300–500°C results in the transfer of the structure of the rapidly quenched Fe–Al alloys to the equilibrium state in complete agreement with the equilibrium diagram (Fig. 1.30). The alloy with 22 at.% Al is characterised by the $B2 \rightarrow DO_3$ phase transition taking place during annealing leading to the formation of the two-component mixture of the phases: A2 (matrix) + DO_3 (ultrafine precipitates). The fact that the rearrangement of the structure takes place in this sequence and not, for example, in the $A2 \rightarrow DO_3 + A2$ sequence, follows from the dimensional correspondence of the ordered regions in accordance with the type B2 in the quenched state with the region, ordered in accordance with the type DO_3 in the annealed condition. In the alloys containing 25–31% Al, the $B2 + DO_3$ phase composition is retained up to the

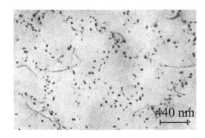

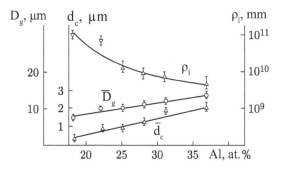

Fig. 1.35. Dislocation loops in the rapidly quenched Fe–25 at.% Al alloy, the bright field, $\mathbf{g} = 220$.

Fig. 1.36. Dependence of the bulk density of the dislocation loops ρ_l, the effective grain size D_g and the cell size d_c on the composition of the Fe–Al alloys.

annealing temperature of 500°C and this is followed by transition to the equilibrium state: $B2 \rightarrow DO_3$.

The spectrum of the investigated compositions of the alloys is characterised, as in the case of the rapidly quenched Fe–Si alloys [27], by the most characteristic feature from the viewpoint of the defects formed in melt quenching: 1) the presence of a high bulk density of the prismatic dislocation loops and the submicropores, whose quantitative characteristics depend greatly on the composition of the alloy; 2) the presence in some sections of the high density of the dislocations subjected to different degrees of relaxation rearrangement, leading to the extensive fragmentation of the initial microcrystals (Fig. 1.35).

The increase of the aluminium content results in a large decrease of the bulk density of the quenching loops (Fig. 1.36). The distribution of the loops and of the submicropores in the grain is non-uniform, and the boundary regions of the microcrystals are enriched with these defects. In addition to this, the distribution of the loops and of the submicropores is of the cellular nature (Fig. 1.35). Prismatic dislocation loops show now distinctive preferential orientation and consist, according to the results of *gb*-analysis, of the dislocations with the Burgers vector $a/2\langle 111 \rangle$ and $a\langle 100 \rangle$. The mean size of the loops in the quenched state away from the grain boundaries for the alloys with 20–22% Al is approximately 25 nm and slightly increases on approach to the boundary volumes and with the increase of the Al content the alloy.

Calculations carried out on the basis of the mean size of the loops and the value of their Burgers vector show [85] that in the Fe–Al alloys the excess vacancies density of the order of 10^{-3}–10^{-6} is quenched; this value depends on the composition. In heat treatment, the internal areas of the grains show a reduction of the dislocation density up to complete removal of defects from the microcrystals at annealing temperatures of the order of 700°C.

The dislocation structure of the rapidly quenched Fe–Al alloys is similar to that the detected previously in the Fe–Si alloys, produced by the same rapid quenching methods [75]. The general tendency for the reduction of the bulk density of the dislocation loops with the increase of the aluminium content is determined, as in the Fe–Si alloys in which the Si concentration was measured, by the reduction of the thermal diffusivity of the material. At the same time, the controlling effect is exerted undoubtedly by the complicated arrangements of the ordered phases in the substitutional solid solution. A similar effect of the superstructures was also reported in the Fe–Si alloys [75] in which the increase of the silicon concentration was accompanied by the transition from the A2 structure to the A2+B2 structure resulting in a large reduction of the bulk density of the prismatic dislocation loops.

The non-uniform distribution of the dislocation loops inside the microcrystals is directly associated [75] with the crystallisation mechanism. In the Fe–Al alloys, crystallisation takes place in the entire investigated concentration range by the dendritic mechanism and the higher density of the loops at the boundaries of the dendritic cells is determined by the 'trap' effect of the free volume in the boundaries of the adjacent dendrites in the processes of movement of the crystallisation front.

In the Fe–Al alloys in the investigated range of the compositions, the crystallisation of the melt at ultrahigh cooling rates, obtained by the spinning method, takes place mainly by the dendritic mechanism without the formation of secondary arms. The increase of the aluminium content results in an increase of the size of the dendritic cells from 0.5 to 2 μm (Fig. 1.36).

The structure of the high-angle grain boundaries in the rapidly quenched Fe–Al alloys is in fact identical with that of the Fe–Si or Fe–Si alloys, produced by the same method [75]. The end sections are characterised by the formation of columnar grains which slightly deviate from the perpendicular line to the surface of the ribbon in the direction of movement of the ribbon on the disc during quenching. In the section, parallel to the ribbon surface, the grains are equiaxed and the effective size of the grains increases with increase of the aluminium content in the alloys (within the range of several microns).

Investigations of the texture, carried out on both sides of the ribbon, show that the crystallographic texture of the alloys is greatly scattered. The texture component (100)[0kl], which is identical for all rapidly quenched alloys, is deflected in the plane through the angle of 10–20°. The dynamics of the variation of the grain size with the increase of the annealing temperature greatly differs from that in the rapidly quenched Fe–Si alloys. The rate of grain growth is lower but is at the same time monotonic. The grain size after annealing for 1 h at 1100°C reaches 40–70 μm which is approximately half the grain size in the Fe–Si alloys. In addition to this, grain growth in the Fe–Al alloys does not result in any increase of the intensity of the crystallographic texture and may even reduce this intensity. In other words, the Fe–Al alloys are characterised by the phenomenon of quasi-secondary recrystallisation, observed in the Fe–Si alloys, and the grain growth processes takes place in the framework of selective recrystallisation. There may be several reasons for the large differences in the nature of the grain growth processes. Evidently, the main reason is the very large scattering of the single (and very weak) component of the (100)[0kl] texture in the Fe–Al alloy. In addition to this, it is possible that the processes of secondary recrystallisation have not progressed sufficiently because of the low density of the quenching loops in the submicropores, playing the role of the inhibitor phase, or also because of their low thermal stability.

1.5. Fe–Ni alloys

The small grain size and the specific special features of the structure of the melt-quenched alloys should obviously have a strong effect on the conditions of martensitic transformations. In a number of investigations [86, 87] it was required to use these methods to produce ultrafine grains in the Fe–Ni alloys and then investigate the effects of the rapidly quenched state of the high-temperature γ-phase on the kinetics of shear transformation and on the morphology of the resultant martensite. This can be carried out most efficiently if the chemical composition of the investigated alloys is selected to ensure that the processes of the formation of the microcrystalline structure of the initial γ-phase in the processes of formation of martensite during cooling takes place in stages: initially, the rapidly quenched structure should form in cooling to room temperature and, subsequently, martensite should form during cooling below room temperature. This would make it possible to analyse in detail the special features of the structure of the initial γ-phase and the effect of these special features on martensitic transformation [88].

Special features of the structure of the low-temperature γ-phase
After melt quenching, the structure of the Fe–Ni alloys with 29, 30 and 32% Ni consists of the microcrystalline γ-phase. Figure 1.37 shows the histograms of distribution of the grains of the initial γ-phase on the basis of the size and the number of the nearest neighbours in the Fe–32% Ni alloy [89]. The histograms for the other alloys were identical. The grain size varied from 0.1 to 12 μm, and the mean grain size varied in the range 2.5–8.7 μm. The number of the nearest neighbours varied from 3 to 12, and the largest number of the grains (30–35% of the total number) was surrounded by five nearest neighbours.

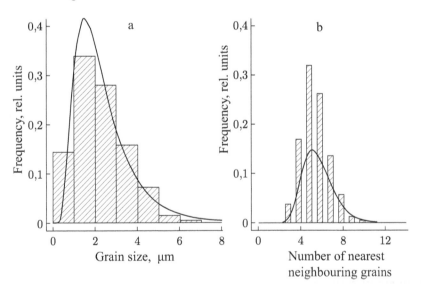

Fig. 1.37. Histogram of the distribution of the size (a) and the number of neighbours (b) of the grains of the γ-phase in the Fe–32% Ni alloy after melt quenching.

The resultant structure is characterised by the presence of dendritic cells with the size from fractions of a micron to several microns having the form close to regular polyhedrons (Fig. 1.38). The cells can be seen more distinctively on the 'free' (not adjacent to the quenching disc) ribbon surface.

The presence of the cellular substructure indicates that in the conditions of melt quenching, the Fe–Ni alloys are characterised by the operation of the dendritic–cellular mechanism of crystallisation and solidification takes place by the movement of the cellular front. Since the cellular nature of the solidification front is determined by the existence of the zone of concentrational supercooling of the melt [45], this indicates some differences in the chemical composition of the cell boundaries where enrichment with the element of the dissolved element should take place, and in the central regions. The nature of electron microscopic contrast of the cell boundaries actually indicates the existence in the rapidly quenched alloys of the chemical heterogeneity on the submicrocrystalline level. As shown in [90], the cell boundaries in the rapidly quenched Fe–Ni (30–40% Ni) alloys are enriched with nickel (by approximately 1% Ni) in comparison with the central zone of the cells.

The rapidly quenched alloys are also characterised by the presence of the dislocation substructure associated with the quenching stresses and the incomplete structure of the grain boundaries (Fig. 1.38). There is a distinctive correlation in the distribution of the cell boundaries and the dislocations: pileup of the dislocations at the cell boundaries so that the dislocation sub-boundaries almost completely coincide with the cell boundaries. Often, there are incomplete low-angle boundaries and also high-angle boundaries containing a high dislocation density.

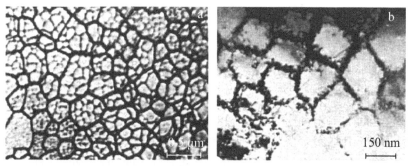

Fig. 1.38. Dendritic cells and quenching defects, segregated at the dendrite cells in the melt-quenched iron–nickel alloys: a) the alloy with 30% Ni, light microscopy; b) alloy with 32% Ni, transmission electron microscopy.

During annealing, the following changes take place in the microstructure of the rapidly quenched alloy. Up to annealing temperatures of 800–900°C there is no large increase in the grain size, and the elements of the cellular structure remain unchanged. A small increase of the grain size is detected after annealing at 950°C, and the rapid grain growth (the grain size increases several times) is detected only after annealing at 1000°C. This is accompanied by the disappearance of the cellular structure indicating the equalisation of the chemical composition of the grain boundaries and in the body of the cells. The dependence of the mean grain size on the annealing temperature is the same for all three investigated alloys and is typical of the rapidly quenched iron-based alloys [85]. It is interesting to note that after high-temperature annealing when the polycrystalline ensemble of the grains becomes evidently considerably more equilibrium, the number of the nearest neighbours remains dominant.

Special features of the course of martensitic transformation
Figure 1.39 shows the effect of the preliminary annealing temperature on the volume fraction of martensite formed in rapid cooling of the investigated rapidly quenched alloys in the regions of the matrix in the immediate vicinity of the contact and free surfaces of the ribbon [86]. In the initial condition (after melt quenching), the volume fraction of martensite on the contact surfaces for all the alloys

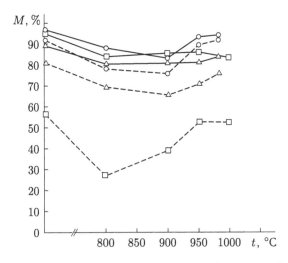

Fig. 1.39. Dependence of the volume fraction of martensite on annealing temperature (1 h) of the alloys with 29% Ni (o), 30% Ni (Δ) and 32% Ni (□). Quenching to martensite at −196°C. The solid line – the contact surface of the ribbon, the broken line – the free surface.

without exception is higher than on the free surface. This difference increases with the increase of the nickel content of the investigated alloys. In addition to this, the volume fraction of martensite on the contact surface changes only slightly with the annealing temperature but changes quite extensively on the free surface. With increase of annealing temperature of the volume fraction of martensite initially decreases and then slowly increases (the minimum amount of martensite corresponds to annealing at 800°C).

Examination of the structure shows that the mean grain size of the initial γ-phase on both surfaces of the ribbon specimens was approximately the same and, in addition to this, the grain size does not change greatly during annealing of the rapidly quenched ribbons up to a temperature of 950°C. As shown by electron microscopic studies, the course of the martensitic transformation was completely the same in the regions adjacent to both the contact and free surfaces of the ribbon (Fig. 1.40). The angular zigzag groups of the martensite crystals on both surfaces of the specimen of the alloys Fe–29% Ni ($M_s = -22$°C), Fe–30% Ni ($M_s = -50 \div -55$°C) and Fe–32% Ni ($M_s = -95$°C) have the same form and are related to the athermal martensite well-known to exist in the coarse-grained-nickel alloys [91]. All the alloys contain partially twinned martensite crystals with straight boundaries, but in the Fe–30% Ni and Fe–29% Ni alloys the twinning was less pronounced. Fe–29% Ni alloy also showed the formation of lath martensite in the regions of the ribbon adjacent to the contact surface. It should be mentioned that in the superfine austenite grains the 'lightning-like' groups of the martensite crystals were distributed

Fig. 1.40. The structure of martensite crystals on the contact (a) and free (b) surface of the ribbon of the rapidly quenched Fe–30% Ni alloy; transmission electron microscopy.

from one grain boundary to another and as a result of the small size of the grains the ratio of the thickness of the plate to the length was close to unity [87]. In addition to this, there were greatly elongated contact planes of the adjacent crystals. Evidently, this increases the relative volume of the martensite phase in every grain, even in small grains, and on approach to the temperature of the finish of the martensitic transformation M_f results in the volume fraction of martensite being almost the same as in the coarse-grained material.

It is well-known that the carbon-free Fe–Ni alloys with 27–33% Ni, quenched from the γ-range, showed the formation of four morphological types of martensite [92]. Cooling to 77 K results in the formation of martensite characteristics of the transformation with the athermal kinetics – the angular zigzag-shaped groups of the plate-shaped partially twinned crystals [93]. Slow cooling below room temperature may result in the formation of lath martensite with the habit plane $\{225\}_A$ by the isothermal kinetics. Usually, the latter forms at temperatures higher than the temperature of the start of athermal transformation [94]. In addition to this, surface martensite forms on the surface of the specimens during electrolytic polishing and its morphology differs from the martensite formed in the volume of the specimens during cooling at temperatures below room temperature [95]. The results indicate that all the rapidly quenched specimens showed the formation of only athermal martensite (with the exception of Fe–29% Ni alloy). During annealing, in the course of equalisation of the composition, the formation of surface or isothermal martensite with the $\{225\}_A$ habit plane was not detected.

The nature of concentration phase separation in the rapidly quenched Fe–Ni alloys

Different dependences of the volume fraction of the transformed volume on the contact and free surfaces of the ribbon specimens in Fig. 1.39 at the same size and kinetics of growth of the austenite grains and also in the case of the identical mechanism of martensitic transformation may be the consequence of either the effect of tensile and compressive quenching stresses or of different contents of the alloying elements on the contact and free surfaces of the ribbon. It is well-known [96] that the external elastic stresses may stimulate the martensitic transformation. Therefore, the partial relaxation of the quenching stresses, formed after melt quenching, in annealing may lead to the reduction of the volume fraction of cooling martensite. Another reason for the reduction of the volume fraction of martensite after annealing at temperatures below 900°C may be the highly non-equilibrium structure of the grain boundaries immediately after melt quenching (Fig. 1.38). These boundaries are less efficient in 'braking' the propagation of the martensitic transformations. Annealing transforms the boundaries to an equilibrium state and the 'transparency' of the boundaries with respect to the propagation of the transformation greatly decreases through the entire volume of the specimens of the alloys.

The main reason for the heterogeneous course of the martensitic transformation in the regions adjacent to the contact and free surfaces of the rapidly quenched ribbons is, according to the authors, the presence of a macrogradient structure determined by the non-uniformity of the chemical composition in the thickness of the ribbon.

The fragment of the equilibrium diagram of the Fe–Ni system, shown in Fig. 1.41, will be investigated. In the process of cooling of the melt from the region L, the alloy containing, for example, 32% Ni, starts to crystallise at a temperature corresponding to the point S in Fig. 1.41. The composition of the first crystals of the γ-phase corresponds to point A. With a further reduction of temperature from point S to point F the volume fraction of the crystalline γ-phase increases, and its composition changes smoothly from the composition corresponding to point A to the composition corresponding to the point B.

If these considerations apply to the case of superfast cooling on the disc-cooler, it may be noted that the crystallisation of the melt, leading to the formation of the ribbon, starts on the contact surface of the melt and the disc (the contact surface of the ribbon) and ends on the opposite surface of the ribbon which is not in contact with the disc (the free surface of the ribbon). If the spinning process

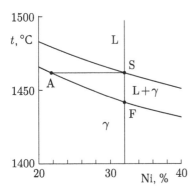

Fig. 1.41. A fragment of the equilibrium diagram of the Fe–Ni system.

is sufficiently slow, the resultant crystals can their composition as a result of the diffusion process in accordance with the curve AF. However, since the quenching rate is very high, the process of diffusion redistribution of the elements is not capable, even at these high solidification temperatures, to ensure the equilibrium concentration of nickel in the rapidly quenched ribbon, especially in the final stage of crystallisation on the free surface when the melt temperature is considerably lower than in the early crystallisation stages. Therefore, we should expect the non-uniform (gradient) distribution of the components of the alloy in the thickness of the rapidly quenched ribbon, especially in the regions adjacent to the free surface. To verify this assumption, the chemical composition of the subsurface regions of the rapidly quenched ribbons was studied by the EDX method [90]. During movement from the contact surface to a thickness of approximately 40–50 µm, the nickel concentration remained approximately constant and then rapidly increased by 1.5–2% in the 10 µm thick layer in the immediate vicinity of the free surface. After annealing to temperatures of 800–900°C, the nickel concentration on both surfaces did not change. Only after annealing at 980°C there was a tendency for the equalisation of the nickel concentration. For example, the nickel concentration on the contact surface in Fe–29% Ni alloy was 29.0%, at a distance of 40–50 µm from the contact surface it was 28.9%, and on the free surface 29.9%.

Figure 1.42 shows the data for the variation of the chemical composition in Fe–32% Ni alloy where the effect of the differences in the volume fraction of martensite on different surfaces was maximum. Since the form of the curves in Fig. 1.39, which contains information on the volume fraction of the resultant martensite for

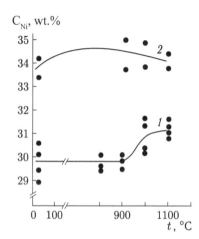

Fig. 1.42. Dependence of the nickel content of the contact (1) and free (2) surfaces of the ribbon of the rapidly quenched alloy with 32% Ni on the annealing temperature (1 h).

the same alloy, and Fig. 1.42 is identical, it may be concluded that the main reason for the occurrence of the detected effect in the alloys produced by melt quenching is the different nickel content in the regions of the alloys adjacent to different surfaces of the ribbon specimens.

Thus, it may be concluded that the process of melt quenching of the Fe–Ni alloys is accompanied by extensive phase separation with respect to nickel and a natural ribbon composite forms.

Effect of melt quenching on the temperature of the start of martensitic transformation

The martensitic point of Fe–32% Ni alloy was measured on different surfaces of the rapidly quenched ribbon specimens [86]. On the contact surface, the martensitic transformations started at –95°C, and on the free surface ($\approx$33.8% Ni) at a considerably lower temperature. To determine whether the martensitic transformation continues on the free surface of the ribbons at temperatures below –196°C (the boiling point of liquid nitrogen), the specimens were cooled in liquid helium to a temperature of –296°C (the boiling point of liquid helium) and this was followed by the construction of the dependences of the volume fraction of the transformed γ-phase on the temperature of preliminary annealing on both surfaces of the ribbon. The identical form of the graphs indicates that the transformation does not continue on the free surface at temperatures lower than 196°C. Comparison of the results with the classic study by O.P. Maksimova et al [97], in which the microcrystalline state was produced by recrystallisation annealing, shows that at the same mean grain size and the nickel content of $\approx$30.3% the value of M_s on the contact side of the ribbon of the Fe–32% Ni alloy in melt quenching is considerably lower than that obtained in [97] (–95°C instead of –50°C). At the same time, the volume fraction of the martensitic phase in the present study proved to be higher (95 instead of 63%). In all likelihood, the difference may be associated with the different methods of formation of the microcrystalline structure of the initial γ-phase. The method of production of the microcrystalline state has a strong effect on the nature of the crystallographic texture and the substructure and also on the special features of the structure of the grain boundaries of the initial phase. The structure of the γ-phase after melt quenching has a number of characteristic features (Fig. 1.38). In particular, as a result of these special features, the martensitic transformation in the rapidly quenched alloys starts at a lower temperature but, after it has started, its rate is considerably higher and this increases the volume fraction of the transformed volume after cooling to a temperature of −196°C.

1.6. Fe–Co alloys

The FeCo alloy with the equiatomic ratio of the components belongs in the group of magnetically soft materials and has the highest Curie point and magnetic saturation induction, and also high values of magnetostriction. The complete realisation of all these properties is associated with considerable difficulties because of the high brittleness of the alloy determined mostly by the formation of the type B2 superstructure [63]. To increase the ductility of the FeCo alloy, it is usually recommended to alloy with vanadium. This slightly reduces the magnetic properties but has a beneficial effect on ductility [98].

It is well-known [63] that cooling of the FeCo alloy is accompanied by two successive phase transformations: the polymorphous transformation $\gamma \rightarrow \alpha$ (FCC→BCC) in the temperature range 960–970°C and the ordering of the BCC solid solution in accordance with the B2 type at a temperature of 730°C. The polymorphous transformation at 'normal' transformation rates takes place to completion, and at room temperature the alloy does not contain any traces of the high-temperature γ-phase. The transition to the ordered state is the phase transition of the second kind and cannot be suppressed by 'conventional' quenching.

Electron microscopic studies show that after melt quenching the alloy does not contain any signs of the martensitic transformation [99]. At the same time, particles of the residual γ-phase were found at the grain boundaries (Fig. 1.43). The volume fraction of the γ-phase is approximately 5% and there is a tendency for a reduction of this value with the increase of the thickness of the quenched ribbon specimens.

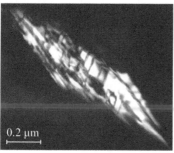

0.2 μm

Fig. 1.43. The dark image of the residual γ-phase in the FeCo alloy produced by melt quenching.

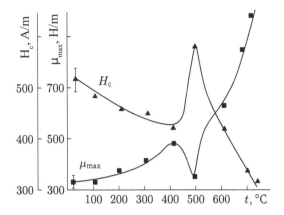

Fig. 1.44. Temperature dependence of the coercive force H_c and the maximum magnetic permittivity μ_{max} for the FeCo microcrystalline alloy.

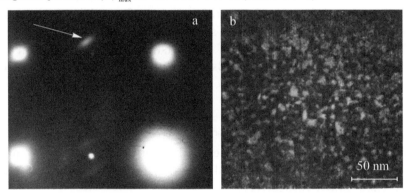

Fig. 1.45. The electron diffraction pattern for the foil orientation (110), where the arrow indicates the additional scattering maximum (a) and the dark field image in the superstructural reflection (b)

The investigations carried out by high-temperature magnetometry show (Fig. 1.44) that at a temperature of 500°C the coercive force greatly increases and the maximum magnetic permittivity decreases. Evidently, this temperature results in the transformation of the residual γ-phase, accompanied by phase hardening. This is confirmed by the fact that the electron microscopic studies of the alloy annealed at 500°C and higher did not show any signs of the existence of the residual γ-phase.

Figure 1.45a shows the electron diffraction micropattern of the FeCo alloy produced by melt quenching. There is a low-intensity superstructural reflection indicating the occurrence of the ordering processes according to the B2 type. Figure 1.45b shows the image of the thermal APBs, obtained under the effect of the superstructural reflection. The morphology of the APBs in this case is typical of the order → disorder transition in the presence of two ordering sublattices. The size of the domains after quenching was 5–10 nm and rapidly increased in isothermal annealing at temperatures higher than 400°C together with the increase of the intensity of the superstructural reflections. As is well-known [30], the intensity of the superstructural reflections cannot be used as a quantitative criterion of the long-range order in the superstructures. In addition to this, the methods of x-ray diffraction analysis cannot be used for the FeCo alloy with similar values of the atomic factors of scattering of the components [63]. For the quantitative evaluation of the degree of ordering by the B2 type, experiments were carried out to determine the degree of splitting of paired superdislocations in the FeCo alloy. Figure 1.46 shows the example of the paired superdislocations formed in the rapidly quenched alloys as a result of slight plastic tensile deformation. In the case of identical types of superdislocations and the plane of the distribution of the shear APBs in the investigated and 'reference' alloys, the degree of ordering can be easily determined from the equation [63]:

$$\eta_{inv} = \eta_{ref} \, (r_{inv}/r_{ref})^{1/2}$$

where η_{inv} and η_{ref} is the degree of the long-range order, r_{inv} and r_{ref} is width of the paired superdislocation for the investigated and 'reference' alloy, respectively.

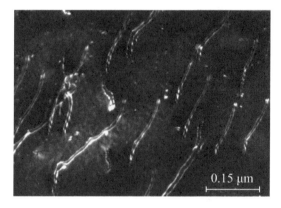

Fig. 1.46. Dislocation structure in the FeCo microcrystalline alloy after slight plastic tensile deformation. The dark field image under the effect of the matrix reflection 110, the low-intensity beam mode.

Statistical analysis of the width of splitting of the paired superdislocations shows that the value η in the alloy produced by melt quenching varies in the range from 0.4–0.5 to 0.75–0.85, depending on the thickness of the ribbon specimens. A larger ribbon thickness corresponds to a higher degree of ordering according to the B2 type. Since the effective cooling rate in spinning of the melt is inversely proportional to the final thickness, in the specimens with the small thickness (higher cooling rate) the degree of ordering is lower than in the specimens with the larger thickness (lower cooling rate). The lowest value of the degree of ordering (0.4–0.5), obtained in melt quenching, is so high that, taking into account previous studies [6, 75], it may be concluded that melt quenching is not capable of suppressing the processes of atomic ordering taking place by the mechanism of phase transformations of the second kind. As regards the polymorphous transformations, it may be mentioned that this transformation also did not change its mechanism and did not acquire any features of the martensitic transformation.

The detailed analysis of the electron diffraction micropatterns of the rapidly quenched alloy shows the existence of the additional maxima of diffusion scattering situated in the nodes of the reciprocal space with the orientation coordinates $2/3\langle 111 \rangle$ and $4/3\langle 111 \rangle$ and slightly displaced from these positions against each other in the direction $\langle 111 \rangle$ (Fig. 1.45a). The dark field images with the maximum diffusion scattering show the characteristic spotty contrast with the individual fragments approximately 5 nm in size (Fig. 1.47). Subsequent annealing did not result in the increase of the intensity of the maximum diffusion scattering nor in the increase of the size of the fragments of the spotty contrast. It is characteristic that no diffusion scattering effects were found in the specimens in which the degree of ordering was minimum but they formed in the specimens in subsequent isothermal annealing.

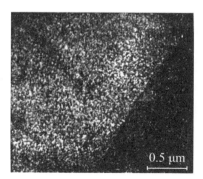

0.5 μm

Fig. 1.47. The dark field image under the effect of the maximum of diffusion scattering of the type $4/3\langle 111 \rangle$. The microcrystalline FeCo alloy, the degree of atomic ordering 0.75–0.85.

The additional maxima of diffusion scattering, recorded by the authors of the book on the electron microdiffraction patterns of the quenched alloys, are in complete agreement with those found previously in the Fe–Al alloys [82]. The authors of [82] showed that these effects are caused by the short-range order of displacements in the distribution of the atoms. In the present case, the diffusion maxima were found only in the melt-quenched specimens in which $\eta = 0.75–0.85$. At the same time, the maxima were not found in the specimens for which $\eta = 0.4–0.5$, but appeared in subsequent annealing. It should be mentioned that the ordered FeCo alloy, produced by the 'conventional' method and having $\eta = 0.75–0.85$, did not show the additional intensity maxima on the electron diffraction patterns. Thus, it is justified to claim that the controlling role in the formation of the short-range order of the displacements in the Fe–Co system is played, on the one side, by the excess vacancies, formed in the solid solution during melt quenching and, on the other side, by the relatively high degree of the long-range order according to the type B2. The absence of at least one of these conditions prevents the effects of statistical displacements from taking place leading to the formation of the diffusion scattering effects on the electron diffraction micropatterns.

In contrast to other ordered iron-based alloys [6, 75], the structure of the FeCo alloy after melt quenching did not contain any prismatic dislocation loops of the quenched origin. There were only individual submicropores distributed mostly at the grain boundaries (Fig. 1.48).

An important special feature of the structure of the rapidly quenched FeCo alloys is the fact that the phase recrystallisation, taking place during fast cooling, changes the nature of the structure formed in the conditions of high cooling rates. In particular, the structure of the FeCo alloy does not contain any prismatic dislocation

0.5 μm

Fig. 1.48. The dark field image of the grain boundary in the FeCo microcrystalline alloy, situated in the foil plane, reflection 200.

loops associated with the supersaturation of the lattice with the quenched vacancies and characterised by the bulk density of 10^{10}–10^{11} mm^{-3} in other iron-based alloys, which do not undergo the polymorphous transformation during quenching [75]. At the same time, the submicropores are not absorbed by the interphase boundaries and are 'inherited' by the structure of the α-phase. The boundaries of the dendritic cells are also 'inherited'; in fact, they represent the regions enriched with one of the components of the alloy. A consequence of the polymorphous transformation is also the unusual shape of the grain boundaries of the α-phase. The interaction of the interphase boundaries during the γ → α transformations with the quenching defects results not only in a large change of the plane of distribution of the grain boundaries but also in a high concentration of the dislocations and submicropores in them. It suffices to say that the rapid grain growth in the FeCo alloy can be 'registered' at annealing temperatures greater than 500°C, whereas in the Fe–Si and Fe–Si–Al alloys this process is detected at temperatures higher than 900°C [6, 75].

The grain size of the FeCo alloy after quenching was approximately 5 μm and did not change up to the annealing temperature of 500°C (Fig. 1.49). This was followed by rapid increase to 20 μm and a subsequent reduction associated with phase recrystallisation. The FeCo microcrystalline alloys are characterised by the grains of arbitrary shapes with a highly developed substructure of different degrees of perfection. In particular, there are flat sections of the grain boundaries, parallel to the crystallographic planes with the low indexes. The investigated alloy is characterised by two special features of the fine structure of the grain boundaries: the presence of 'screw-like' boundaries with 180° rotation of the boundary plane at distances of the order of several microns (Fig. 1.50) and the presence of a high density of defects (dislocations and submicropores) in the

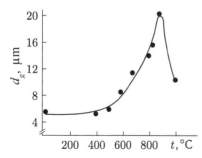

Fig. 1.49. Dependence of the grain size on annealing temperature in the melt-quenched FeCo alloy.

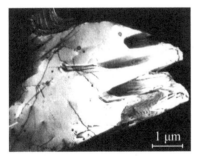

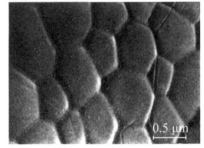

Fig. 1.50. The fine structure of the grain boundaries of the FeCo alloy, subjected to melt quenching. The dark field image in the 110 matrix reflection.

Fig. 1.51. The solidification structure of the FeCo melt-quenched alloy, on the side of the free surface, scanning electron microscopy; the image in secondary electrons.

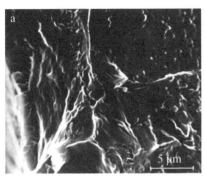

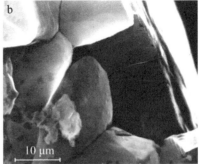

Fig. 1.52. Fracture surface of the rapidly quenched FeCo alloys with the degree of atomic ordering of 0.4–0.5 (a) and 0.75–0.85 (b); scanning electron microscopy; the image in secondary electrons.

boundary plane (Fig. 1.48). Annealing results in a large reduction of the dislocation density.

The free surface of the ribbon specimens examined by scanning electron microscopy shows dendritic crystallisation cells with the size of 0.5–1 µm (Fig. 1.51). The cells disappear after annealing at temperatures above 800°C.

The ductility of the alloy is sufficiently high. The ribbons were wound without problems on a bar with a diameter of 5 mm. Nevertheless, the brittle fracture susceptibility of the rapidly quenched alloys increases with the increase of the degree of ordering and this is characteristic of the 'massive' state produced by the conventional method. Examination of the fracture surface of the specimens, fractured by uniaxial tensile loading at room temperature, showed changes in the nature of fracture from transcrystalline with the traces of the dimpled relief at $\eta = 0.4-0.5$ (Fig. 1.52a) to intercrystalline at $\eta = 0.75-0.85$ (Fig. 1.52b).

It may be seen that melt quenching results in a large increase of the plasticity of the FeCo alloy. The susceptibility to brittle grain boundary fracture increases with increase of the degree of ordering in the melt-quenched specimens. This fact was determined by the authors of the book previously [100] for the FeCo alloy, produced by 'standard' technology of melting and heat treatment. The experiments carried out with melt quenching again confirm the conclusion made in [100] according to which this circumstance is not associated with the susceptibility to the formation of the grain boundary segregations (the impurities are displaced to the dendritic cell boundaries), and is caused by the reduction of the effective energy of the grain boundary APBs.

In addition to the measurement of the temperature dependence of μ_{max} and H_c (see Fig. 1.44), the experiments carried out in [101] were used to determine the values of the saturation induction and magnetostriction at room temperature. These values were 2.36 T and $70 \cdot 10^{-6}$, respectively.

It may be assumed that melt quenching produces a relatively ductile vanadium-free Permendur (FeCo) alloy which cannot be obtained by the 'standard' methods of melting, processing and heat treatment. This should also increase the important service parameter – magnetic saturation induction, since it is well-known that vanadium reduces this parameter. The saturation induction of 2.36 T obtained in the study cannot obviously be regarded as limiting because the preliminary heat treatment is evidently not optimum from the viewpoint of complete removal of the residual γ-phase.

1.7. Ni–Fe–Nb alloys

Alloying the alloys of the Ni–Fe systems with elements such as Nb, Mo, and production of these alloys by melt spinning in the form of microcrystalline ribbons greatly improves their mechanical properties and makes it possible to obtain a higher level of effective magnetic permittivity in the frequency range 0.5–1.0 MHz [102, 103]. The optimum combination of the physical–mechanical properties of these alloys is determined to a large extent by the processes of variation of the phase composition and evolution of the grain structure, taking place under thermal effects.

An isothermal section of the Ni–Fe–Nb ternary metastable diagram was published in [104]. In the vicinity of the composition corresponding to the stoichiometric composition $Ni_3(Fe,Nb)$, in addition to the disordered γ-phase there is also the β-phase with the composition Ni_3Nb ordered in accordance with the DO_{22} type [105, 106]. According to the results obtained in [107–109], in heat treatment in the concentration range similar to the ranges studied previously there are ordered metastable phases: γ' – with the type LI_2 superstructure, and γ'' with the DO_{22} superstructure [105].

Investigations of the rapidly quenched alloys show that high cooling rates may result in the formation of new metastable phases. Electron microscopic studies of the fine structure of the $Ni_{81}Fe_{15}Nb_4$ alloy showed that in the initial condition, the alloy contained mostly equiaxed grains [110] (Fig. 1.53). Quenching defects in the form fine dislocation loops with the size of 10–15 nm and micropores were found in the body of the grains and at the grain boundaries. However, the high density of the quenching defects, which is characteristic of a number of microcrystalline alloys in the initial condition, for example in the alloys based on Fe and Ni [6, 102], is not achieved.

In addition to this, studies of the structure of the alloy found a cellular dislocation structure, especially in the regions close to the free surface of the ribbon. On the contact side of the ribbon there were areas with the fine equiaxed grains with the size of 0.5–1.5 μm [111].

The dendritic–cellular structure is characteristic of the alloy with the higher niobium content: $Ni_{79}Fe_{15}Nb_6$. The mean cell size was approximately 100 nm. The cells may correspond to a network of high-angle boundaries or they may form an independent network. The dislocation density at the boundaries of the dendritic cells was

670 nm

Fig. 1.53. The structure of the rapidly quenched $Ni_{81}Fe_{15}Nb_4$ alloy in the quenched condition. The bright field image.

Fig. 1.54. Dislocations at the boundaries of the dendritic cells of the $Ni_{79}Fe_{15}Nb_6$ alloy in the initial condition. The dark field image, matrix reflection $g = 220$.

higher (Fig. 1.54) and there were also nanoparticles of a new phase. However, electron microscopic analysis does not make it possible to detect reflections from any phases, with the exception of the matrix phases.

The alloy with the maximum Nb content ($Ni_{77}Fe_{15}Nb_8$) contained precipitates of the phase at the cell boundaries and weak reflections on the diffraction patterns which indicate the presence of the $β-Ni_3Nb$ phase in the structure.

The variation of the structure and phase composition of the microcrystalline alloys of the Ni–Fe–Nb system in heat treatment depends on the Nb content. For example, in annealing the $Ni_{81}Fe_{15}Nb_4$ alloy in the temperature range 400–500°C processes of grain fragmentation start to take place and the rate of these processes is especially high after annealing at 700–800°C. The fragment boundaries are represented by the type $a/2$ dislocations [110]. The disorientation of the individual fragments of the grains is ~5° (Fig. 1.55a, b). Equidistant flat clusters of grain boundary dislocations are also found.

Annealing of the $Ni_{79}Fe_{15}Nb_6$ alloy is characterised by a high rate of the process of precipitation of the phases. At temperatures of 600–800°C the metastable phase $γ''$, ordered in accordance with the type DO_2 (Fig. 1.56a) forms in the area of dislocation clusters at the cell boundaries. Its morphology has the form of thin discs (Fig. 1.56b), oriented in relation to the matrix as follows: $(100)_{γ''}||(100)_γ[010]_{γ''}||(010)_γ$. In the same annealing temperature range the $γ$-matrix contains tetrahedral stacking faults (Fig. 1.57) and annealing twins. After heat treatment at 900°C no dislocation clusters

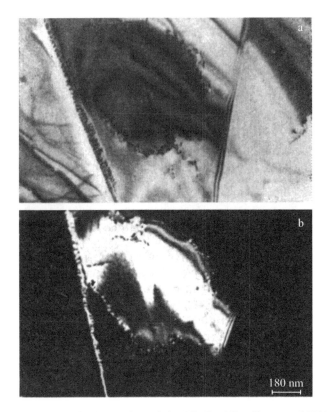

Fig. 1.55. Fragmentation in the grains of the $Ni_{81}Fe_{15}Nb_4$ alloy, $t = 800°C$: a) the dark field; b) the bright field, matrix reflection $\mathbf{g} = 200$.

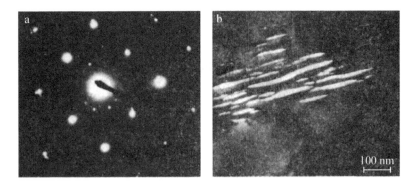

Fig. 1.56. Precipitates of the γ'' phase in the $Ni_{79}Fe_{15}Nb_6$ alloy, $t = 800°C$: a) the electron diffraction pattern, the axis of the zone [100]; b) the dark field, the reflection of the phase $\mathbf{g} = 001$.

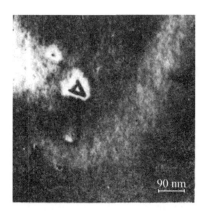

Fig. 1.57. Tetrahedral stacking faults in the γ-matrix of $Ni_{79}Fe_{15}Nb_6$ alloy, $t = 800°C$, dark field, matrix reflection $\mathbf{g} = (220)$.

90 nm

are found in the alloy. The γ''-phase is distributed in the body of the grain in the form of island-shaped colonies. In addition to this phase, the equilibrium β-modification Ni_3Nb with the orthorhombic lattice also precipitates. The lattice parameters of this phase were calculated on the basis of the results of x-ray diffraction analysis and equal $a = 0.5128$ nm, $b = 0.4248$ nm, $c = 0.4530$ nm which, in general, is in agreement with the values obtained for the nickel-based alloys where this phase also precipitates. The β-phase is located mainly at the grain boundaries with the specific orientation in relation to the matrix: $(111)_\gamma \| (010)_\beta [110]_\gamma \| [100]_\beta$. Its morphology may differ depending on the heat treatment temperature. For example, at 800°C this phase usually precipitates in the form of bars or plates, growing from the grain boundaries in the form of a wedge. The base of the wedge is located at the grain boundary. The growth of the β-phase precipitates may also be associated with the migration of the grain boundaries during annealing. In this case, several precipitates with the same orientation have common areas with the grain boundary which assumes a unique tooth-like form (Fig. 1.58). Areas of the matrix free from the γ''-particles are found around the precipitates of the β-phase. With increasing annealing temperature the form of the precipitates of the γ-phase becomes 'coarser'. The precipitates may have the form of large particles of irregular shapes. The γ''-phase may have the form of plates with the internal areas showing the contrast from the stacking faults in the form of transverse bands observed in the reflections of the β-Ni_3Nb phase. The nature of the processes of the precipitation of the γ'' and β phases in the alloy with 8 at.% Nb is the same but the rate of the processes is higher.

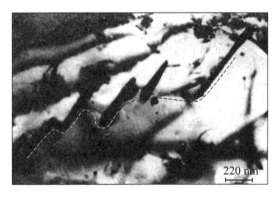

Fig. 1.58. Grain boundary precipitates of β-Ni$_3$Nb in the Ni$_{79}$Fe$_{15}$Nb$_6$ alloy after annealing at t = 800°C, the bright field.

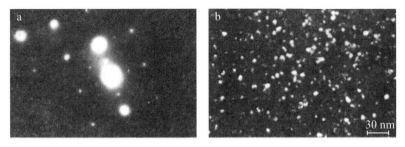

Fig. 1.59. Precipitates of the γ'-phase in the microcrystalline alloys of the Ni–Fe–Nb system: a) electron diffraction pattern of the Ni$_{77}$Fe$_{50}$Nb$_8$ alloys after annealing at t = 900°C. The axis of the zone [100]; b) Ni$_{79}$Fe$_{15}$Nb$_6$ alloy, t = 1100°C, the dark field, the reflection of the phase **g** = 010.

The presence of the γ'' and β phases after high-temperature annealing at 1100°C was not confirmed by the methods of x-ray diffraction analysis and electron microscopic studies in any of the alloys.

In addition to the previously described phases, formed in the course of heat treatment, the dark field images for the alloys Ni$_{79}$Fe$_{15}$Nb$_6$ and Ni$_{77}$Fe$_{15}$Nb$_6$, starting at t = 600°C, show nanoparticles uniformly distributed in the body of the grain and visible in certain superstructural reflections of the γ''-phase. Evidently, the lattice parameters of these phases differ only slightly. More detailed analysis of the resultant diffractions and additional investigations produced the images of the reciprocal lattice, characteristic of the phase ordered in accordance with the type L1$_2$ (Fig. 1.59a). After high-temperature annealing at 1100°C, the mean size of the particles of this phase greatly increased and equalled 5–7 nm (Fig. 1.59b) and at 1200°C the largest particles reached 100 nm. In the alloy with 8 at.% Ni, the superstructural reflections from the γ'-phase on the electron diffraction patterns are characterised by higher intensity, and the mean size of the particles at $t = 1100°C$ is approximately the same as in the alloy with 6 at.% Nb and equals 5–10 nm. As a result of the non-uniform nature of the structure already in the initial condition, in some areas of the ribbon the rate of growth of the particles was higher because there was a tendency for the ordered distribution of the particles on the cube faces.

According to the equilibrium diagram of the Ni–Nb system [112], the solubility of Nb and Ni at temperatures lower than 1000°C is approximately 4 at.%. When the solubility limit of Nb and Ni is exceeded, the stable β-Ni_3Nb phase with the orthorhombic lattice appears. The structure of the ternary alloys of the Ni–Nb–Fe system, produced by the conventional methods, in the vicinity of the stoichiometric composition $Ni_3(Nb, Fe)$ is also characterised by the presence of this phase. However, detailed investigations of the equilibrium diagrams of the alloys of the system in the region of the compositions with a high nickel content have not been carried out. According to the results, in the initial condition the $Ni_{81}Fe_{15}Nb_4$ alloy is a single-phase γ-solid solution based on nickel [102]. After melt quenching the $Ni_{77}Fe_{15}Nb_8$ alloy contains the β-Ni_3Nb phase. Evidently, the fine precipitates on the dislocation clusters in the $Ni_{79}Fe_{15}Nb_6$ alloy are the β-Ni_3Nb phase. Thus, as a result of rapid quenching by the spinning method the supersaturation of the γ-solid solution with niobium in the microcrystalline alloys is not large at a ribbon thickness of 50–70 μm. In heat treatment, the variation of the phase composition of the rapidly quenched alloys has a number of special features. The precipitation of the γ''- and γ'-phases in heat treatment in these alloys precedes the formation of the β-Ni_3Nb phase. The investigated alloys with 6 and 8% Nb showed after annealing fine precipitates of the phase ordered in accordance with the type $L1_2$ which was uniformly distributed in the matrix and as a result of the processes of ordering in the nickel-based γ-solid solution. The rate of these processes was higher in the alloys in heat treatment at 900–1100°C after suppressing the long-range order in the initial state by rapid quenching.

It is well-known that the electron microscopic studies of the Ni_3Fe superstructure (type $L1_2$) in the Ni–Fe alloys are associated with difficulties because of the similar atomic scattering factors of these elements. The observed intensification of the superstructural reflections can evidently be explained by the fact that niobium occupies the positions in the Fe sublattice so that the development of ordering in the γ-matrix can be studied.

In contrast to the γ'-phase, the precipitation of γ'' takes place preferentially by the heterogeneous mechanism. The nuclei of the γ''-phase are found after heat treatment, especially at the dislocations of the type $a/2\langle 110\rangle$. Therefore, the distribution of the particles of the γ'-phase is determined by the distribution of the dislocation clusters within the grain. The dendritic crystallisation mechanism leads to the formation of a higher density of dislocations in the interaxial space of the dendritic cells and to the formation of fragmented grains. The precipitation of the γ''-phase starts first of all in the regions in the immediate vicinity of the cell boundaries. The experimental results show that the morphology of the γ''-phase and its orientation relationships $\gamma''-\gamma$ remain the same as in the stable alloys.

At higher annealing temperatures the alloys with 6 and 8 at.% Ni are characterised by the high rate of the process of formation of the stable β-Ni$_3$Nb phase. It was reported in [106] that there are two mechanism of nucleation of the β-phase in the stable alloys: heterogeneous and homogeneous. Our results confirm that the first nucleation mechanism operates in the investigated alloys, i.e. the β-phase forms mainly at the boundaries of the grains andtwins or at the stacking faults, formed in the particles of the metastable γ''-phase during passage of partial dislocations of the type $C\delta$ through them. In [102] we found that the microcrystalline alloys of the Ni–Nb–Fe system are characterised by the extensive refining of the crystals during heat treatment, with the degree of refinement in other rapidly quenched alloys considerably smaller or not detected at all. On the basis of the results it may be assumed that the observed refining is associated with the intensification of the orientation of the subgrains during formation of the particles of the β-phase at their boundaries.

The upper limit of the existence of the γ''- and β-phases with respect to temperature can be assumed to be $t = 1000°C$. The results obtained in x-ray diffraction electron microscopic analysis showed that the γ''- and β-phases do not form in the alloys with 6 and 8 at.% Nb at 1100°C. The increase of the lattice spacing of the matrix from 0.357 to 0.360 nm in the annealing temperature range 900–1100°C for the Ni$_{77}$Fe$_{15}$Nb$_8$ alloy indicates that during dissolution of the γ''- and β-phases the atoms are transferred to the ordered γ-matrix. The dissolution of the phases is also indirectly indicated by the data on the magnetic properties. For example, the coercive force H_c of the Ni$_{79}$Fe$_{50}$Nb$_6$ alloy rapidly decreases from 220 to 55 A/m, and the peak of the effective magnetic permittivity μ_{ef} at a frequency of 1 MHz appears after annealing at 1000°C. The large increase of microhardness HV in the temperature range 600 and 800°C for this alloy, observed in [104], may be explained by the ordering processes in the γ-matrix with extensive precipitation of the hardening γ″- and β-phases.

The characteristic feature of the phase transformations in the investigated alloys is the acceleration of these processes in comparison with the alloys produced by the conventional methods in which the annealing time at the same temperatures required to complete the breakdown processes is expressed in tens of hours [107–109].

Regardless of the differences in the mechanism of formation of the nuclei of the ordered γ″- and γ′-phases, the rate of formation and growth of the phases in both cases is high. The fact that the density of the quenched defects, characteristic of the rapidly quenched alloys, in the investigated alloys is not high, is evidently explained by the processes of their relaxation and displacement of the areas of excess Nb concentration which is an effective factor accelerating the phase formation and ordering processes in the γ-matrix. The formation of the tetrahedral stacking faults in the matrix during annealing also indicates the relaxation of quenching defects. Thus, the main factor, accelerating the process of phase transformations in the γ–γ′ case, is the excess vacancy density, and in the γ–γ″ case there are a number of additional factors:

–the concentrational heterogeneity, i.e. the higher content of Nb and Fe in the interaxial spaces of the dendritic cell;

–the stresses in melt quenching leading to the development of the grain fragmentation processes.

As regards the mechanism of acceleration of the formation of the β-phase, it can be described as follows. Since in the initial state the β-phase is distributed at the grain boundaries, i.e. in the areas of intensive sink of the quenching defects, in heat treatment as a result of the development of the processes of grain fragmentation and primary recrystallisation the β-phase appears in the areas of formation of the high-angle boundaries. The acceleration of the diffusion processes in the conditions of the excess vacancy concentration also results in the accelerated growth of the phase.

At the same time, it has been reported that autocatalysis is one of the possible models of the formation of the β-phase [106]. In growth of a plate dislocations form in the matrix as a result of the stresses and they dissociate into partial dislocations ($C\delta = A\delta + B\delta$). Passage of the partial dislocation of the type $C\delta$ through the precipitates of the β-phase results in the formation of a stacking fault, and the stacking order ABCABC, characteristic of the DO_{22} lattice, is disrupted and becomes ABAB. This corresponds to the nucleius of the β-phase.

Taking these considerations into account, the sequence of the processes of phase transformations for the $Ni_{79}Fe_{15}Nb_6$ and $Ni_{77}Fe_{15}Nb_8$ alloys can be described by the following scheme:

– stage I (annealing 600–700°C). The formation at the dislocations of the dispersed precipitates of the γ''-phase, the ordering processes in the γ-matrix, fragmentation of the grains;

– stage II (annealing at 800–900°C). The growth of the particles of the γ''-phase, intensive precipitation of the β-phase at the cell boundaries, refining of the grain structure, formation of the recrystallisation nuclei, development of ordering in the matrix;

– stage III (annealing 1000–1100°C). Dissolution of the phases γ'' and β, high rate of the ordering processes in the γ-matrix, growth and orientation of the γ'-particles on the cube planes, increase of the mobility of the high angle boundaries.

1.8. Fe–Cr–Al alloys

The alloys based on the Fe–Cr–Al system are used widely as materials for electric heaters, precision resistors and resistance strain gauges. At the present time, the Fe–Cr–Al three-component system is often used as a base for the development of new alloys with a certain level of the properties. For example, changing the chromium and aluminium concentration, and also introducing new alloying additions, such as gallium, germanium, molybdenum and vanadium, it is possible to produce materials characterised not only by the high specific electrical resistivity and strain sensitivity, but also by the given value of the temperature coefficient of electrical resistivity. A significant shortcoming of all the alloys of the Fe–Cr–Al system (including multi-component alloys) is their susceptibility to embrittlement referred to as 475 brittleness in the literature. At the same time, it is well-known that melt quenching greatly improves the ductility of these low-deformability alloys, including Fe–Si, Fe–Si–Al and Fe–Co [6, 99]. The positive effect of melt quenching on the mechanical properties in these cases is associated mainly with the refining of the grains and formation of the fragmented structure and also with the effect on the atomic ordering processes.

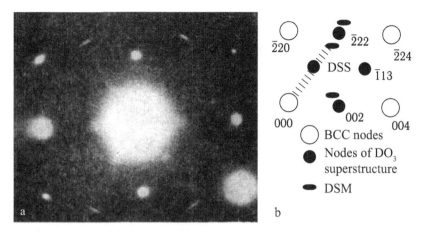

Fig. 1.60. Electron diffraction micropattern of Fe–14% Cr–10% Al alloy (a) and the distribution of the diffusion scattering effects on the pattern (b); the axis of the zone $\langle 110 \rangle$.

Electron microscopic studies of the fine structure of the quenched Fe–Cr–Al alloys [8] showed three additional diffraction effects (Fig. 1.60) on the electron microdiffraction patterns, in addition to the reflections typical of the BCC solid solution: scattering, associated with the formation of the atomic order of the DO_3 type (only in Fe–14% Cr–10% Al alloy); diffusion scattering strands (DSS) in the $\langle 111 \rangle$ direction, passing through the zero reflection and ending in the nodes of the superstructure with the even indexes of the type 222 (only in the Fe–14% Cr–10% Al alloy); maxima of diffusion scattering (MDS), not associated with the main and superstructural reflections of the BCC lattice, distributed in the nodes of the reciprocal space with the orientation coordinates 2/3 (222) and 4/3 (222) and slightly displaced from these positions in the $\langle 111 \rangle$ direction to the node with the indexes 222 [8].

The dark field images, obtained under the effect of the MDS and the 200 superstructural reflection show small regions with the size 2–4 nm (Fig. 1.61). The spotty contrast detected under the effect of the DSS is qualitatively similar to that shown in Fig. 1.61. The size of the glowing regions in this case is almost the same. The variation of the chromium content in the alloys has no effect on the intensity of the MDS and the size of the visible regions in them. Heat treatment also does not cause any significant changes in the overall pattern of diffusion scattering. Only the increase of the intensity of the superstructural reflections 111 and 222 with the increase of the annealing time (470°C, 1 h) is detected and indicates the development in the Fe–14% Cr–10% Al alloy of the

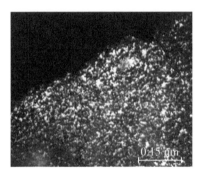

Fig. 1.61. Dark field image in the diffusion scattering maxima; Fe–14% Cr–10% Al alloy.

processes of atomic ordering of the type DO_3. Thus, the structural state, responsible for the appearance of DSS and MDS on the electron diffraction patterns of the quenched materials is characterised by high stability in a wide annealing temperature range.

The additional DSS found on the electron diffraction micropatterns in [8] were detected for the first time because they were not reported in a previous study concerned with the examination of the structure of the Fe–Cr–Al alloy [83]. The distribution of the MDS in the investigated alloys was in agreement with the one published in [82, 83]. The authors of these studies, investigating the scattering of the x-rays by the single crystals of the Fe–Al and Fe–Cr–Al alloys, showed that the formation of the MDS is determined by the static displacement of the atoms from the nodes of the BCC lattice and is associated with the formation in the solid solution of the local atomic configurations of the anti-ω-phase type (the short-range order of displacement). In the present case, the MDS were detected in all alloys irrespective of the chromium content and also of the presence or absence of superstructural reflections, determined by the substitutional atomic order.

The formation of DSS is determined by the simultaneous existence of two types of order in the alloy – displacement and substitution. The absence of any of these orders results in the absence of strands of diffusion scattering on the electron diffraction patterns. It is also evident that there are no reasons to assume that the DSS are the superposition of the maxima of two scattering effects, determined by the order of displacements and substitution since the development of atomic ordering in annealing does not cause any significant changes in the DSS.

The joint application of the scanning and transmission electron microscopy showed a number of special structural features determined

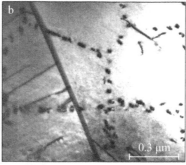

Fig. 1.62. Crystallisation substructure; the image was produced by scanning electron microscopy (a) and transmission electron microscopy in the bright field (b).

by the processes of crystallisation of the melt. The free surface of the ribbon specimens, examined by scanning electron microscopy (SEM), showed crystallisation cells with the form similar to the regular polyhedrons (Fig. 1.62a). Examination of the structure by transmission electron microscopy (TEM) revealed a distinctive correlation in the distribution of the crystal lattice defects of the quenched origin – submicropores and dislocation loops – with the boundaries of the crystallisation cells (Fig. 1.62b). This correlation is found up to annealing temperatures of 700°C.

A similar pattern was previously detected in the microcrystalline alloys of the Fe–Si–Al system [68], but the Fe–Cr–Al alloys showed [8] another interesting special structural feature which may be described as follows. Detailed examination of the nature of the diffraction contrast on the defects showed that in addition to the prismatic dislocation loops of the vacancy origin, resulting in the 'pure' dislocation contrast, the structure of the alloys also contains loops with plate-shaped precipitates inside them. A distinguishing feature of these loops is that the images of the loops in the matrix reflection show the banded contrast of the type of 'displacement bands' [30, 113] (Fig. 1.63a). In a number of cases, it was possible to obtain from the precipitates the primary contrast in the dark field images in the reflection of the phase. The total number of the precipitates in the alloys containing 14% Cr was not large (2–3%) and this caused difficulties in identification. Therefore, the crystal structure of the phase was determined by the methods of x-ray diffraction analysis using Fe–23% Cr–5% Al alloy which contains a large number of plate-shaped precipitates distributed strictly at the boundaries of the crystallisation cells (Fig. 1.63b, c). The detected phase had a body-centered tetragonal lattice with the parameters $a = 0.3096$ nm and $c =$

Fig. 1.63. Dark field (a, b) and bright field (c) images of the precipitates of the Cr₂Al phase in the Fe–14% Cr–10% Al (a) and Fe–23% Cr–5% Al (b, c) alloys: a) in the 220 matrix reflection; b) in the reflection of the phase.

1.0288 nm and the expected composition Cr_2Al. In the alloys, produced by conventional technology, this phase was detected at the grain boundaries after long-term annealing. The authors of [114] reported on the precipitates of Cr_2Al in the microcrystalline alloys of the Fe–Cr–Al system containing a large amount of Cr and Al in comparison with the alloys investigated in this study, respectively 35 and 23 at.%.

The grain size in the ribbon plane after quenching was 2–4 μm and did not change after annealing at temperatures below 600°C. Higher temperatures resulted in an anomalous reduction of the grain size of the Fe–23% Cr–5% Al alloy (Fig. 1.64a). This is associated with the formation of a network of low-angle boundaries coinciding with the boundaries of the crystallisation cells (Fig. 1.64b). The disorientation between the adjacent grains was 1.5–2°. Rapid grain growth took place at annealing temperatures of 700–800°C. It is interesting to note that in the quenched alloys the grain boundaries coincided only rarely with the cell boundaries: in fact, there were two networks of defects with different scales. This may be associated with a strong stimulus to migration and the high mobility of the grain boundaries during cooling from quenching temperature.

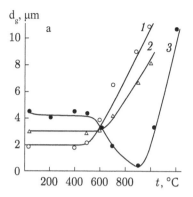
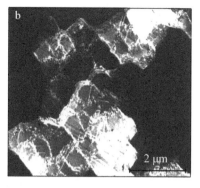

Fig. 1.64. Dependence of the grain size d_g on annealing temperature (a) and the dark field image in the matrix reflection of the dislocation structure of Fe–23% Cr–5% Al alloy after annealing at 900°C, 1 h (b): 1) Fe–14% Cr–5% Al; 2) Fe–14% Cr–10% Al; 3) Fe–23% Cr–5% Al.

It may be concluded that high-speed melt quenching does not suppress the formation of the short-range order of the displacements. In this case, in contrast to the substitutional order, annealing did not result in any significant changes in both the intensity of the MDS and the nature of electron microscopic contrast, detected under the effect of the MDS. This athermal mechanism of formation of the short-range order of the displacements assumes the realisation of different mechanisms, formed by the kinetics of the processes of development of the substitutional order and the displacement order [83]. Atomic ordering takes place by the diffusion mechanism with the increase of the degree of the order in annealing. The short-range order of displacements is formed by the diffusionless mechanism and, therefore, is not affected by the cooling rate and subsequent date treatment. However, in the present study, the short-range order of the displacements in the Fe–14% Cr–10 % Al alloy was also produced by the isothermal mechanism. The results show that severe plastic deformation suppresses both types of ordering. The kinetics of restoration of the substitutional order and the displacement order depends on the annealing temperature and is evidently controlled by a single mechanism – the diffusion mechanism. This is indicated by the fact that the appearance and dynamics of the increase of the intensity of the scattering effects on the electron diffraction patterns take place synchronously. Complete restoration (the initial quenched state) of the scattering pattern of the electrons and of the nature of the diffraction electron microscopic contrast in the superstructural reflection take place after annealing for 1 h at 700°C, which corresponds to the recrystallisation temperature of the alloy or is slightly higher than this temperature. Evidently, the short range order of displacement cannot be regarded as responsible for the increase of electrical resistivity in annealing in the temperature range 200–350°C (see Fig. 1.67). Usually, this temperature instability of the electrical properties is linked with the formation of the K-state. However, as shown by the experiments, annealing of the quenched specimens has no effect on the short-range order of the displacements; on the other hand, restoration of the order after deformation takes place at higher temperatures (500–700°C).

Since the coefficient of distribution in iron for chromium and aluminium is $K > 1$ [115], these elements segregate along the cell boundaries. As a result of the local increase of the concentration of Cr and Al, the thermodynamic probability of the formation of the Cr_2Al phase at the cell boundaries is higher than in the volume. The amount of the phase increases with the increase of the chromium content of the alloy. At the same time, the kinetics of the process of precipitation of Cr_2Al is evidently determined by the presence of nucleation centres (as in the Fe–Cr–Al microcrystalline alloys) of dislocation loops characterised by the preferential distribution of the cell boundaries (Fig. 1.63b). This also explains the observed precipitation of the phase particles at the dislocation loops in the process of cooling of the solid solution. It should be stressed that the detected phase is not metastable and also appears in the 'massive' materials of the same composition after long-term annealing, and the grain boundaries are the areas of nucleation and growth.

Thus, an interesting feature of the Fe–Cr–Al microcrystalline alloys has been found: regardless of the high cooling rate, quenching from the liquid state does not suppress and in fact accelerates the processes of breakdown of the solid solution. It may be assumed that in this case this is caused mainly by the operation of the cellular crystallisation mechanism and also by the high density of the crystal lattice defects of the vacancy origin. In addition to this, the operation of the cellular crystallisation mechanism makes it possible to explain the anomalous reduction of the grain size in annealing of Fe–23% Cr–5% Al alloy (Fig. 1.64a). This is caused by the formation of a network of low-angle boundaries coinciding with the cell boundaries. Evidently, the nature of this phenomenon is associated with the difference in the thermal expansion coefficients of the matrix and the volumes adjacent to the cells and by the formation as a result of the dislocation structure determined by the thermal stresses.

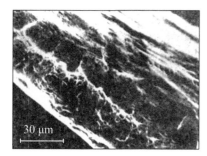

Fig. 1.65. Fractograph of the specimen fractured by uniaxial compression, alloy Fe–14% Cr–10% Al.

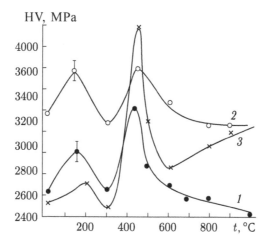

Fig. 1.66. Dependence of microhardness HV on annealing temperature for the alloys Kh14Yu5 (1), Fe–14% Cr–10% Al (2) and Fe–23% Cr–5% Al (3).

All the alloys in the quenched state are characterised by high ductility in bending. Fractographic analysis of the specimens, fractured by the uniaxial tensile loading method, showed the ductile dimpled appearance of the fracture surface. Fracture is preceded by high deformation causing local narrowing of the specimen in the fracture area (Fig. 1.65).

The 475° brittleness, typical of these alloys, was not detected after heat treatment in the given conditions. Even after annealing at 470°C, 10 h (cooling with the furnace) the alloys showed high ductility in bending and ductile failure. The strength properties of the alloys were estimated by the value of microhardness HV. All the compositions are characterised by two maxima and the dependence of HV on the annealing temperature (Fig. 1.66).

The high-temperature peak (400–500°C) is associated with the phase separation of the solid solution with respect to chromium

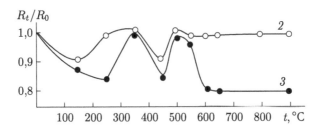

Fig. 1.67. Dependence of electrical resistivity R_t/R_0 on the annealing temperature for Fe–14% Cr–10% Al (2), Fe–23% Cr–5% Al (3) alloy; R_t is the electrical resistivity after annealing at temperatures t, R_0 is the electrical resistivity of the quenched alloy.

which is also detected for the alloys produced by the conventional technology, and the low-temperature peak (150–to 50°C) is associated with the formation of quenching defects. It is characteristic that both hardening peaks are found in the annealing temperature ranges in which the specific electrical resistivity decreases (Fig. 1.67). It is interesting to note that the microhardness of Fe–23% Cr–5% Al alloy increases after annealing at 600°C which in all likelihood is caused by grain refining.

Because of the high ductility properties, the Fe–14% Cr–10% Al microcrystalline alloy could be rolled to a thickness of 5 µm with the total reduction of 85% without intermediate annealing. Investigation of the deformed material by the TEM method showed the formation of a developed dislocation structure ($d_c \approx 0.3$ µm) with a high dislocation density at the cell boundaries.

Attention should be given to the fact that the electron diffraction patterns of the deformed specimens did not contain any of the previously investigated diffusion scattering effects. Evidently, deformation disrupts both the atomic order and also the structural state which results in the formation of the maxima and diffusion scattering strands on the electron diffraction patterns. Annealing at 500–700°C restores the scattering pattern of the electrons and the increase of temperature results in the simultaneous increase of the intensity of both the superstructural reflections and diffusion scattering. After annealing at 700°C for 1 h (this is higher than the recrystallisation temperature), the diffusion scattering pattern becomes the same as in the quenched materials. The dark field images in the superstructural reflections, MDS and DSS in this case are identical with those shown in Fig. 1.61.

The method of high-temperature resistometry was used to measure the temperature dependence of electrical resistivity ($\Delta R/R$) of the

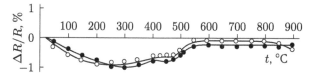

Fig. 1.68. Temperature dependence of electrical resistivity $\Delta R/R$ for the Fe–14% Cr–5% Al alloy, alloyed with Ga, V, Mo, produced by the conventional technology (●) and melt quenched (○).

Fe–14% Cr–5% Al microcrystalline alloy, alloyed with a small amount of gallium, vanadium and molybdenum. The alloy, produced by the conventional methods, has the temperature coefficient of electrical resistivity (TCER) close to the zero in a wide temperature range. The tests showed that the $\Delta R/R$ value of the material after melt quenching is almost identical with that for the conventional alloy, and the microcrystalline material has higher ductility (Fig. 1.68).

1.9. $Ni_3(Mn, V)$ alloys

In [116] it was shown that the mechanism of formation of the long-range atomic order of the type $L1_2$ in the microcrystalline alloys of the $Ni_3Mn–Ni_3Al$ system, produced by melt quenching, is determined by the proximity of the composition of the γ'-phase to the two-phase $(\gamma+\gamma')$ range to the intermetallic compound Ni_3Al, which forms as a result of melt crystallisation. In this case, the values of T_c of the γ'-phase are close to the crystallisation temperature and this ensures the establishment of a high degree of the long-range order S in the alloys produced by melt quenching. In the γ'-phase of the $Ni_3Mn–Ni_3Al$ system, the values of T_c are considerably lower than the crystallisation temperature [117], and, consequently, it is possible to analyse the effect of the cooling rate from the disordered state on the extent of the processes of the long-range order by the $L1_2$ type in the multicomponent metallic systems.

In [118, 119] the possible effect of the method of quenching (from the temperature range of the melt or from the temperature range of existence of the high-temperature disordered γ-phase) on the degree of atomic ordering by the type $L1_2$ in the three-component ordering alloys $Ni_3 (Mn_{1-x}V_x)$, where $x = 0.3$ and 0.4, was investigated. The alloys were selected to ensure that they are in the temperature–concentration range of the phase transition to the ordered state A1 (FCC)→$L1_2$ at slightly different values of the critical temperature T_c. The melt quenching temperature or the temperature of quenching

Table 1.2. The main parameters of the investigated alloys

Specimen No.	Content of elements, at.%			d_m, µm	T_c, K	T_q, K	S	ε, nm
	Ni	Mn	V					
1	75.0	17.5	7.5	151.8	1010	1373	0.44 ± 0.04	4.5
2	75.0	17.5	7.5	1.1	1010	1600	0.58 ± 0.07	5.8
3	75.0	15.0	10.0	145.4	1060	1373	0.64 ± 0.06	6.2
4	75.0	15.0	10.0	1.3	1060	1600	0.80 ± 0.08	9.4

from the solid state in both alloys were identical. The compositions of the investigated alloys and the values of the quenching temperature T_q and the ordering temperatures T_c are presented in Table 1.2.

Halves of the specimens of each alloy were quenched in water from a temperature of 1373 K from the range of the γ-phase (specimens No. 1 and No. 3 in accordance with the notation in Table 1.2). The other half of the specimens was melt-quenched from the temperature of 1600 K in an argon atmosphere (specimens No. 2 and No. 4) by the spinning method. The quantitative evaluation of the degree of the long-range order was carried out by the method of neutron x-ray diffraction analysis in a neutron diffractometer with the wavelength of the neutrons of $\lambda = 0.128$ nm [117]. The application of neutron diffraction analysis for the investigation of the atomic order in this system is based on the favourable ratios of the values of the amplitudes of the nuclear scattering of the neutrons of the components of the alloy ensuring the unique sensitivity of this method.

Figure 1.69 shows the fragments of the neutron diffraction patterns of the specimens No. 1–4 with the superstructural (I^{100} and I^{110}) and main (I^{111} and I^{200}) reflections of the FCC lattice ordered in accordance with the L1$_2$ superstructure. The superstructural reflections I^{100} and I^{110} indicate the presence of the long-range atomic order, as in the specimen No. 1 and No. 3, quenched from the solid phase, and also in the specimens No. 2 and No. 4 melt-quenched from the higher temperatures.

The parameters of the long-range order S were calculated for the Ni$_3$(Mn, V) ternary system in the quasi-binary approximation using the standard equation [120] and the relationships I^{100}/I^{200} and I^{110}/I^{220} of the γ'-phase. It was assumed that the mutual substitution of the atoms of Mn and V takes place in the crystal lattice of the L1$_2$ superstructure while retaining the stoichiometry Ni$_3$(Mn, V) [121]. The mean values of S of the γ'-phase, calculated from the paired ratios $I^{100}/$

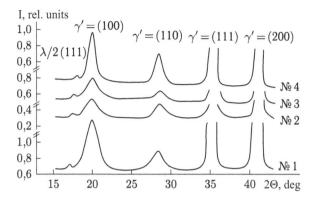

Fig. 1.69. Fragments of the neutron diffraction patterns of the specimens No. 1–4 of the Ni$_3$(Mn, V) alloys, ordered in accordance with the type L1$_2$.

I^{200} and I^{110}/I^{220} are presented in Table 1.2. In addition to this, the values of the mean grain size, measured by metallographic analysis, and of the antiphase ordering domains ε, obtained by the analysis of the half width of the superstructural reflections I^{100} and I^{011}, are also presented.

Detailed analysis of the results, presented in Table 1.2, can be used to make a number of important conclusions:

1. The values of the parameter S of the investigated specimens increase with the increase of the appropriate values of the temperatures T_c and with the approach of these values to the crystallisation temperature;

2. Values of the parameter S for both alloys after melt quenching (specimens No. 2 and No. 4) are higher than after quenching from the solid state and these differences are well outside the range of the measurement error.

If the first conclusion is assumed to be logical and corresponds to the results obtained in [119], the second conclusion should be regarded as paradoxical (the Glezer–Goman'kov paradox in accordance with the terminology proposed in [118]), since the melt quenching rate ($\approx 10^6$ deg/s) greatly exceeds the quenching rate from the solid phase ($\approx 10^3$ deg/s). In other words, it can *apriori* be expected that as the rate of quenching from the disordered state increases the degree of atomic ordering, taking place in the solid phase, should decrease.

In particular, this evident relationship was found when varying the melt quenching rate by changing the speed of rotation of the quenching disc in spinning. The quantitative criterion of the quenching rate from the melt can be evidently represented by the mean grain size, formed in the material [1]. The experimental results show that with the increase of the melt quenching rate and the appropriate reduction of the mean grain size from 3.2 μm to 1.2 μm, the degree of the order S changed from 0.68 to 0.58, i.e. the increase of the melt quenching rate resulted in the evident decrease of the degree of the long-range order.

Thus, the Glezer–Goman'kov paradox in the investigated alloys is evidently associated with the fact that quenching is carried out from different aggregate states: quenching from the liquid state carried out at a considerably higher rate nevertheless stimulates the establishment of the long-range order in the solid solution. This fact can be explained if it is remembered that in melt quenching the very high density of the excess vacancies in the solid solution is retained. The difference between the equilibrium concentration of the vacancies at the melting point and at room temperature in the nickel-based alloys may reach 4–5 orders of magnitude [39]. The excess vacancies can partially leave the crystal to go to the external and internal sinks but most of them remain in the temperature range T_c in the solid solution and greatly accelerate the diffusion processes, including the extent of atomic ordering. In a theoretical study [122], M. Hillert et al proposed a mechanism of capture of the vacancies by the rapidly growing solidification front, with the point defects capable of playing the role of an additional component in the solid solution. Consequently, the mobility of the atoms in the crystal rapidly increases. For example, according to the calculations carried out in [122], the mobility of the copper atoms in melt quenching becomes independent of temperature and equals $M = 5 \cdot 10^{-7}$ m⁴/J·s, which is considerably higher than the mean mobility, calculated without taking the capture of the excess vacancies into account.

On the basis of the electron microscopic studies of the melt-quenched Ni–Mn–V alloys it may be assumed that the crystal lattice in the temperature range of transition to the ordered state contained a very high concentration of point defects accelerating the course of the atomic ordering process.

1.10. Fe–C–Si alloys

It has been shown that melt quenching increases the solubility limit of carbon in austenite, greatly refines the grain structure and results in the formation of metastable crystal and amorphous phases [36]. The authors of [3] analyse in detail the structure and properties of the Fe–3.5% C–2.5% Si eutectic alloy (rapidly quenched cast iron). Depending on the melt quenching conditions, the structure consisted of austenite, metastable HCP phases, martensite and cementite. The highest mechanical properties were shown by the alloy with the structure of non-equilibrium austenite and the metastable ε-phase with the HCP lattices. Light microscopy showed two varieties of the structures: dendritic and lath. In fact, the laths are dendrites with indistinctive second-order axes.

It should be stressed that the structure of the alloy after melt quenching contained two different ε-phases with the HCP structure with different carbon content (2 and 4% C). The former filled the interdendritic space, and the latter was distributed inside the austenite dendrites. Investigations by electron microscopic experiments and computer modelling showed also that the regions of the ε-phase with the size of approximately 10 nm are characterised by the ordering of the carbon atoms with the formation of a superstructure, with the most probable composition of the superstructure corresponding to the formula $Fe_{2.4}C$.

References

1.   Ultra-fast hardening of liquid alloys, Ed. G. Herman, Moscow, Metallurgiya, 1986. -
2.   Metastable and non-equilibrium alloys, Ed. Y. Efimova, Moscow, Metallurgiya, 1988.
3.   Baguzin, S.V., Suyazov, A.V., Crystalline iron alloys obtained by melt quenching, Moscow, Chermetinformatsiya, 1988, Obz. Inform., Ser. Metallurgy and heat treatment, No. 1..
4.   Herlach, D., Galenko, P., Horland-Morits, D., Metastable materials from supercooled melts, Moscow–Izhevsk, Institute of Computer Science, 2010.
5.   Glezer, A.M., Aldokhin, D.V., Amorphous alloys, structure, properties, Applications, Advanced Materials, Tolyatti, MISiS, 2006, 65–88.
6.   Glezer, A.M., Molotilov B.V., Sosnin V., Fiz. Met. Metalloved., 1984, V. 58, No. 2. 370-376.
7.   Zhigalina, O.M., Sosnin, V.V., Glezer, A.M., Fiz. Met. Metalloved., 1993, No. 2, 132-139.
8.   Glezer, A.M., Maleeva I.V., Novoselov N., Fiz. Met. Metalloved., 1990, V. 69, No. 1, 122-130.
9.   Takahashi, M., Nushiro, K., Physica Status Solidi A, 1985, V. 89, No. 1, K27-K29.
10.   Glezer, A.M., Sosnin,V.V., Izv. AN SSSR, Ser. fiz., 1989, V. 53, No. 4, 671-678.
11.   Kozakai, Y., Takabatake, J., Spinodal decomposition in Fe-Mo and Fe-W binary alloys prepared by liquid-quenching, Rapidly Quenched Metals, 4RQM, Proc. of the 4th Intern. Conf., Ed. T.Masumoto, K. Suzuki, Sendai, The Japan Inst. of Metals, 1982, V. 2, 1573-1576.

12.   Inokuti, V., Cantor, B. , Acta Metall. 1982, V. 30, No. 2, 343-356.
13.   Rayment, J. J., Ashiru, O., Cantor, B., The as-quenched microstructures of rapidly solidified Fe-25% Ni, Proc. of Intern. Conf. on Solid to Solid Phase Transformation, Ed. H. I. Aaronson, et al., TMS-AIME, Warrendale, 1982, 1385-1389.
14.   Glezer, A. M., Pankova, M.N., J. Phys. IV. 1995, V. 5, 299-303.
15.   Pushin, V.G., et al., Fiz. Met. Metalloved., 1997, V. 83, No. 6, 149-156.
16.   Pushin, V.G., Kourov, N.I., Kuntsevich, T.I., Matveeva, N.M., Popov, V.V., Fiz. Met. Metalloved., 2001, V. 92, No. 1, 68-74.
17.   Glezer, A.M., Sosnin, V.V., Molotilov, B.V., Pluchek, B.Ya., Structural stability and physico-mechanical properties of microcrystalline alloy Sendast, in: Problems of studying the structure of amorphous metallic alloys, Moscow, MISiS, 1984, 171-175.
18.   Valiev, R.Z., Rossiiskie nanotekhnologii, 2006, V. 1, No. 1-2, 208-216.
19.   Tushinsky, L.I., Classification and evolution of structures in modern materials science, Abstracts Second All-Russian Conference on Nanomaterials, Novosibirosk, IKhTTM Russian Academy of Sciences, 2007, 431-432.
20.   Glezer, A.M., Pozdnyakov, V.A. Zh. Tekhn. Fiz., 1995, V. 21, No. 1, 21-36.
21.   Yagodkin, Yu.D., Lubina, Yu.V., MiTOM, 2009, No. 1, 27-34.
22.   Hadjipanayis, G.C., J. Magn. and Magn Mater., 1999, V. 200, 373-391.
23.   Kekalo, I.B., Vvedensky, V.Yu., Nuzhdin, G.A., Microcrystalline soft magnetic materials, Moscow, MISiS, 1999.
24.   Polishchuk, V.E., Selissky, Ya.P., Ukr. Fiz. Zh., 1969, V. 14, No. 10, 1722–1724.
25.   Chang, Y., Acta Met., 1982, V. 30, 1185–1192.
26.   Arai, K., Tsuya, N., Ohmuri, K., et al., J. Magn. and Magn. Mater., 1980, V. 19, 83–87.
27.   Glezer, A.M., Molotilov, B.V., Fiz. Met. Metalloved., 1973, V. 36, no. 1, 162–168.
28.   Marsinkovsky, M.G., Theory and direct observation of antiphase boundaries and dislocations in the superstructures, Electron microscopy and strength of crystals, Moscow, Metallurgiya, 1968, 212–320.
29.   Saburi, T., Yamauchi, T., Nenno, S., J. Phys. Soc. Japan, 1972, V. 32, No. 3, 694–701.
30.   Hirsch, P., Howie, A., Nicholson, R., Pashley, D., Whelan, M., Electron microscopy of thin crystals, Moscow, Mir, 1968.
31.   Glezer, A.M., Zolotarev, S.N., Molotilov, B.B., Pikus, E.A., Kristallografiya, 1978, V. 23, No. 1, 128–137.
32.   Nesterenko, E.G., Osipenko, I.A., Firstov, S.A., Fiz. Met. Metalloved., 1969, V. 28, No. 6m 987–992.
33.   Gregory, D.P., Acta Met., 1963, V. 11, No. 6, 623–624.
34.   Mikin, G., et al., Vacancy loops in quenched molybdenum, in: Defects in quenched metals, Moscow, Atomizdat, 1969, 131–133.
35.   Shults, G., Quenching vacancies in tungsten, in: Defects in quenched metals, Moscow, Atomizdat, 1969, 58–62.
36.   Honeycombe, R.W., Rapidly quenched crystalline alloys, in: Rapidly quenched metals, Moscow, Metallurgiya, 1983, 58–67.
37.   Glezer, A.M., Molotilov, B.V., Prokoshin, A.F., Sosnin, V., Fiz. Met. Metalloved., 1983, V. 56, No. 4, 750–757.
38.   Glezer, A.M., Molotilov, B.V., Fiz. Met. Metalloved., 1973, V. 35, No. 1, 176–187.
39.   Zasimchuk, I.K., Metallofizika, 1981, V. 3, No. 1, 57–71.
40.   Amelinks, S. Methods of direct observation of dislocations, Academic Press, 1968.
41.   Glezer, A.M., Molotilov, B.V., Izv. AN SSSR, Ser. fiz., 1979, V. 43, No. 7, 1426–

1433.

42.   Glezer, A.M., Molotilov, B.V., Pluchek, B.Ya., Sosnin, V.V., Izv. RAN, Ser. Fiz, 1982, V. 46, No. 4, 701–710.

43.   Arai, K., Tsuya, N., Ohmori, K., IEEE Trans. Magn., 1982, V. Mag-18, No. 6. 1418–1420.

44.   Masumoto, H., Yamamoto, T., On the new alloy Sendust and ternary alloys, containing FeSiAl, and magnetic and electrical properties, Sendust, Tokyo, 1980, 218–236.

45.   Fleming M., Solidification process. Academic Press, 1977.

46.   Glezer, A.M.,The main regularities of the structure and mechanical properties of the alloys of iron with varying degrees of atomic and crystalline ordering, in: Ordering of the atoms and its influence on properties of alloys, Sverdlovsk, IMP Ural Branch of RAS, 1983 Part 1, 12–14.

47.   Miroshnichenko, C., Quenching from the liquid state, Moscow, Metallurgiya, 1982.

48.   Lee, E., Koch, C., Liu, S. Rapidly solidified long range ordered alloys, in: Rapidly solidified amorphous and cryst. alloys, Proc. Mater. Res. Soc. Annu. Meet., Boston, Mass., USA, Nov. 1982, N.Y., 1982, 375–379.

49.   Glezer, A.M.,Nanocrystals, quenched from the melt, structure, properties, and application, Proceedings of the IV All-Russian Conference 'Physical chemistry of ultrafine systems', Moscow, 1998, 27–28.

50.   Wakamiua, M., Norita, Y., Senno, H., Hirota, E., A study on crystallographic textures affected by cold rolling and heat treatment of rapidly quenched high silicon iron alloys, in: Rapidly Quenched Metals 4RQM, Proc. of the 4th Intern. Conf., Ed. T. Masumoto, K. Suzuki. Sendai, The Japan Inst. of Metals, 1982, 1577–1580.

51.   Glezer A. M. Melt-quenched nanocrystals, in: Nanostructured materials, Science and technology, Dordrecht, Kluver Acad. Publish., 1998, 143–162.

52.   Chuistov, K.V., Aging of metal alloys, Kiev, Akademperiodika, 2003.

53.   Glezer, A.M., Molotilov, B.V., Matveev, Yu.A., Zakharov, A.I., Izv. RAN, Ser. fiz., 1979, V. 43, 1415–1421.

54.   Precision alloys, Ed. B.V. Molotilov, Moscow, Metallurgiya, 1983.

55.   Glezer, A.M., Maleeva, I.V., Zakharov, A.I., Atomic ordering and mechanical properties of doped high-silicon iron, Izv. RAN, Ser. fiz., 1985, V. 49, No. 8, 1633–1644.

56.   Glezer, A.M., Molotilov, B.V., Pogosov, V.Z., Izv. RAN, Ser. fiz., 1982, V. 46, No. 4, 696–697.

57.   Arai, K., Tsuya, N., Ohmori, K., Matsuoka, V., IEEE Trans. Mag.. 1982, V. 18, No. 6, 1418–1420.

58.   Tsuya, N., Arai, K., Ohmori, K., Shimanaka, H., Kan, V., IEEE Trans. Mag., 1980, V. 16, No. 5, 728–733.

59.   Swann, P.R., Granas, L., Lehtinen, B., Met. Science, 1975, V. 9. 90–96.

60.   Glezer, A.M., Usikov, M.P., Utevsky, L.M., Analysis of electron images of ordered and heterophase alloys, in: Diffraction electron microscopy in metallurgy, Moscow, Metallurgiya, 1973, 392–429.

61.   Glezer, A.M., Molotilov, B.V., Sosnin, V.V., Fine ordering structure in iron–silicon alloys, in: Atomic and magnetic ordering in precision alloys, Moscow, Metallurgiya, 1985, 21–25.

62.   Enokizono, M., Teshima, N., Narita, K., IEEE Trans. Mag., 1982, V. 18, No. 5. 1007–1013.

63.   Glezer, A.M., Molotilov, B.V., Ordering and deformation of iron alloys, Moscow, Metallurgiya, 1984.

64. Glezer, A.M., Pozdnyakov, V.A., Kirienko, V.I., Zhigalina, O.M., Mater. Sci. Forum, 1996, V. 225–227, 781786.
65. Murakami, K., Shiraishi, A., Okamoto, T., Acta Met., 1983, V. 31., 1417–1424.
66. Tiller, W.A., Solidification, Physical Metallurgy, Ed. R. Kahn, Academic Press, 1968, V. 2, .
67. Glezer, A.M., Defects and phase transformations in melt-quenched nanocrystalline and microcrystalline iron-based alloy, Abstracts of the 11 Intern. conf. on Rapidly Quenched and Metastable Materials, Oxford, Oxford University, 2002.
68. Sosnin, V.V., Glezer, A.M., Molotilov, B.V., Fiz. Met. Metalloved., 1985, No. 3, V. 59, 507–516.
69. Glezer, A.M., Sosnin, V.V., Molotilov, B.V., Izv. AN SSSR, Ser. fiz. 1985, V. 49, No. 8, 1593–1605.
70. Bulycheva, Z.N., Mironov, L., et al., Izv. AN SSSR, 1985, V. 49, No. 8, 1588–1592.
71. Sosnin, V.V., Prospects for the production of electrical anisotropic steel for highfrequency magnetic cores, IN: I.P.Bardin and metals science, Moscow, Metallurgizdat, 2003, 241–245.
72. Reynaud, F., Phys. stat. sol., 1982, V. A72, 1159.
73. Swann, P.R., Duff, W.R., Fisher, R.M., Met. Trans., 1972, V. 3, No. 2, 409–419.
74. Polishchuk, V.E., Selissky, Ya.P., Fiz. Met. Metalloved., 1970, V. 29, No. 5, 1101–1104.
75. Holland-Moritz, D., Schenk, T., et al., Mater. Sci. Engin., 2004, V. A375–377, 98–103.
76. Junqua, N., Grilhe, J., Scr. Met., 1983, V. 17, No. 4, 441–444.
77. Yamashiro, Y., Teshima, N., Narita, K., J. Appl. Phys. 1985, V. 57, No. 8, Part 2B, 4249–4251.
78. Gadieu, F. J., Russak, M.A., Pirich, J. Magn. and Magn. Mater., 1986, V. 54–57. Part 3, 1598–1600.
79. Sagane, H., Oki, K., Eguchi, T., Trans. Japan Inst. Metals, 1977, V. 18, No. 6, 488–496.
80. Krizanowski, J. E., Allen, S.M., Acta Met., 1986, V. 34, 1035–1050.
81. Nakamura, F., Takamura, J., Proc. of V Yamada Conf. on Point defects and defect interactions in metals, Eds. J. Takamura, M. Doyama and M. Kiritani, Kyoto, 1981. 627–633.
82. Vlasov, E.Y., Dyakonov, N.B., Fiz. Met. Metalloved., 1986, V. 61, No. 3, 569–574.
83. Dyakonov, N.B., Vlasova, E.N., Belov, A.A., Gavrilova. A.V., Fiz. Met. Metalloved., 1987, V. 64, 533–539.
84. Naumova, M.M., Semenovskaya, S.V., Fiz. Tverd. Tela, 1971, V. 13, 371–379.
85. Honeycombe, R.W.K., Rapidly quenched crystalline alloys, in: Rapidly quenched metals, translated from English, Moscow, Metallurgiya, 1983, 58–66.
86. Blinov, E.N., Glezer, A., Pankov, M.N., Krotkina, E.L., Fiz. Met. Metalloved., 1999, V. 87, No. 4, 49–54.
87. Blinova, E.N., Glezer, A.M., Pankova, M.N., J. Mater. Sci. Technol., 2000, V. 16, No. 1, 33–36.
88. Blinov, E.N., Glezer, A.M., Zhorin, V.A., D'yakonova, N.B., Izv. RAN, Ser. fiz., 2001, V. 65, No. 10, 1444–1449.
89. Kovalenko, V.V., Glezer, A.M., Blinov, E., Izv. RAN, Ser. fiz., 2003, V. 67, No. 10, 1408–1411.
90. Blinov, E.N., Glezer, A.M., Vestn. Tambovsk. Univ., Ser. Estestv. Tekh. Nauki, 2000, V. 5, No. 2, 163–165.
91. Izotov, V.I., Khandarov, P.A., Fiz. Met. Metalloved., 1972, V. 34, No. 2, 332–337.

92.   Kurdyumov, G., Entin, R.I., Utevsky, L.M., Transformations in iron and steel, Moscow, Nauka, 1977.
93.   Lobodyuk, V.A., Estrin, E.I., Martensitic transformation, Moscow, Fizmatlit, 2009.
94.   Roitburd, A.L., Present state of the theory of martensitic transformations, in: The imperfections of the crystal structure and martensitic transformation, Moscow, Nauka, 1972.
95.   Schastlivtsev, V.M., Rodionov, D.P., et al., Structure and crystallography of lath martensite, Phase and structural transformations in steels, Magnitogorsk, MDP, 2001, Issue 1, 53–71.
96.   Pankova, M.N., Utevsky, L.M., Crystallography of martensitic transformation with small plastic tension and compression of high-nickel steels, in: Martensitic transformations, Kiev, Naukova Dumka, 1978, 79–83.
97.   Maximova, O.P., Zambrzhitsky, V.N., Fiz. Met. Metalloved., 1986, V. 62, No. 5, 974–984.
98.   Glazyrina, M.I., Glezer, A.M., Molotilov, B.V., Fiz. Met. Metalloved., 1983, V. 56, No. 4, 733–740.
99.   Glezer, A.M., Maleeva, I.V., Fiz. Met. Metalloved., 1989, V. 68, No. 1, 170–178.
100. Glezer, A.M., Maleeva, I.V., Fiz. Met. Metalloved., 1988, V. 66, No. 6, 1228– 6
101. Herman, R., Loser, W., et al., Mater. Sci. Engin., 2004, V.A 375–377, 507–512.
102. Sosnin, V., Glezer, A.M., Zhigalina, O.M., Metalloved. Term. Obrab. Met., 1992, No. 3, 28–32.
103. Sadchikov, V.V., Sosnin, V.V., Zhigalina, O.M., A process for preparing microcrystalline ribbons of high hard magnetic alloy of the permalloy grade, Auth. Cert. No. 4943474/02, Bull. izobr. 1991.
104. Savin, V.V., Fiz. Met. Metalloved., 1989, V. 68, No. 1, 143–149.
105. Matveeva, N.M., Kozlov, E.V., Ordered phases in metal systems, Moscow, Nauka, 1989.
106. Sundaraman, M., Muknopadhyay, P., Banerdjee, S., Met. Trans. A, 1988, V. 19A, 453–465.
107. Kirman, I., J. Iron Steel Inst., 1969, V. 207, 1612–1618.
108. Kirman, I., Warrington, D.H., Met. Trans., 1970, V. 1, 2667–2675.
109. Qoist, W.E., Taggart, R., Polonis, D.H., Met. Trans., 1971, V. 2, No. 3, 825–832.
110. Zhigalina, O.M., Sosnin, V.V., Glezer, A.M., Fiz. Met. Metalloved., 1993, No. 2, 132–139.
111. Sosnin, V.V., Glezer, A.M., Zhigalina, O.M., The structure of melt-quenched Ni–Fe-based alloy, Abstracts of the 7th Intern. Conf. RQM, Stockholm, 1990, PC063.
112. Nash, P., Nash, A. The NiNb (niobium–nickel) system, Bull. of Alloy Phase Diagrams, 1986, V. 7, No. 2, 124–130.
113. Electron microscopic images of dislocations and stacking faults, Ed. V.M. Kosevich and L.S. Palatkina, Moscow, Nauka, 1976.
114. Naohara, T., Inoue, A., Minemura, T. et al., Met. Trans. A, 1982, V. 13A, 337–343.
115. Crystallization from the melt, translated from German, I. Bartel, et al., Moscow, Metallurgiya, 1987, 320.
116. Gomankov, V.I., et al., Dokl. RAN, 2004, V. 396, No. 2, 183–186.
117. Gomankov, V.I., et al., Fiz. Met. Metalloved., 1990, No. 10, 80–84.
118. Glezer, A.M., et al., Dokl. RAN, 2006 V. 407, No. 4, 478–480.
119. Gomankov, V.I., et al., Fiz. Met. Metalloved., 2006, V. 102, No. 6, 630–635.
120. Gomankov, V.I., et al., Fiz. Met. Metalloved., 2000, V. 90, No. 4, 91–97.
121. Ruban, A.V., Skriver H.L., Phys. Rev. B, 1997, V. 55, No. 2, 856–874.
122. Hillert, M., Schwind, M., Selleby, M., Acta Mater., 2002, V. 50, No. 12, 3285–3293.

2

Nanocrystals produced by melt quenching at rates close to critical (type II nanocrystals)

The process of transition from the amorphous or liquid to crystalline state can be considered as a transition of the order–disorder type. In principle, this can be done either by heating the amorphous state (see chapter 3), or in the cooling process from the melt at a rate close to the critical. In the first case, the process of crystallization takes place under constant heat input (either at constant or continuously increasing temperature) and the additional effects of heat released during solidification. As a result, the system is characterised in most cases by the formation, at a specific stage of heat treatment, of a structure consisting of two distinctive structural components: amorphous and crystalline [1]. The nature of the structure in this case in particular depends on the rate of heating and cooling, annealing temperature and environment.

A completely different morphology of the structure can be implemented in the early stages of crystallization in rapid cooling of the melt, with effective heat removal from the crystallizing system. Such amorphous–crystalline formations have been studied very little, but the mechanical properties implemented at the same time, can be classified as unique.

Finally, there is another way of forming amorphous–crystalline structures in which dispersed crystalline particles (typically refractory metal carbides) are blown into the 'puddle' of molten metal formed on the quenching disc–cooler. As a result, the amorphised melt and then the solidified amorphous matrix contain particles of the crystalline phase uniformly distributed in the volume

The specifics of the amorphous–crystalline state and the mechanical properties obtained by melt quenching at a rate close to critical, when a sharp reduction in the temperature of the melt allows only the formation of submicroscopic crystals, have been studied very little to date. We return to the results obtained in [2, Introduction Ref. 44] on the Fe–Cr–B alloys.

Fig. 2.1 shows the change of microhardness HV as a result of decreasing the effective cooling rate in melt quenching in the range of the critical cooling rate v_{cr} for the $Fe_{70}Cr_{15}B_{15}$ alloy. At cooling rates close to v_{cr}, a sharp peak corresponding to transition of the alloy to the crystalline state appears. When $v > v_{cr}$ the alloy is in an amorphous state, which has the specific diffuse rings on electron diffraction micropatterns, and at $v < v_{cr}$ it is in the crystalline state formed by several phases. Characteristically, the HV values in all states obtained by melt quenching are significantly higher than the values corresponding to this alloy obtained by 'standard' technology of melting and heat treatment (8 GPa). In the region of the HV(v) curve related to the amorphous state, there is an increase of the strength of the electron microscopic effects due to the presence of regions of high correlation in the position of the atoms (see chapter 3). This also determines a smooth increase of HV with decreasing v closer to v_{cr}.

In the region of the maximum on the HV(v) curve (21–22 GPa), corresponding to the transition from the amorphous to crystalline

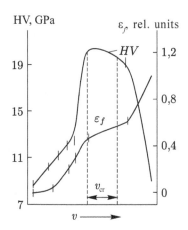

Fig. 2.1. Dependence of microhardness HV and ductility in the bend test ε_f on the effective melt quenching rate for an Fe–Cr–B alloy.

state, in addition to the small volume fraction of the amorphous phases there is also the ultrafine crystalline phase (Fig. 2.2a). Attempts for the identification of the electron diffraction micropattern, similar to that shown in Fig. 2.2a, resulted in the assumption by the authors of the study [introduction Ref. 44] that the structure contains mainly the submicrocrystals with the BCC or similar lattice with the parameter close to 0.285 nm (in the case of the BCC lattice) and the large scatter of the lattice spacing, reaching several percent. The presence of the scatter in the lattice spacing results in the formation of a 'cloud' of the point reflections on the electron diffraction micropatterns (Fig. 2.2a), with the evident variation of the length of the vector of the acting reflection for each system of the reflections of the azimuthally disoriented particles. The mean grain size of the individual crystals was 8–10 nm (minimum size 1–2 nm). The morphology of the submicrocrystals can be clearly seen on the dark field images under the effect of one or several points reflections (Fig. 2.2b). The broadening of the reflections, often detected only electron diffraction micropatterns, and the reduction of the intensity of the primary diffraction contrast in the peripheral areas of the dark field images of the individual crystals indicate that the parameter of the crystal lattice is capable in this case to change not only from a single microcrystal to another one but also from the central part of each crystal to the peripheral part. In addition to this, it is likely that the boundary regions at the very early stages of the formation of the ultrafine structure ($v \approx v_{cr}$) are partially amorphous. In particular, this is indicated by the fact that in the transition

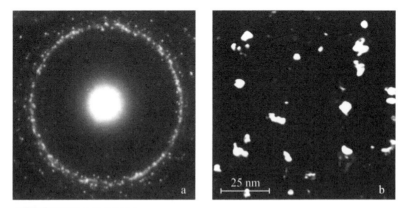

Fig. 2.2. Electron diffraction micropattern (a) and the dark field image (b) in one of the reflections of the first ring, corresponding to the Fe–Cr–B alloy after quenching from the melt at a rate of $v \approx v_{cr}$.

amorphous–crystalline state there is no characteristic banded contrast from the grain boundaries which obviously does not form.

On the basis of these results we can propose the following structural model of the anomalously strong amorphous–crystalline state, produced by melt quenching. The microcrystals, forming a uniform conglomerate, are characterised by the smallest change in the degree of the crystalline order: in the central part of each crystal, formed at a high temperature, there is the ideal crystalline structure which on approach to the periphery (i.e. with a large reduction of the solidification temperature) gradually changes to the amorphous structure (Fig. 2.3). The observed interlayers of the amorphous phases are enriched with the atoms of the metalloid (boron) and show now distinctive boundaries with the crystalline phase. The same structure can be regarded as microcrystalline in which the grain boundaries between the individual crystals are 'smeared' to such an extent that they appear as relatively long areas of the amorphous phases.

Attention should be given to a very important circumstance typical of the investigated amorphous–crystalline state: the boron concentration in the microcrystals is higher than the equilibrium concentration for α-iron. This is indicated by the dependence of the values of coercive force H_c and saturation induction B_s of the $Fe_{70}Cr_{15}B_{15}$ alloy in the quenched state on the effective quenching rate v (Fig. 2.4). The point is that in the amorphous state the alloy is paramagnetic: saturation induction and high values of v is close to 0. The formation of the crystalline phase with the reduction of v results in an increase of both saturation induction and coercive force. The peak of H_c coincides, as shown by the experiments, with the peak of HV, i.e. corresponds to the relevant amorphous crystalline

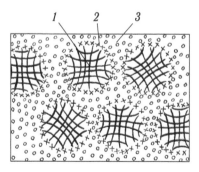

Fig. 2.3. Structural model of the transition amorphous–crystalline state formed in melt cooling with the rate $v \approx v_{cr}$: 1) the region of the microcrystal with the variable parameter of the crystal lattice; 2) the region of transition from the crystalline to amorphous state; 3) the thin interlayers of the amorphous phases.

state. It may be seen that the maximum of H_c corresponds to the value of B_s which is still below the maximum value equal to the saturation induction of α-Fe–Cr. This can be explained only by the fact that in the amorphous–crystalline state the crystals contain a large amount of boron. As a result of the subsequent reduction, the crystals of α-Fe–Cr with no boron form in the structure (saturation induction reaches a maximum value), but the hardness and coercive force, corresponding to such a structure, are no longer optimum.

The results of the static bend tests show [Introduction Ref. 44] that the ductility of the alloy in the transition state is lower than in the amorphous state but at the same time it is far away from the zero ductility (Fig. 2.1). Thus, it may be concluded that the investigated state of the Fe–Cr–B alloys, produced by melt quenching, has not only the uniquely high strength but also sufficiently high plasticity. Investigations of the specifics of the structure of the plastically deformed ultrafine alloys show that the process of plastic yielding has the features typical of deformation of the amorphous alloys. Thus, for example, investigations were carried out into shear bands by transmission scanning electron microscopy (Fig. 2.5). Systems of highly localised bands (the height of the slip steps reached 0.3–0.4 μm) indicated the deformation equal to several hundreds of percent. This is typical of the amorphous materials. Since electron microscopic studies of the shear bands in [2] did not reveal any features of the existence of the dislocations, the authors of [2] assume that the process of plastic deformation in the ultrafine state is localised in the amorphous intercrystalline interlayers and resembles to a certain extent the process of gliding of the grain boundaries.

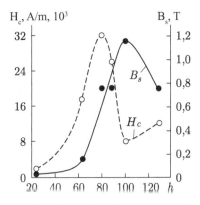

Fig. 2.4. Dependence of coercive force H_c and saturation induction B_s on the effective melt quenching rate (ribbon thickness h).

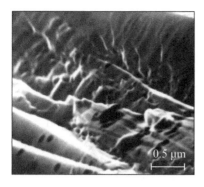

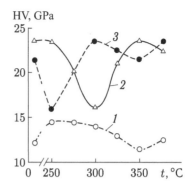

HV, GPa

Fig. 2.5. Slip lines in a bent ribbon specimen of $Fe_{70}Cr_{15}B_{15}$ alloy produced by melt quenching at a rate of $v \approx v_{cr}$. Scanning electron microscopy in secondary electrons.

Fig. 2.6. Dependence of microhardness HV on annealing temperature t for the annealing time of 1 h for the $Fe_{70}Cr_{15}B_{15}$ alloy, produced by melt quenching at a rate of $v>v_{cr}$ (1), $v\approx v_{cr}$ (2) and $v<v_{cr}$ (3).

It is of considerable interest to investigate the question of the thermal stability of the ultrafine state produced by melt quenching. This question is answered to some degree by the dependence of HV on the annealing temperature for the annealing time of 1 h for the $Fe_{70}Cr_{15}B_{15}$ alloy, melt quenched at different rates (Fig. 2.6). Curve 1 describes the alloy quenched at a rate of $v > v_{cr}$ which is consequently in the amorphous condition. The alloy is characterised by a small increase of HV, determined by the low-temperature relaxation hardening which is typical of almost all amorphous alloys of the metal–metalloid type (chapter 3). Curve 2 corresponds to the transitional amorphous–crystalline state produced by melt quenching at a rate of $v \approx v_{cr}$. In this case, at annealing temperatures slightly higher than 200°C the values of HV rapidly decreased with subsequent recovery but already in the absolutely brittle state. As shown by the investigations of the structure, a similar dependence is associated with the simultaneous occurrence of two processes: reduction of the boron concentration in the matrix and high rate of boride formation. The precipitation of boron takes place mostly in the amorphous interlayers at the crystal boundaries and blocks the propagation of plastic shear at the boundaries. Curve 3 corresponds to the possible crystalline state produced by melt quenching at a rate of $v < v_{cr}$. This state is characterised by the lowest stability, although the processes taking place in the alloys during tempering appear to be shifted by the phase in relation to the those which take place in the amorphous–crystalline state. A paradox which requires explanation is that as the initial structure becomes more non-equilibrium, its stability in relation to the thermal effect increases: the amorphous state is characterised by the high stability, and the lowest thermal stability is recorded for the crystalline state.

Summarising, it should be mentioned that there are several reasons for the anomalously high strength and hardness of the amorphous–crystalline state produced by melt quenching and having the structure shown schematically in Fig. 2.3. Firstly, the alloys in this condition are characterised by the very small size of the microcrystals, supersaturated with the boron atoms. Secondly, the amorphous interlayer at the crystal boundaries fully prevents the operation of the dislocation mechanism of transfer of deformation from one crystal to another. Finally, the high boron concentration in the amorphous interlayer creates additional conditions for the realisation of high stresses of the start of plastic yielding.

In addition to this, it should be mentioned that the structural state of the Fe–Cr–B alloys, melt quenched at the critical rate and described in the studies [Introduction Refs. 2, 44] may be realised also in amorphising systems. Actually, such a state, accompanied by the very high strength (σ_p = 6 GPa), was detected recently in Co_4B alloy.

Blowing particles of refractory carbides into an amorphizing melt
Another method of producing the amorphous–crystalline structure and correspondingly very high mechanical properties may be described as follows. In the production of ribbons of the amorphous alloy by the spinning method, a nozzle is placed in the contact zone of the melt with the spinning disc–cooler and is used for blowing disperse particles of a refractory carbide into the 'puddle' of the melt on the disc (2–3 μm). Similar experiments were carried out in [3] where the particles of WC or TiC were blown into the amorphizing melt of $Ni_{75}Si_8B_{17}$ in an inert gas jet (He). This operation resulted in the formation of a ribbon containing the uniformly distributed carb ide particles in the amorphous matrix. The technology of production of similar amorphous–crystalline composites makes it possible not only to vary the type of carbide but also its volume fraction in the structure of the alloy.

The presence of the carbide phase results in a large increase of the Young modulus of the amorphous alloy, and the dependence of this parameter on the volume fraction of the carbide phase is linear (Fig. 2.7).

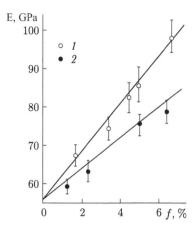

Fig. 2.7. Dependence of the Young modulus of the volume fraction of the particles of the carbide phase in the amorphous matrix of the $Ni_{75}Si_8B_{17}$ alloy. The points correspond to the experimental data, the solid lines to the dependence resulting from the equation (2.1): 1) WC; 2) TiC.

The yield limit of the alloy also increases with increase of the volume fraction of the carbide phase particles in accordance with the rule of additive composition of the Young modulus of the structural components. In accordance with the results published in [3], the dependence has the following form:

$$\sigma_T = \sigma_T^m \left[1 + f \left\{ \left(\frac{E_c}{E_m} \right) - 1 \right\} \right],$$

(2.1)

where E_m and E_c is the Young modulus of the matrix and the carbide particles, respectively; σ_T^m is the yield limit of the amorphous matrix.

Figure 2.8 shows how the value σ_T varies in this case in relation to the value $x = 1/p^{1/2}$ or $x = 1/s^{1/2}$, where p is the mean free path between the particles of the carbide phase; s is the mean distance between the particles. The values of p and s are determined by the following expressions:

$$p = \left(\frac{2d}{3} \right) \frac{1-f}{f}, \quad \frac{s}{d} = 0.6 \left(\frac{2\pi}{3f} \right)^{1/2} - 1,$$

(2.2)

where p is the mean distance from the surface of one particle along the selected direction to the surface of another particle; d is the mean particle diameter; v is the volume fraction of the particles.

It may be seen that the variation of σ_T in relation to the parameter x resembles strongly the situation existing in the crystals in the processes of dispersion hardening. On the basis of this analogy the authors of [4] assume that the effect of hardening of the amorphous matrix can be described by the model of flat clusters.

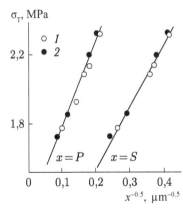

Fig. 2.8. Dependence of the yield limit σ_T on the value of parameters $x = 1/p^{1/2}$ and $x = 1/s^{1/2}$ for the $Ni_{75}Si_8B_{17}$ amorphous alloy, containing carbide phase particles: 1) WC; 2) TiC.

The very brief description of the defects, associated with the simultaneous presence of the amorphous and crystalline phases in the structure leads to a very important consequence: at the optimum combination of the volume fraction, structure and the morphology of the components, the amorphous–crystalline composite may have higher mechanical properties in comparison with the crystalline or even amorphous materials. The problem is only to obtain the optimum combination of strength and ductility using the easily reproducible technology of production of the material and its subsequent heat treatment. Undoubtedly, the production of fully microcrystalline materials through the amorphous state is very useful and highly promising. A macroscopically homogeneous ultrafine structure may form in this case on the structure which has, under specific conditions, not only high strength but also high ductility and in some cases shows the superplastic effects. Undoubtedly, this direction of structural metals science is only in the initial stage of development and offers great possibilities for producing alloys with high mechanical properties.

References

1.   Skakov Yu.A., Phase transitions in heating and isothermal holding in metallic glasses, Itogi Nauki i Tekhniki, Metalloved. Term. Obrab., VINITI, Moscow, 1987, vol. 21, 53–96.
2.   Glezer A.M., Chicherin Yu.E., Structural model and mechanism of plastic yielding of ultrafine alloys produced by melt quenching, in: Problems of investigation of the structure of amorphous alloys, Moscow Institute of Steel and Alloys, Moscow, 1988, 455–456.
3.   Ziclinski P.G., Aat D.G., Aota Met., vol. 32, No. 3, 1984, 397–405

Nanocrystals produced by controlled annealing of the amorphous state (type III nanocrystals)

This method of production of the nanocrystalline structures using melt quenching is the most widely used method and, consequently, the crystallisation processes, taking place as a result of the thermal effect on the amorphous state, have been studied most extensively. The nanocrystals form in two stages: initially, melt quenching produces the amorphous state and, subsequently, annealing under special conditions stimulates the processes of crystallisation leading to the formation of nanocrystals. In heating the amorphous state, the crystallisation process takes place as a first order phase transition in the conditions of constantly supply (either at a constant or continuously increasing temperature) and under the additional effect of heat generated in the crystallisation process. Consequently, structures consisting of two distinctive structural components: amorphous and crystalline [chapter 2, Ref. 1], form in a certain stage of heat treatment. Single-phase multiphase nanocrystalline states of the material can form in later stages of heat treatment. The nature of the structure depends strongly on the heating and subsequent cooling rate and on the annealing temperature and medium.

We should initially simulate the situations associated with partial transition under the effect of the temperature of the highly non-equilibrium amorphous state to a thermodynamically more stable state. The amorphous–nanocrystalline and nanocrystalline states formed in this case may show the qualitatively new level of the physical–mechanical properties. For example, the amorphous state is characterised by the low values of the elasticity modulus [1], and evidently the partial crystallisation greatly increases the elasticity modulus and creates suitable conditions for more efficient application of amorphous alloys in the cases in which the material should have not only high strength and ductility but also sufficiently high elasticity moduli. Evidently, important factors in this case are the melt quenching and subsequent annealing conditions resulting in the formation of the nanocrystalline phase because these parameters determine the morphology, phase composition and amount of the structural components in the amorphous–nanocrystalline and nanocrystalline conditions.

The amorphous state, formed in melt quenching, is itself of considerable scientific importance because it has unique properties [1]. However, in this book we restrict ourselves only to the aspects of the amorphous state which are important in investigating this state as the precursor for the formation of nanocrystalline phases. It should be remembered that in accordance with the classification discussing the introduction, the amorphous state itself can be classified in the group of nanocluster materials.

3.1. The amorphous state of the solid and its genetic relationship with nanocrystals

3.1.1. Amorphous metallic alloys

The amorphous state of the solid is one of the least investigated areas of the physics of the condensed state. The amorphous state can be defined as the state with the absence of a long-range order (absence of correlations between the atoms at large distances) whilst retaining the short-range order (the presence of such correlations in several (no more than two or three) coordination spheres [2]. Subsequently, we will discuss mainly the so-called metallic glasses, i.e. amorphous metals or alloys produced by melt supercooling. In this case, the amorphous state of the solid reflects efficiently the structure of the liquid, and the description of its structure should be based on taking into account the fluctuations of density, the local environment and chemical composition. This introduces the probability and statistical aspects into the description of the structure.

The concept of the similarity of the liquid and the condensed amorphous state was introduced by Ya.I. Frenkel' [3] who assumed that the melting process can be characterised as amorphization with no changes in hardness. The latter means that the purely elastic behaviour of a substance in accordance with the principle of the solid or purely ductile behaviour of a substance in accordance with the liquid does not result from its natural properties and is of a relative nature, which depends on the external loading rate. In principle, there are two limiting cases of the behaviour of the solid under external loading. In the first case, which corresponds to the ideal solid, deformation is proportional to applied stress. The second limiting case is characterised by the viscous liquid for which the rate of deformation is equal to the applied load divided by the viscosity coefficient. The amorphous solids, including metallic solids, are neither ideally elastic nor ideally viscous and combine the elastic and viscous properties. This means that the total deformation of the amorphous solid appears to consist of two parts: elastic and viscous, which Ya.I. Frenkel' termed 'solid' and 'liquid', respectively. The amorphous alloys, which were of course unknown to him, have confirmed completely the productivity of a similar 'dual' approach to the mechanical behaviour of the amorphous solids.

An important special feature, typical of the structure of all amorphous alloys without exception, is that the atomic ensemble has a specific short-range order. If the topological short-range order, describing the degree of local ordering in accordance with the type of crystal, has no analogue in the conventional crystals, then the chemical (composite) short-range order, describing the tendency of the atoms to surround themselves by the atoms of a specific type, is very similar to that which exists almost always in the multicomponent crystals. In addition to this, the methods of quantitative description of the composite short-range order in the amorphous systems [4] and short-range order in the crystals [5] are identical in principle. Therefore, an interesting situation forms in this case: the structural state, which appears to exist on different bands of the atomic–crystalline order, can be described by the general relationships, and the assumptions characteristic of the short-range atomic order in the ordered crystals can be efficiently applied for describing the atomic correlation in the disordered systems.

Although the existence of a very distinctive chemical short-range order in the amorphous alloys has been clearly confirmed, its quantitative characteristics are very difficult to determine. In [6], the authors propose the determination of the order parameter η_{AB}, based on the partial coordination number Z_{ij}, and

$$Z_A = Z_{AA} + Z_{AB} \neq Z_B = Z_{BB} + Z_{BA} \qquad (3.1)$$

where $Z_{AB} = (x_B/x_A)Z_{BA}$; x_i is the fraction of the atoms of the corresponding type.

Taking into account (3.1), the expression for the degree of the short-range composite order has the form

$$\eta_{AB} = \frac{Z_{AB}(x_A Z_A + x_B Z_B)}{x_B Z_A Z_B} - 1. \qquad (3.2)$$

It may easily be seen that the equation (3.2) is identical to the classic equation for the determination of the short-range atomic order in the crystalline and binary systems [5]. For the completely disordered amorphous alloys $\eta = 0$. The values $\eta < 0$ and $\eta > 0$ corresponds to the tendency for phase separation and ordering, respectively. The maximum value of η_{AB} has not as yet been calculated, but the determination of this parameter of the short-range composite order in the experiments and in the models for which the values of the partial coordination numbers are available, shows that certain alloys are completely ordered [7].

The problem of the topological short-range order in the amorphous alloys is far more complicated than the problem of the composite (chemical) ordering because the selection of the order parameter is associated with difficulties. The topological order in the amorphous alloys is exclusively polytetrahedral. This order cannot be compared with the short-range order spatially de-concentrated in the three measurements which exists in the crystals characterised by translational symmetry. Nevertheless, it has been shown [8] that a similar complete polytetrahedral packing is possible in the distorted three-dimensional space (i.e. on the surface of the four-dimensional polytype). To image this structure in the three-dimensional space, it is necessary to introduce defects. Defects are represented by, for example, rows of dislocation lines.

A number of structural models of amorphous alloys has been proposed so far and they can be divided into two large groups [9]: the first group of the models is based on the quasi-liquid description of the structure using a continuous network of the randomly distributed close-packed atoms; the second group of the models is based on the description of the structure by means of the crystals containing a high density of defects of different type (in particular, intergranular boundaries).

The distribution of the atoms in the amorphous state can be described by the atomic function of the radial distribution. A partial structural factor is also introduced; this factor is the Fourier transformant of the partial atomic function of the paired distributions. The latter describes a specific number of atoms of j-th type in the unit volume at distance r from the atom of the type i. In the n-component system there are $n(n + 1)/2$ partial distribution functions. For example, in a binary amorphous alloy there are three partial functions of the paired distributions and, correspondingly, three partial structural factors which can be determined as a result of three independent diffraction experiments (for example by EXAFS spectroscopy) [10].

Initially, the structure of the one-component amorphous systems using the 'quasi-liquid' model was described by the Bernal model proposed at the time for the description of the structure of simple liquids.

This model is based on the random dense packing of rigid spheres. However, the methods of computer modelling (either the methods of successive annexations or the methods of collective rearrangement), actively used for the formation of the quasi-liquid structure, did not make it possible to produce the structure of randomly distributed close-packed rigid spheres of the same density as that detected in the experiments. Further modernisation of the model (application of the 'soft' spheres governed by the Lennard–Jones paired interatomic potential, instead of the 'rigid' spheres) resulted in considerable improvement of the correspondence between theory and experiment. In addition to Bernal polyhedrons, the application of Voronoi polyhedrons also proved to be productive. It is not yet known whether the specific type of polyhedron is linked by an unambiguous relationship with the type selected for the description of the atomic structure. In addition to this, any displacement of the atoms or distortion of local areas of the structure results in changes in the type of Voronoi polyhedrons [9]. In the final analysis, the amorphous one-component structure should be investigated in the form of the ensemble of distorted octahedrons and tetrahedrons, existing in a simple close-packed structure; the sequence of alternation of the configurations of the tetrahedrons and octahedrons can again be described within the framework of the paired correlation functions [10].

The attempts made to solve the problem of two-component amorphous systems using the quasi-liquid model led the researchers to the definite conclusion regarding the correctness of stereochemical considerations, presented for the first time in [11]. The stereochemical approach is based on describing the amorphous structure by specific structural elements consisting of the central atom A and the atoms B surrounding the central atom and forming together some coordination cell (for example, in the form of a trigonal prism) (Fig. 3.1). The most accurate description of the structure can be obtained using a coordination cell whose symmetry is identical to that which forms in the crystalline phases, formed in the same binary system. It is natural to assume that the structural elements existed in the initial melt in the form of associates and in supercooling of the melt were inherited by the metallic glass.

Here it is necessary to mention that the previously described structural elements, correctly describing the structure of amorphous alloys, were referred to as 'clusters'. In this case, this name is not completely correct. In fact, the Physical Encyclopaedia [2, volume 2] describes this term as follows: 'the cluster – a system consisting of a large number of weakly bonded atoms or molecules. The clusters occupy an intermediate position between the van-der-Waals molecules, containing several atoms or molecules, and fine dispersion particles'. It can be seen that in the present case the structural element contains a small number (no more than 10) of the atoms distributed into two or three coordination spheres and

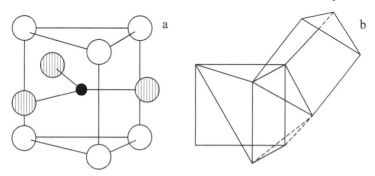

Fig. 3.1. Trigonal–prismatic nanocluster used for describing the structure of a two-component amorphous alloy of the metal–metalloid type (a): the black circle – the metalloid atom; open circles – metal atoms; circles with crosshatching – metal atoms distributed in the second coordination sphere; the scheme of 'coupling' of nanoclusters (b).

characterised by very strong (metallic and covalent) interatomic bonds. In this connection, the cluster models of the structure of the binary (and multicomponent) amorphous metallic alloys determine the presence in the structure of metallic nanoclusters with the metallic and covalent nature of the interatomic bonds and the nanoscaled dimensions (2–3 coordination spheres, i.e. less than 10 nm).

For the majority of the amorphous systems of the metal–metal type and, in particular, the metal–metalloid type, the nanocluster model provides a correct description of the structure. In this case, the binary alloys of different composition are often regarded as the 'two-phase mixture' of the close-packed regions of the pure metal and the regions with the close-packed structure typical of the basic metallic nanocluster. In more complicated cases (for example, for multicomponent systems) we can use the polycluster models [12] in which the amorphous matrix is formed by several types of nanoclusters each of which is locally ordered and separated from the neighbours by intercluster boundaries represented by flat defects and consisting of two-dimensional monolayers with imperfect local ordering of the atoms.

We now describe briefly another approach to the description of the structure of the amorphous state – the pseudocrystalline model. It can be said with a high degree of reliability that a defective crystal, containing dislocations, even if the dislocation density exceeds 10^{12} cm^{-2}, is not adequate to the structure of metallic glass [13]. This does not apply if these defects are represented by disclinations. The regular filling of the crystal with these defects is capable of transferring the crystal to a structural state close to amorphous. The disclination theory shows [14] that dislocations cause only Cartan torsion in the crystal without changing its metrics. However, if a crystal contains disclinations, the latter have the controlling effect on its metrics, i.e. the crystal should be treated as a crystal in space characterised by the Riemann–Christoffel curvature as a function of the tensor of disclination density, with the disclinations being the linear defects of such a structure. A similar method of construction of the models of the structure of the amorphous alloys from polytypes in the distorted space has been used widely, especially for calculating the electronic properties. For example, it was shown [15] that the close-packed non-crystalline structure can be produced by displaying the polytype figures from the distorted space in the Euclidean space. This imaging is achieved by introducing a network of disclination lines, converting the space curvature to 0. As an example, the structures generated by imaging the polytype {3, 3, 5} on the three-dimensional sphere in the Euclidean space were investigated in [15]. The structures obtained in this case can be classified as networks of disclination lines. As regards the metallic glasses, produced by melt quenching, the disclination model cannot be used in this case because the mechanism of formation of crystals with a very high disclination density in the process of superfast cooling of the liquid phase is not known. Evidently, the disclination considerations are suitable for describing the amorphous state formed as a result of high plastic deformation of the crystalline intermetallics [16].

Although the ensemble of the randomly oriented microcrystals or nanocrystals has no translational symmetry over large distances, it has been shown that its radial distribution function differs in principle from the identical characteristic of the amorphous state. The fact that the microcrystalline model cannot be used in this case reflects the fundamental difference in the nature of the topological short-range order of the amorphous and crystalline phases: the polytetrahedral in the first case and the crystalline (with the elements of translational symmetry) in the second case. At the same time, the microcrystalline approaches to the description of the amorphous state are very tenacious. This is associated primarily with the fact that the x-ray diffraction diagrams of the nanocrystalline objects are very similar to those of the amorphous alloys. Even the term 'x-ray amorphous state' has been proposed, i.e. the state regarded as amorphous on the x-ray diffraction level but in principle is microcrystalline or nanocrystalline. Additional confirmation of the model has been obtained by electron microscopy experiments (of course, not always accurate) in the regime of direct resolution of the atomic structure. Considerable methodical difficulties, arising in visualisation of the crystallographic planes and individual atoms as a result of the formation of the electron microscopic phase contrast, can be overcome only by the application of very thin (of the order of 1–5 nm) objects and by carrying out parallel computing experiments of the resultant images under the identical conditions of electron scattering [17]. Figure 3.2 shows the image of the atomic structure of $Fe_{74}B_{26}$ metallic glass in the direct resolution mode [18]. According to the author [18], the structure of the alloy can be described by the ensemble of nanocrystals with the size of approximately 1 nm. Later, other studies were published in which the authors describe the structure of metallic glasses either by nanocrystals or nanocrystals distributed in a liquid-like amorphous matrix [19].

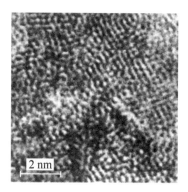

2 nm

Fig. 3.2. Electron microscopic image of the atomic structure of $Fe_{74}B_{26}$ in the direct resolution mode.

Here, it is necessary to make another important detour and say that the pattern, formed in the description of the structure of metallic glasses, is surprisingly similar to the situation formed in the description of the structure of the short-range atomic order in certain ordering solid solutions. As an example we mention the so-called K-state in the Fe–Al, Ni–Mo and Ni–Cr alloys, well known in the literature, in which the anomalies of the short-range order could be explained only on the basis of the structural model of the local long-range order: the ensemble of the stable microdomains (more accurately, nanodomains) of the long-range order, distributed in the disordered crystalline matrix [20]. Another type of the short-range atomic order is a homogeneous long-range order over the entire crystal with a very small (nanoscale) size of the domains and correspondingly high volume density of the anti-phase boundaries [5] which is evidently the complete analogy of the nanocrystalline model of the amorphous state.

3.1.2. Criteria of transition from the amorphous to nanocrystalline state

Obviously, there is a purely methodological aspect of the analysis of the submicroscopic structure of the amorphous state. It is associated in particular with the current confusion in the scientific literature regarding the terms 'nanocrystal' and the 'cluster' (or, more accurately, 'nanocluster'). As mentioned previously, the nanoclusters are the inseparable structural elements of the liquid-like model of the structure of metallic glasses. Their size can be comparable with the size of the nanocrystals used for describing the structure of metallic glasses using the quasicrystalline model. However, there is a principal difference: the nanocrystals are always characterised by the translational elements of symmetry. Usually, the nanoclusters do not have these elements and genetically determine the non-crystalline symmetry of the amorphous state. There is another methodological question: what is the limit to which the size of the nanocrystal can be reduced when describing them using the structure of the atomic ensemble?

We believe, that there is a precise criterion which determines the lower limit of the length of the nanocrystal (as is well-known, the upper limit is 100 nm [introduction Ref. 4]).

When the size of the nanocrystal, characterised by the strict set of the symmetry elements, is reduced, a moment arises when the reduction of the size of the crystal results in the loss of certain symmetry elements. This is the critical size of the crystal at which the crystal still retains the symmetry elements typical of the given type of crystal, and should be regarded as the lower limit of the nanocrystal size. For example, if the crystal has the BCC or FCC lattice, found widely in the nature, it is easy to show that the critical size using our approach is equal to 3 coordination spheres. For example, the critical size of the nanocrystal for α-Fe is approximately 0.5 nm, for Ni it is approximately 0.6 nm, and so on. The fact that the proposed criterion is correct can be confirmed by examining Fig. 3.3 which shows the high-resolution electron microscopic image of a nanocrystal 1.5 nm in size, produced by D.V. Shtanskii [21].

Within the framework of the nanocrystalline model of the structure of the metallic glasses, the nanocrystals are structural elements filling the entire volume of the solid (the amorphous state is in fact identical with the nanocrystalline state) or filling it only partially together with the liquid-like state. The latter structural state appears to be two-phased and in our view is of special interest.

Previously, we drew reader's attention to the differences between the amorphous and nanocrystalline states of the solid. However, there are also important coincidences between them. We shall discuss

Fig. 3.3. A nanocrystal with the size of 1.5 nm in the Ti–Al–B–N film.

two most important ones, in our view, which reduce the difference between the amorphous and nanocrystalline materials.

The absence of the long-range order in the distribution of the atoms in the amorphous metallic materials appears to exclude the existence of anisotropy of any physical properties. It should be mentioned that the absence of the anisotropy of the properties of the offers alloys is erroneously mentioned by the latest edition of the Physical Encyclopaedia [2, volume 1]. In fact, the microvolumes of amorphous ferromagnetics are characterised by magnetic anisotropy and by the associated ordered distribution of the vectors of spontaneous magnetisation which has been reliably confirmed by experiments [22]. Initially, the long-range magnetic order in the atomically disordered medium was attributed to the existence of some nanocrystalline regions in the material. This was done assuming that ferromagnetism cannot exist without the crystal lattice. However, in the 60s of the previous century, A.I. Gubanov theoretically justified the possibility of existence of the amorphous ferromagnetics [23] which was subsequently unambiguously confirmed by the experiments. The overall physical pattern of the macroscopic anisotropy of the magnetic properties has not as yet been established. The experiments show that the main contribution to the magnetic anisotropy of the amorphous ferromagnetics is provided by magnetic–elastic anisotropy and the anisotropy of orientation-ordered atomic pairs (the so-called 'directional ordering'). Since the new methods of practical application of amorphous and amorphous-nanocrystalline alloys, associated with their unique magnetic properties, are being developed intensively at the present time, the nature of the ferromagnetism of the disordered and low-dimensional systems is of considerable importance.

The capacity for magnetic reversal of the ferromagnetics (and, consequently, high magnetic permittivity) is controlled by the inhibition of the moving domain boundary by the structure (Bloch walls). In the amorphous alloys, this inhibition is very small and is determined mainly by the previously mentioned effect of local magnetic anisotropy. Since the size of the ferromagnetic nanocrystals is considerably smaller than the Bloch wall thickness (100–200 nm), the inhibition of the moving Bloch wall is also not significant and the magnetic permittivity of the ferromagnetic materials rapidly increases with the reduction of the grain size in the nanometer range [24].

Thus, the amorphous and nanocrystalline ferromagnetics are in principle high-permittivity materials. The addition to the amorphous matrix of nanocrystals capable of reducing the effect of local magnetic anisotropy and the magnetostriction constant of the two-phase system leads to an additional increase of the magnetic permittivity of the already high-permittivity amorphous ferromagnetics. This principle was used recently for the development of a new generation of magnetically soft amorphous–nanocrystalline iron-based alloys – Finemet, Nanoperm and Thermoperm [25] whose unique magnetic properties are higher than those of the amorphous ferromagnetics.

There is another common feature of the amorphous nanocrystalline alloys – the same mechanism of plastic deformation taking place under external loading. It is well-known that the process of plastic deformation of crystals takes place by the nucleation, interaction and annihilation of dislocations. At a small crystal size the presence of imaging forces, associated with the elastic fields of stresses at the dislocations, causes that the introduction of the dislocations to the crystal becomes undesirable from the energy viewpoint. Consequently, the process of plastic yielding of the nanocrystals is of the non-dislocation nature [26]. It should be mentioned that for this reason the nanocrystals demonstrate the anomalous reduction of the deformation stresses with the reduction of the grain size in the nanocrystalline range [Introduction Ref. 5]. The same situation forms in the metallic glasses: the absence of translational symmetry exclude the existence of conventional dislocations and plastic deformation is also of the non-dislocation nature. Omitting the details, it should be said that the plastic yielding of the nanocrystalline and amorphous solids has common features and, consequently, the common pattern of mechanical behaviour.

3.2. Main relationships of crystallisation

3.2.1. Crystallisation thermodynamics

Conventional crystallisation of the AMS is regarded as a solid-phase transformation, governed by the classical crystallisation thermodynamics of the supercooled liquid (Fig. 3.4). The variation of enthalpy ΔH, entropy ΔS, and free energy ΔG in crystallisation of the AMS [Introduction Ref. 16] is equal to:

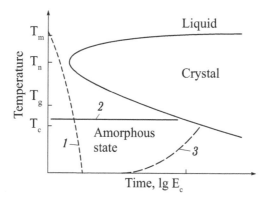

Fig. 3.4. Schematic representation of the temperature–time–transformation diagram for the processes of formation of the crystals in the supercooled melt; 1) quenching with the formation of the amorphous state, 2) isothermal annealing of the amorphous state, 3) slow heating of the amorphous state, leading to crystallisation at temperature T_c.

$$\Delta H^{c-am}(T) = -\Delta H_m + \int_{T}^{T_m}(C_p^{am} - C_p^c)\,dT,$$

$$\Delta S^{c-am}(T) = -\frac{\Delta S_m}{T_m} + \int_{T}^{T_m}\left(\frac{C_p^{am} - C_p^c}{T}\right)dT,$$

$$\Delta G^{c-am} = \Delta H^{c-am}(T) - T\Delta S^{c-am}(T), \tag{3.3}$$

were ΔH_m is the difference of the enthalpies of the amorphous and crystalline states at the melting point T_m, C_p^{am} and C_p^c are the specific volume (or molar) heat capacities of the amorphous and crystalline phase, respectively. The thermodynamic parameters $\Delta H^{n-am}(T)$, $\Delta S^{n-am}(T)$ and ΔG^{n-am}, identical to (3.3), can be determined [Introduction 16], [27] for the transformation from the amorphous the nanocrystalline state:

$$\Delta H^{n-am}(T) = -\Delta H_m(T_c) + \int_{T}^{T_m}(C_p^n - C_p^{am})\,dT,$$

$$\Delta S^{n-am}(T) = -\Delta S_0^{n-am} + \int_{T}^{T_m}\left(\frac{C_p^n - C_p^{am}}{T}\right)dT,$$

$$\Delta G^{n-am} = \Delta H^{n-am}(T) - T\Delta S^{n-am}(T). \tag{3.4}$$

Here $\Delta H_m(T_c)$ is the difference of the enthalpies of the amorphous and nanocrystalline states at the crystallisation temperature T_c. The enthalpy difference can be measured by calorimetric measurements. ΔS^{n-am} is the entropy difference of the nanocrystalline and amorphous state at 0 K which can be estimated from the concentration of the free volume in two states [28], C_p^n is the heat capacity of the nanocrystalline state.

The processes of crystallisation and nanocrystallisation have considerable thermodynamic differences. For the crystallisation process, the entropy variation is negative, $\Delta S < 0$. With increase of temperature the value ΔS decreases and ΔG increases. For nanocrystallisation $\Delta S > 0$, i.e. entropy increases. With increase of temperature ΔS and ΔH increase and ΔG at the given grain size decreases.

3.2.2. Crystallisation mechanisms

There are four crystallisation mechanisms [27].

1. *Polymorphous crystallisation* in which the composition of the transformation products is the same as that of the amorphous matrix. This type of crystallisation can take place only in pure elements or compounds on the equilibrium diagrams and takes place by single 'jumps' of the atoms through the crystallisation front, i.e. by the diffusionless mechanism. The shape of the crystals is determined by the anisotropy of the growth rate in different crystallographic directions. Since the amorphous alloys usually correspond to the eutectic compositions characterised by the large difference of the composition of the crystalline phases and the amorphous matrix, this type of crystallisation is encountered only in a small number of cases, especially in the formation of metastable crystalline phases which then transforms to the equilibrium transformation with considerable concentrational supersaturation.

2. *Eutectic crystallisation* in which the amorphous matrix crystallises with the simultaneous formation of two phases in a close structural relationship. In this case, the components may be redistributed at the solidification front but in a number of cases this distribution is not sufficiently complete. Although the transformation takes place in the solid state, the resultant colonies are referred to as eutectics and not eutectoids because the amorphous matrix is a supercooled liquid. As an example, Fig. 3.5 shows the eutectic colonies of α-iron and Fe_3B boride uniformly distributed in the amorphous matrix in the melt-quenched Fe–Cr–B alloy. Structural components of the eutectic have the form of very thin (no more than 10 nm thick) alternating plates distributed in a strict orientation relationship. Since the solubility of boron in α-iron is extremely small,

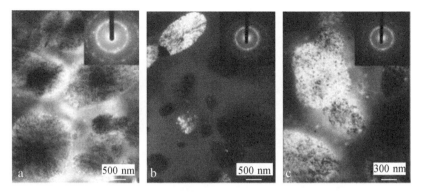

Fig. 3.5. Eutectic colonies of Fe (Cr)–(Fe, Cr)$_3$B, distributed in the amorphous matrix of the Fe$_{70}$Cr$_{15}$B$_{15}$ alloy; transmission electron microscopy, a) the bright field, b, c) the dark field.

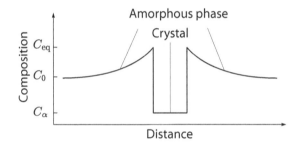

Fig. 3.6. The concentration profile, formed in the presence of the primary crystal. C_0 is the mean concentration in the amorphous phase, C_α is the concentration in the solid phase, C_{eq} is the concentration in the liquid phase.

oscillations in the boron concentration between the structural components of the detected exceed 30 at.%.

3. *Primary crystallisation* in which the crystals with the composition different from the composition of the amorphous matrix form in the initial stage (Fig. 3.6). The growth rate of these crystals is controlled by diffusion of one or several components in the initial matrix. After enrichment of the amorphous matrix with these components polymorphous or eutectic transformation takes place in the matrix with the formation of different or the same crystalline phases. This crystallisation mechanism is characteristic of the alloys of the hypoeutectic or hypereutectic composition, and also of the alloys whose composition slightly differs from the composition of the alloys undergoing polymorphous transformation. In most cases, the primary crystals grow in the form of dendrites (Fig.

Fig. 3.7. Dendritic growth of primary α-Fe crystals in the Fe B amorphous alloy.

    3.7), and the formation of the characteristic dendrite axes depends strongly on the heating rate and annealing temperature of the amorphous alloy. Measurements of the isothermal growth rates of the primary crystals are taken to evaluate the diffusion rate of the atoms of the metalloid in the amorphous alloy.

4. *Crystallisation of the amorphous matrix after preliminary phase separation* takes place in two stages. If the amorphous matrix is susceptible to phase separation already prior to crystallisation, the resultant amorphous 'phases' crystallise independently and also at different temperatures. The phase separation processes of the amorphous state have not as yet been completely explained and, therefore, the concept of the amorphous 'phase' and, correspondingly, specific features of the behaviour of the phase in transition to the crystalline state are not yet completely accurate.

Nevertheless, regardless of the dependence on the specific crystallisation mechanism, it is possible, under the optimum treatment conditions, to produce the amorphous–crystalline mixture and analyse its mechanical behaviour in dependence on the volume fraction and the type and morphology of the crystalline phases distributed in the amorphous matrix.

It is important to note that the first phase, formed in the amorphous matrix during heating, is the phase which is the first to form during slow cooling of the melt. This means that the amorphous matrix contains a large number of the nuclei of the crystalline phase inherited from the liquid. It is fully possible that such a nucleus is a cluster incorporated in some manner in the structure of the amorphous alloy or it may even be an internal element of the structure.

3.2.3. Crystallisation kinetics

The crystallisation kinetics of the amorphous alloys is the result of the effect of thermodynamic factors and kinetic parameters [29]. The kinetics of this process depends on a large number of parameters, in particular, on the crystallisation mechanism, the number of 'frozen-in' crystallisation centres, the activation energy of diffusion, and on the driving force – the difference of the free energies of the amorphous and possible crystalline phases. In addition to this, the nature of crystallisation is influenced by the quality of the surface and the effect of external factors (radiation, pressure, deformation).

For the accurate description of crystallisation it is important to know the kinetic equation of every elementary process taking place during crystallisation: the nucleation of crystals of the new phase, diffusion, growth and coalescence of nuclei, etc. The mathematical description of different stages of crystallisation with time and with the variation of temperature taking into account the contribution of each elementary process is quite difficult.

Phase transformations in the amorphous alloys are studied using the Kissinger or Johnson–Mehl–Avrami equation [30-32].

In the first case, at a constant heating rate, the crystallisation rate is described by the first order equation:

$$\left(\frac{\partial x}{\partial \tau}\right)_T = k(1-x),$$

$$(3.5)$$

where x is the amount of the material crystallised during the period τ at a temperature T, k is the rate constant. The constant can be determined from the equation $K = A \exp\left[-\dfrac{\Delta E}{RT}\right]$, where E is the activation energy, R is the gas constant. If the temperature changes with time, the reaction rate is described by the equation of the type

$$\frac{dx}{d\tau} = \left(\frac{\partial x}{\partial \tau}\right)_T + \left(\frac{\partial x}{\partial T}\right)_\tau \frac{dT}{d\tau}. \tag{3.6}$$

Since both the number of nuclei and position of the components in the system are constant at a fixed time, then $(\partial x/\partial T)_\tau = 0$ and, combining (3.4) and (3.6), we obtain the equation which can be used for any temperature:

$$\frac{dx}{d\tau} = A(1-x)\exp\left[-\frac{\Delta E}{RT}\right]. \tag{3.7}$$

Differentiating (3.7) with respect to time we have

$$\frac{d}{d\tau}\left(\frac{dx}{d\tau}\right) = \left[\left(\frac{\Delta E}{RT^2}\right)\left(\frac{dT}{d\tau}\right) - A\exp\left(-\frac{\Delta E}{RT}\right)\right]\left(\frac{dx}{d\tau}\right). \tag{3.8}$$

At the very beginning of the crystallisation process, the rate of the process is close to 0, the rate at the maximum point of the exothermic peak on the DTA curves is maximum and, consequently, at these temperatures $dx/d\tau = 0$ and

$$A\exp\left(-\frac{\Delta E}{RT_m}\right) = \frac{\Delta E}{RT_m^2}\left(\frac{dT}{d\tau}\right), \tag{3.9}$$

where T_m is the temperature corresponding either to the maximum peak or the start of the transformation. Denoting the heating rate $dT/d\tau = \beta$, we obtain

$$-\left(\frac{\Delta E}{R}\right)\left(\frac{1}{T_m}\right) = \ln\left(\frac{\beta}{T_m^2}\right) + const. \tag{3.10}$$

The angle of inclination of the straight-line in the $\ln\left(\dfrac{\beta}{T_m^2}\right) - \dfrac{1}{T_m}$

coordinates can be used to determine the activation energy of the crystallisation process ΔE and the pre-exponential multiplier A.

In the case of isothermal holding, the fraction of the crystallised material can be described by the classic Johnson–Mehl–Avrami equation

$$X(\tau) = 1 - \exp\left[-b\tau^n\right] \tag{3.11}$$

where $x(\tau)$ is the volume fraction of the crystalline phase, formed during the time τ, b is the rate constant, n is the exponent which can be used to indicate the mechanism of the process. According to Christian, each stage of crystallisation has its own specific value of n [33]. This may be interpreted as follows. In crystallisation of amorphous alloys, the value of n can change from 1.5 to 4. If

n = 1+0.5, the grain growth is controlled by diffusion, there is no nucleation; at n = 2.5+0.5 nucleation takes place with the constantly increasing rate of nucleation; the value n = 3.5–4 indicates that the crystallisation rate is controlled by the growth rate of the crystals at a constant rate of formation of the crystallisation centres, or corresponds to the eutectic breakdown [34].

The dependence of the rate constant b on the temperature is described by the Arrhenius equation:

$$b=b_0 \exp\left(-\frac{\Delta E}{RT}\right),$$

$$(3.12)$$

where b_0 is the pre-exponential multiplier, ΔE is the activation energy, R is the gas constant.

The crystallisation temperature T_c is not strictly defined – the transformation can take place, although at a lower rate, in the process of isothermal holding at a temperature several degrees (in the case of pure amorphous metals) or tens of degrees (for amorphous alloys) lower than the value regarded as T_c. Crystallisation temperature T_c is usually the temperature at which the rate of transformation is quite high (10^{-3}–10^{-1} of the volume of the specimen per minute).

Activation energy E_a and the latent heat of transformation Q play the controlling role in the crystallisation kinetics of the amorphous films. The relationship is unambiguous: as the purity of the specimen increases, the values of T_c and E_a decrease and Q increases [35, 36]. The stability of the amorphous alloys increases with the reduction of the thermodynamic stimulus and with the increase of the kinetic barrier of their transformation (lower temperature and higher activation energy of crystallisation). The activation energy of the crystallisation process, as shown by the experiments, changes in a wide range: from 40 to 400 kJ/mole [chapter 2: 1].

The value T_c equals 13–23 K for the amorphous alloys of pure metals with a thickness D > 20 nm and may reach several hundreds of degrees for the amorphous alloys. For the majority of glasses T_c is in the range $(0.4–0.65)T_m$ for the normal heating rates used in DSC investigations, equal to 10–100 K/s (here T_m is the equilibrium melting point of the crystalline alloy). The absolute values of the range at the present time are 40–1500 K (for the amorphous alloys based on refractory metals).

The crystallisation of the amorphous alloys and metals in heating takes place by the formation of crystalline nuclei in the amorphous matrix followed by their growth. The kinetics of nucleation of the crystalline phases have been discussed extensively. There are different views regarding the relative role of true nucleation in the amorphous matrix and the athermal growth of the pre-precipitates or frozen-in nuclei. Additional difficulties with interpretation are associated with the measurements of the nucleation rates. The volume density of the crystallisation centres, depending on time, can be calculated directly from micrographs. The nucleation rate should be determined on the basis of the measurement of the growth rate and the general transformation kinetics. In the first case, a serious error may be made if the effect of the small cross-section of the thin foil is not accurately estimated. In addition to this, in both cases, the results are incorrect in the early stages of crystallisation when the number and dimensions of the crystals are small.

In many cases, nucleation takes place homogeneously and the nucleation rate is approximately constant with time at the given temperature. However, homogeneous nucleation is only one of the possibilities onto which the heterogeneous nucleation, athermal nucleation or even nucleation, determined by the 'frozen-in' crystallisation centres, is often superposed [37]. The surfaces or interphase boundaries may act as a catalyst for the nucleation of centres since the new crystalline phase replaces part of the surface, reducing the total surface energy. However, heterogeneous nucleation on the surface and the interphase boundaries in the amorphous alloys is not observed so widely as in the case of transformations in the crystalline solids. The 'frozen-in' centres are sufficiently large to act as crystallisation centres at typical annealing temperatures. They do not always have the regular structure and become the effective crystallisation centres only after some rearrangement which requires a certain period of time. At high temperatures, these centres dissolve during annealing.

3.2.4. Crystallisation stages

In most cases, the crystallisation of amorphous alloys is a complicated process taking place in several stages, including the formation of intermediate metastable crystalline phases. Taking into account the data for different amorphous alloys, the process can be described as follows: initially, the high-dispersion metastable phase $MS - I$ forms in the amorphous phase, this is followed by the formation of a mixed structure, gradually, bypassing the intermediate phases, transferring to the crystalline metastable phase $MS–N$, which transforms at high temperatures into a stable equilibrium structure.

Table 3.1. Variation of the phase composition of several amorphous alloys after annealing

$Fe_{83}B_{17}$ [38]	$Co_{70}Fe_5Si_{15}B_{10}$ [39]	$Fe_{64}Co_{21}B_{15}$ [40]	$Pd_{81}Cu_7Si_{12}$ [41]	$Fe_{90}Zr_{10}$ [42]
A	A	A	A	A
α-Fe+A	A-2	A+α–Fe+ +α-Co+β-Co	A+Pd+ γ-(Pd–Cu)	α'-Fe(Zr)+ +α''-Fe(Zr)
α-Fe+ +Fe_3B^T+A	A+FCC Co	α–Fe+α-Co+ β-Co+Fe_2B+ +Fe_3B+CoB	Pd+Pd_3Si+ Pd_9Si_2+γ- (Pd–Cu)	α-Fe+ Fe_3Zr
α-Fe+ +Fe_3B^T	FCC Co+ +Co_3B+Fe_5Si_3			
α-Fe+ Fe_3B^T+ +Fe_3B^0	FCC Co+Co_3B+ +HCP Co			
α-Fe+ +Fe_3B^T+ +Fe_3B^0+ +Fe_2B	FCC Co+ HCP Co+ +Co_3B+ +Co_2B+Fe_5Si_3			
α-Fe+Fe_2B	HCP Co+Co_2Si+ +Co_2B+Fe_5Si_3			

Comment. Fe_3B^T is the metastable phase with the BCC lattice, Fe_3B^0 is the metastable phase with the orthorhombic lattice; A-2 is the additional diffusion maximum on the x-ray diffraction diagrams of the amorphous alloys; α'-Fe(Zr) and α''-Fe(Zr) are the solid solutions of Zr and Fe in the cubic and slightly tetragonal lattice, respectively, in isothermal annealing.

Table 3.1 shows the sequence of transformations in heating several amorphous alloys in accordance with the above scheme, with the exception of $Fe_{90}Zr_{10}$ phase in which phase separation takes place.

As a result of systematisation of the data on the initial stages of crystallisation and phase transformations, taking place during the breakdown of the amorphous state, it may be concluded that the structure, phase composition and morphology of the crystalline phases, formed during annealing of the amorphous alloys, strongly depend on the production conditions, preliminary heat treatment and the ratio of the chemical elements of the investigated alloy. The complicated nature of crystallisation can be determined by the calorimetric method showing several distinctive exothermic peaks [43]. The total crystallisation heat is usually ~40% of the melting heat of the alloy; the remaining part of enthalpy corresponds to the

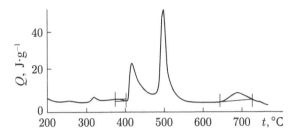

Fig. 3.8. The DSC curve, produced in heating the amorphous alloy $Ni_{44}Fe_{29}Co_{15}Si_2B_{10}$.

heat generated from the specimen during melt quenching because the heat capacity of the melt is considerably higher.

In many cases, the DSC curve of glass contains two crystallisation peaks [40, 44] (Fig. 3.8). This situation is not unusual: the individual peaks, if they are detected, are always far away from each other. The double crystallisation peak may be determined by the initial primary crystallisation during which the remaining amorphous matrix changes its composition and subsequently solidifies at a higher temperature with the formation of another phase or in principle may be associated with crystallisation without any change of the composition followed by a crystalline polymorphous transformation (however, no such case has as yet been unambiguously confirmed). The temperature of the first crystallisation peak depends not only on the heating rate but also on the free volume in the glasses. In the multistage process, the first peak, corresponding to the crystallisation of the amorphous phase to a metastable crystalline phase, may be lower than the subsequent peaks as a result of the development of further processes of recrystallisation and phase transformations. The presence of a small amount of dissolved oxygen in an amorphous alloy also results in the transition from single stage to two-stage crystallisation and, therefore, in analysis of the information obtained using DSC it is important to take into account the possible role of the impurities.

3.2.5. The size of crystallisation products

In recent years, together with the beginning of the era of nano-technology and nanomaterials, the incorrect term 'nanocrystallisation' appeared in the literature [45]. In most cases, the meaning of this term is that of the crystallisation products having the nanosized dimensions. However, this crystallisation stage occurs in all amorphous alloys if, after all, the critical size of the nucleus of the crystal phase does not exceed 100 nm which is very rare! Evidently, the term 'nanocrystallisation' refers to the phenomenon in which the products of breakdown of the amorphous matrix are very small in a relatively wide temperature–time range and the coalescence of these products is for some reason difficult or completely blocked.

The transformation of the amorphous state of the material to the nanocrystalline state in heating can be regarded as the breakdown of the amorphous phase into intracrystalline and grain boundary components. Consequently, the molar excess energy of the grain boundary phase ΔE_b in relation to the perfect crystalline state may be described by the equation [Introduction Ref. 16]

$$\Delta E_g = \frac{\Delta H_{n-am}(T) - \Delta H_{c-am}(T)}{x},$$
(3.13)

where ΔH_{n-am} and $\Delta H_{c-am}(T)$ is the variation of the enthalpy of nanocrystallisation and crystallisation of the amorphous phase, measured by differential scanning calorimetry.

The types and morphology of the products of nanocrystallisation are determined by the crystallisation mechanism linked with the chemical composition and thermodynamic characteristics of the resultant crystalline phases [Introduction Ref. 16]. The size of the grains of the nanocrystalline structure, formed during crystallisation of the amorphous alloy, is strongly influenced by the heat treatment conditions and the chemical composition of the metallic glass. One of the most important factors which determine the grain size in isothermal annealing is the annealing temperature. The annealing time is usually determined by the time to completion of the transformation of the isothermal phase to nanocrystalline phase. In most cases, an increase of annealing temperature increases the mean grain size of the structure. This was found in the systems Ni–P [46], Fe–Co–Zr [47], Fe–Ni–P–B [48], Si [49], and others. However, some alloys show the opposite tendency: the size of nanocrystals decreases with increase of the isothermal annealing temperature, for example for the Co–Zr [50] and Fe–B [51] system. The grain size of the Fe-Si–B system [52] decreases with the increase of temperature in the range 450–500°C, reaches the minimum value of approximately 25 nm at around 500°C and greatly increases with a further increase of annealing temperature. In the polymorphous nanocrystallisation of the $NiZr_2$ metallic glass, the mean grain size of the resultant nanostructure was constant in the wide annealing temperature range [52]. Figure 3.9 shows the dependence of the grain size of the nanocrystalline phase on the heat treatment temperature for different alloy systems

[Introduction Ref. 16]. It may be seen that the minimum grain size is obtained during heat treatment at a temperature close to $0.5T_m$, where T_m is the melting point of the alloy. Evidently, this is determined by the nanocrystallisation mechanism.

At the moment, there is only a small number of experimental data for the effect of alloying elements on the grain size of the nanocrystalline phase, formed by crystallisation of the metallic glasses. In particular, it has been established that the additions of C and Si in the amorphous iron-based alloys increase the diffusion mobility of the metalloids and, consequently, increase the growth rate of the products of primary crystallisation [53]. These additions can also reduce the concentration of nuclei and, at the same time, support the formation of a structure with larger grains. The introduction of the additions of Cu or Au to glasses based on Fe increases the cooling rate of the α-Fe crystals by several orders of magnitude. The addition of the elements which reduce the diffusion rate, such as Nb, Zr or Mo, reduces the gross rate of the crystals and increases the dispersion of the structure [54]. The microadditions of Cr, Cu, Ni or Pd have no significant effect on the primary crystallisation of the Fe-based metallic glasses [55].

The experimental results show that the grain size of the nanocrystalline phase is influenced by the oxygen content of the metallic glass. The zirconium-based amorphous alloys strongly absorb oxygen in preparation by the melt spinning method. Metastable

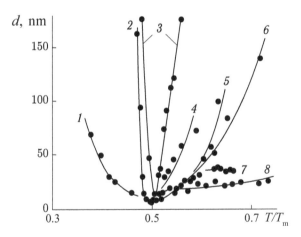

Fig. 3.9. Dependence of the mean size of the nanocrystals on annealing temperature normalised with respect to the appropriate melting point T_m for the processes of crystallisation of different systems of metallic glasses: 1) Fe–B; 2) Co–Zr; 3) Fe–Si–B; 4) Ni–P; 5) Si; 6) Fe–Co–Zr; 7) Pd–Cu–Si; 8) Fe–Ni–P–B.

compounds, stabilised with oxygen, formed in the process of primary crystallisation of zirconium-based alloys [54]. It has been assumed that the presence of oxygen reduces the interfacial energy and also reduces (by approximately an order of magnitude) the growth rate of the crystals in the amorphous matrix at an oxygen content of 1% and results in the formation of a highly dispersed structure as a result [54].

The mean size of the nanocrystals, produced by the crystallisation of the amorphous state, and also the nature of the size distribution of the nanocrystals may greatly differ depending on the annealing conditions. Annealing at a superhigh heating rate (pulsed annealing) results in the formation of highly dispersed nanostructures in comparison with conventional furnace and annealing. In the case of the Fe–Cu–Si–B metallic glass [56], pulsed annealing results in the formation of α-Fe(Si) nanocrystals with the mean size of approximately 20 nm, whereas conventional annealing leads to the formation of nanocrystals 80 nm in size. In metallic glasses based on Fe–Nb and Fe–Zr, crystallisation under the effect of pulsed annealing results in the formation of the nanocrystals with the size of 8–20 nm which is considerably smaller than in conventional annealing [57].

The effect of the initial structure of the amorphous state on the grain size of the nanocrystalline phase was investigated on Fe–B–Si metallic glasses [ntroduction Ref. 16]. Four specimens were prepared by the spinning method with different quenching rates as a result of changing the speed of rotation of the quenching disc-cooler. The experimental results show that with the reduction of the linear speed of rotation of the disc (i.e. with the reduction of the quenching rate) from 41.5 to 17.0 m/s the minimum size of the crystallisation products increased from 25 to 70 nm at the same morphology and crystal structure (solid solution Fe(Si) and boride Fe_3B). The increase of the quenching rate increases the 'degree of amorphicity', i.e. the degree of disorder which, evidently, results in a reduction of the size of the crystallisation products.

The experimental results confirm that the minimum possible size of the nanocrystals in the polymorphous and eutectic nanocrystallisation is several nanometres, and in primary crystallisation it is considerably larger (15–30 nm). The minimum grain size for a specific crystallisation variant does not depend on the number of elements in the alloy. Different values of the size of the nanocrystals for different solidification mechanisms indicate that the limiting size is determined by the mechanism of nucleation and the structure of the interphase boundaries and also by the structure of the amorphous matrix.

3.2.6. Surface crystallisation

Surface crystallisation is a special form of crystallisation of the amorphous alloys produced by melt quenching. In this case, the nanocrystalline phase forms, as is well-known, in the surface layers of the rapidly quenched amorphous ribbon. This process is characterised by different states of the surfaces (internal, adjacent to the disc, and the external, which is in contact with the atmosphere). The accelerated crystallisation phenomena of the surface can be detected on both sides of the ribbon but there are cases of preferential crystallisation on one of the sides [chapter 1 Ref. 2] [58]. The reasons for this phenomenon have not as yet been completely established. At the same time, the practical value of surface crystallisation is very high for practical application of amorphous alloys. For example, crystallisation of the surface layers is reflected in their catalytic activity and corrosion resistance [50, 59], and in the properties of magnetically soft materials [60, 61].

The experimental results show that the crystallisation of both surfaces of the amorphous alloys based on iron and cobalt starts earlier than in the volume [62, 63]. However, the activation energy of this process on the outer surface is the same as in the volume, and on the inner surface it is considerably lower (in particular, this is evident at low temperatures). Taking into account the data obtained by the methods of low-angle x-ray scattering and electron Auger spectroscopy [62, 63], the authors have concluded that accelerated crystallisation of the outer surface of the amorphous alloys is determined to a large degree by the presence of an excess free volume in the surface layer, high concentration of the submicropores, whereas the acceleration of crystallisation of the inner side of the alloy is determined by changes of the chemical composition of these layers.

3.2.7. Structure of nanocrystals

The most important element of the structure of the nanocrystalline materials are the grain boundaries (GB) and interphase boundaries (IB). The state and length of these boundaries determine to a large extent the macroscopic properties of the specimen. A large number of investigations have been devoted to explaining the structural characteristics of the boundaries in the nanocrystals. Investigations of the structure [Introduction Ref. 6] [64, 65] confirm that the width of the grain boundaries does not exceed 1 nm, and the structure of the grain boundaries does not differ greatly from the structure of the grain boundaries of the conventional polycrystals. The main difference of the special features of the nanocrystals of the third type, produced by melt quenching, is the presence of an intermediate amorphous or amorphous– crystalline state and the extent of completion of crystallisation from this state. In a number of studies, high-resolution electron microscopy was used to investigate the structure of the grain boundaries and interphase boundaries in the alloys produced by complete crystallisation of the amorphous metallic alloys. The experimental results show that as regards the structure and morphology, the boundaries do not differ from the boundaries of the coarse-grained materials. In [66], investigations were carried out into the structure of α-Fe (Si) GB and IB of Fe_2B crystals in the specimens produced by crystallisation of the $(Fe_{99}Mo_1)_{78}Si_9B_{13}$ amorphous alloy. It has been established that the grain boundaries are sufficiently flat and have the form of the conventional high-angle grain boundaries. The interphase boundaries consist of the facets separated by steps. The orientation of the crystals is usually random, and the IB are characterised by complete incoherence.

The nanocrystals, produced by polymorphous crystallisation of the $Ni_{33}Zr_{67}$ alloy, which consists only of one intermetallic phase $NiZr_2$, usually have straight, perfect grain boundaries. The only type of the interphase boundaries are the coherent twin boundaries with a specific structure. It may be expected that these boundaries are distributed in the low-energy configuration with a very small excess volume.

In the Ni–B alloys, the nanocrystals are the product of eutectic nanocrystallisation and the majority of the inner boundaries are the interphase boundaries $Ni–Ni_3B$ or $P–Ni_3P$. Examination by high-resolution electron microscopy showed flat IBs with the low-energy configuration [67]. As a result of the presence of the orientation relationship between the two phases, the structure of the IB should be quite specific.

The energy of the grain boundaries is a thermodynamic parameter reflecting the nature of the boundary. Calorimetric measurements taken during grain growth yielded the values of the energy of the interphase boundaries of $Ni–Ni_3P$ of 0.155 $J \cdot m^{-2}$ and the Ni–Ni grain boundaries of 0.111 $J \cdot m^{-2}$ [Introduction Ref. 16] [67], corresponding to the typical energies of the low-angle grain boundaries. This means that the state of the interphase boundaries in the investigated nanocrystals is characterised by low-energy and this contradicts the conclusions and the orientation relationships of the boundaries. The possible reason for this contradiction is that the lattice parameters of the crystals in the nanometre range of the grain size differ from the parameters of the perfect crystal lattice.

The nanocrystals of the first type may show the formation of different types of interphase boundaries: the coherent, semi-coherent and non-coherent, depending on the crystallisation mechanism. The non-coherent interphase boundaries are often detected in the equiaxed grains, formed, for example, in the processes of primary crystallisation. The coherent interphase boundaries form in high-energy processes, such as polymorphous and eutectic crystallisation. In the study [Introduction Ref. 16], the nanocrystals were classified in accordance with the morphology of their interphase boundaries (Table 3.2).

The defective structure was investigated efficiently by the method of electron-positron spectroscopy [Introduction Ref. 16]. Analysis of the spectra of the lifetime of the positrons for the crystals of the third type showed a number of common features of these materials [Introduction Ref. 16]. For the majority of systems of the alloys, the mean lifetime of the positron is longer than in the appropriate amorphous alloys. However, no differences were found in the lifetime of the positrons for the Fe–Zr and Fe–Cu–Nb–B alloys [64].

Table 3.2. Classification of nanocrystals in accordance with the morphology of their interphase boundaries

	Crystals with orientation relationships				Randomly oriented crystals
	Type of boundaries				
Coherence	Coherent without coherent deformation	Coherent with coherent deformation	Incoherent with interphase boundaries	Fully incoherent	Fully incoherent
Specific energy of IB	Very low	Increases	Increases	High	High, depends on orientation
Type III	Exist (NiZr$_2$)	Exist (Ni–P)	Exist	Ti–Ni–Si	Fe–Si–B, Se, etc
Nanocrystals produced by other methods	No	No	No	No	Yes

The short component of the lifetime τ_1 for the nanocrystals is greater than the appropriate component for the de-localised states of the volume crystals, but shorter than the lifetime of the positrons on single vacancies. This may be regarded as the fact of localisation and annihilation of the positrons of the elements of the free volume of the interphase boundaries which are smaller than for the vacancies. The intermediate component of the lifetime of the positrons τ_2 corresponds to the annihilation at the intersections (contacts) of two or three interphase boundaries and/or grain boundaries, and also on the nanopores whose volume corresponds to the volume of 10–15 vacancies [65]. The large pores which may be associated with the long component of the lifetime of the positrons τ_3 were not detected in the nanocrystals of the first type. Some nanocrystalline materials did not even contain the intermediate component of the lifetime of the positrons which indicates that they do not contain the nanopores. Consequently, the nanocrystals of the third time are characterised by the higher density than those obtained by the methods. The results of electron microscopic studies show that they also contain a smaller number of defects [Introduction Ref. 16].

It has also been established that the nature of distribution of the defects in the nanocrystals of the third type changes with the change of the grain size because the values of τ_1 and τ_2 increase with the reduction of the size of the crystals. This means that the mean size of the units of the free volume increases with the reduction of the grain size, and the size of the nanopores decreases or they are capable of absorbing a small number of 'vacancies'. Thus, as the grain size of the nanocrystals decreases the number of units of the free volume per unit area of the IB increases and the number of nanopores decreases. In other words, the density of the material of the interphase boundaries increases with the reduction of the grain size.

Another important question should be answered: are there any changes in the structure of the lattice of the nanocrystals in relation to the perfect crystal lattice? In the crystals, formed in crystallisation of the amorphous alloys, there are two types of structural changes – formation of the supersaturated solid solutions and the formation of the distortion of the lattice in pure elements and stoichiometric compounds.

Taking into account the Thomson–Gibbs effect [2], the solubility of the impurity elements in the solid solution should increase with the reduction of the size of the crystals to the nanosized dimensions.

For example, the results published in [68] showed the increase of the solubility of hydrogen in Pd nanocrystals (at a concentration of 10^{-3} and lower) in comparison with the nanocrystals by many orders of magnitude. The identical effect was reported in [69] for the solubility of Bi in the Cu nanocrystals, reaching approximately 4% at $100°C$, whereas the equilibrium solubility of Bi in the Cu nanocrystals is less than 10^{-4}. Another confirmation of the increase of solubility in the nanocrystals is the report on the formation of Cu–Fe in Cu–W solid solutions [70, 71].

The increase of solubility was also detected in the nanocrystals of the first type NiP. In the nanocrystalline condition, the alloy consists of two phases – the Ni(P) solid solution and the Ni_3P compound. The concentration in the FCC Ni was measured as the function of the grain size. The results show that the P concentration in the Ni(P) solid solution is in the range 1.86–2.81% which is 10–15 times higher than the equilibrium solubility [72]. It was also shown that the measured P concentration in the Ni nanophases is considerably higher than the concentration calculated using the Gibbs–Thomson equation. With the reduction of the mean grain size the P concentration of the nanoparticles of Ni decreases instead of increasing as indicated by the ratio. This behaviour, which is in agreement with the variation of magnetic susceptibility with the grain size, can be explained by investigating the effect of annealing temperature on the formation of the Ni(P) nanocrystalline solid solution from the amorphous state. It may be concluded that the formation of the supersaturated structure of the nanomaterials is caused not only by the effect of the superfine grains but also by the history of formation of the nanostructure.

The problem associated with the thermal stability of the nanocrystals is of considerable importance [73]. It is strongly influenced by the processes of normal and anomalous grain growth, phase transformations, relaxation, diffusion and a number of other processes. Below, we consider two important examples of the evolution of the nanostructures under the thermal effect.

3.2.8. Special features of nanocrystallisation in Finemet alloys

An important group of rapidly quenched materials is formed by the amorphous–nanocrystalline materials whose properties may be higher than those of the amorphous and nanocrystalline alloys. For example, the alloys of the Fe–Si–B system with the small additions of Cu and Nb, referred to as the Finemet alloys [74], have very high magnetically soft properties, higher than the properties of the appropriate amorphous materials.

The Finemet alloys were described for the first time in [75] and subsequently in [76]. It was shown that the size of the crystals formed in the $Fe_{76.5-x}Si_{13.5}B_9M_xCu_1$ alloys can be reduced mainly by adding niobium (Nb = Ta > W >V >Cr) resulting in the unique magnetic properties. The optimum magnetic characteristics were obtained for the composition $Fe_{73.5}Si_{13.5}B_9Nb_3Cu_1$ which was also referred to as Finemet [77–79].

The nanocrystalline phase formed under controlled heat treatment of the amorphous state of this and similar compositions has the form of the solid solution of silicon in iron α-Fe(Si), ordered in accordance with the type DO_3 with the size of the nanocrystals of 9–20 nm and the silicon content of 16–23 at.% [74–78]. The volume fraction of the nanocrystalline phase is 75–80%, the remainder is the amorphous phase (Fig. 3.10).

The kinetics of the structural phase transformations in the nanocrystals based on the Fe–Si–B–Cu–Nb system has been studied quite extensively at annealing temperatures from t = 480°C to 900-950°C [79–82]. Special attention in the studies was given to the range t = 480–550°C. It has been assumed that no significant structural changes take place in the amorphous matrix at temperatures below 480°C [82] but in [83] studies by pro-angle scattering of x-rays revealed structural–phase changes also in the low temperature range.

The published literature data on the kinetics of transition from the amorphous to nanocrystalline state obtained using the Johnson-Mehl–Avrami equation by different authors greatly differ both in the values of effective crystallisation energy E^* and the values of the exponent n which characterises the special features of the transformation by the mechanism of nucleation and growth [84, 85].

In [86] X-ray diffraction studies were carried out (including the method of low-angle X-ray scattering) of the special features

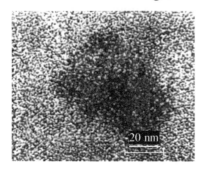

Fig. 3.10. The nanocrystalline phase formed after annealing of the amorphous state in Finemet alloy. High-resolution electron microscopy.

and kinetics of formation of the nanocrystalline structure in the $Fe_{73.5}Si_{13.5}B_9Cu_1Nb_3$ alloy in isochronous and isothermal (430 and 550°C) annealing. Data for the low-angle scattering of the x-rays showed that the transition from the amorphous to nanocrystalline state in the given heat treatment conditions is characterised by the same non-monotonic dependence of the intensity of low-angle scattering on annealing temperature (time). This non-monotonic relationship shows that in the initial stages of the phase transformation, the contribution of the processes controlling the phase transformation differs. The number of the scattering centres of electronic density increases, i.e. the number of nanoscale phase formations formed in the amorphous matrix increases. Although the relaxation of the free volume characteristic of the amorphous alloys (in particular, with the formation of the micropores [19]) can contribute to the intensity of low-angle scattering, nevertheless the resultant dependences can be linked to a certain extent with different ratios of the contributions of the stages of nucleation and growth of the new phase to the transformation.

According to a large number of electron microscopic data, the number and size of the nanocrystals in the investigated and similar alloys increases with increase of annealing temperature (time). This is in agreement with the x-ray data. As indicated by the data published in [86], the values of the Avrami parameter are considerably lower in comparison with the theoretical value $n = 25$. When explaining these differences, it is assumed that the presence of Cu in the alloy increases the number of the 'frozen-in' crystallisation centres, formed directly in the process of preparation of the initial specimens, and followed by the athermal transformation of these centres during heating [87].

This process may also be combined with thermally activated nucleation [88]. In addition to this, the growth rate of the nanoparticles may also be reduced by the factors such as the difficult diffusion of Si and B due to the presence of niobium which imposes restrictions on the growth of α-Fe (Si) crystals, the presence of local heterogeneities of the chemical composition in the thickness of the ribbon determined by high temperature gradients in melt quenching, etc. [88, 89].

It may be assumed that in low-temperature (430°C) annealing when the diffusion processes are still associated with difficulties ($D \approx 7 \cdot 10^{-22}$ m²/s), the main contribution to the transition from the amorphous nanocrystalline state comes from the athermal transformation on the 'frozen-in' centres. The increase of the intensity of low-angle scattering in this stage is logically associated with the increase of the formed 'growth points', and its subsequent reduction with the simultaneous increase of the dimensions of the nanocrystals from 38 to 41 nm – with the coalescence process.

The experimental results confirm the dominance of the growth factor of the nucleation process in the transition of the Finemet alloy to the nanocrystalline state. The results of calorimetric investigations are described more adequately by the kinetic processes, controlled by the growth of nanoparticles [90], and not by the processes including the stages of both nucleation and growth.

Special features of the processes of nanocrystallisation of Finemet alloy are attributed by many authors to their initial micro-heterogeneous (nanocluster) structure. A large amount of experimental data have been collected which shows that the amorphous alloys, like the melts, are characterised by the presence of regions with a short range (topological and/or composition) order, and the polycluster model of the structure [91] of the amorphous alloy adequately describes their structure-sensitive properties. On the basis of the experimental data obtained by high-resolution advanced methods (for example [92]), it is assumed that the structure of the alloy consists of the regions enriched with Fe–B, and also the clusters containing mostly atoms of Fe and Si [84]. This microheterogeneity of the initial alloy determines the formation of the magnetically soft nanocrystalline α-phase during heating.

It is widely believed that the structure of the melt of the Fe–Si–B–Cu–Nb system is described more than adequately by the microheterogeneous model, and the structural components in the melt are the clusters of the atoms of iron, silicon and intermetallic compounds of iron with niobium and boron [93]. The microheterogeneous (nanocluster) state, formed in the process of amorphisation of the alloy, results in the situation in which the processes of crystallisation of the $Fe_{73.5}Si_{13.5}B_9Cu_1Nb_3$ amorphous alloy takes place almost completely without nucleation by the athermal transformation of the frozen-in crystallisation centres and by the formation of the amorphous–nanocrystalline structural state in the initial stage of the crystallisation process [93].

To explain the role of the Cu atoms as the areas of nucleation in the process of nanocrystallisation of Finemet alloys, excellent results have been obtained using the EXAFS method. The method was used for the first time in [94] for detecting the fact that the copper-enriched particles have the FCC structure and form at temperatures considerably lower than the crystallisation temperature with the precipitation of α-FeSi. The results obtained in [75, 76, 95] show the reduction of the crystallisation temperature of α-FeSi, caused by the additions of copper, but it was not possible to confirm directly the hypothesis according to which the copper atoms support heterogeneous nucleation. Later, Homo et al [96] showed by experiments with the atomic probe that the copper clusters formed in these alloys already after short annealing times and concluded that the nanocrystals based on Fe form in the areas outside these clusters.

The study [97] confirm the formation in the early stages of a copper-enriched phase and confirmed also the FCC structure of the particles in the melt-quenched alloys. These investigations were followed by the study [98] in which the structural model of the formation of nanocrystals in Finemet alloys was developed. In particular, it was shown that even in the melt quenched amorphous state there are copper clusters with the size of one elementary cell of the FCC lattice. After annealing clusters grow and new clusters already form in the early stages of heat treatment (less than 8 min). If this situation is compared with the alloy with no niobium (for example, $Fe_{76.5}Si_{13.5}B_9C_4$), it may be seen that the Finemet alloy has considerably smaller clusters. The alloy with no niobium does not contain copper clusters after melt quenching but annealing of this alloy results in the formation of larger copper clusters.

The proposed model may be described as follows. Melt quenching results in the formation of the amorphous state with copper nanoclusters. In early stages of annealing new copper nanoclusters form at the optimum temperature together with the growth of the clusters already formed after quenching. In the formation of α-FeSi nanocrystals, the copper nanoclusters with the FCC lattice are the areas of heterogeneous nucleation. The concept of the heterogeneous nucleation of the α-FeSi nanoclusters on the copper clusters is based on the minimisation of the energy of the interphase boundaries [99]. Finally, in the stage of growth of α-FeSi, the copper clusters no longer play the role of the potential areas fur nucleation, and grow to a size of approximately 5 nm.

A significant role in understanding the nature of structural transformations in the Finemet-type alloy has been played by the study [96] performed by atomic-probe microscopy. The results show that all five components of the alloy (Fe, Si, B, Cu, Nb) are uniformly distributed in the initial amorphous matrix. Nanocrystallisation resulted in the formation of the two-phase state: α-Fe–Si solid solution in the form of the nanocrystalline phase and the amorphous phase enriched with B and Nb and depleted in Si. Secondary crystallisation involved the amorphous phase, distributed between the nanoparticles. Since the Cu atoms are not soluble in Fe, the precipitation of Cu from the amorphous matrix should take place in the early stages of crystallisation and should inhibit the nucleation of α-Fe–Si nanocrystals. The following sequence of phenomena was observed in the process of nanocrystallisation at 550°C [100]:

1) the primary matrix is fully amorphous;
2) in the initial stages of crystallisation, the Cu atoms form clusters with the size of several nanometres;
3) the α-FeSi nuclei form in the areas in which clusters formed;
4) the α-FeSi nanocrystals grow, displacing the Nb and B atoms into the surrounding amorphous matrix.

After nanocrystallisation, three phases should exist for the formation of the optimum structure. It is the BCC solid solution (~20 at.% Si) Fe–Si, the remaining amorphous phase with ~5 at.% Si and 10–15 at.% Nb and B, and also the third phase with ~60% Cu, < 5 at.% Si, B and Nb and ~30% Fe. These observations indicate that the formation of the Cu clusters stimulates the chemical aggregation and nucleation of the primary phase. Nb and B, characterised by low solubility in Fe, are displaced from the BCC Fe–Si nanocrystals. The amorphous matrix, enriched with Nb and B, surrounds the nanocrystals. This sequence of the process is shown schematically in Fig. 3.11.

The Finemet effect, i.e. formation of a very high volume density of the nanocrystals in the amorphous matrix in the process of crystallisation in annealing, which do not subsequently grow, is typical not only of the Fe–Si–B–Nb–Cu alloys, but also a number of other amorphous alloys. For example, patents were granted for the Fe–M–B–C nanocrystalline alloys (M = Zr, Nb, Hf) [101] (under the name Nanoperm), with the composition of this alloys selected to ensure the nanosize of the BCC of the α-Fe particles, distributed in the amorphous matrix. The Thermoperm alloys [102–104] are characterised by the formation of the α- and α'-nanocrystalline phases with the BCC lattice and the B2 superstructure (FeCo) respectively with greatly improved high-temperature magnetic properties in comparison with the first two alloys. The Finemet effect was also detected in the production of the nanocrystals of the third type in the amorphous alloys Ni–Fe–Cu–Si–B and Fe–Ni–B alloys [105].

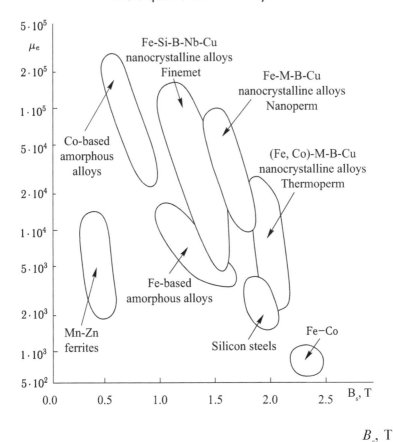

Fig. 3.11. Sequence of the structural-phase transformations in heating of the Finemet alloy.

In [106] a structural model of the inhibition of growing crystals was proposed for the alloys with the Finemet effect. The model is based on fulfilling three conditions:

1) the presence in the amorphous alloy of at least one active alloying element which increases the crystallisation temperature of the amorphous matrix;
2) the active alloying element must be characterised by low solubility in the lattice of the resultant nanocrystal;
3) the amorphous matrix should have the conditions for the nucleation of a large number of nanocrystals.

All the three conditions were realised by the authors of [106] in the Fe–B–Y–Nb–Cu bulk amorphous alloys. Crystallisation resulted in the precipitation of α-Fe nanocrystals with the size of 11–30 nm, and 'barrier' the regions, enriched with Y and B atoms, formed around them in the amorphous matrix. On the one hand, these atoms

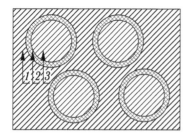

Fig. 3.12. The scheme illustrating the inhibition of growth of an α-Fe nanoparticle by a 'cloud' of the highly stable amorphous phase enriched with boron; 1) amorphous matrix, 2) the 'cloud', 3) the nanocrystal.

did not dissolve in the nanocrystals and, on the other hand, increased the thermal stability of the amorphous state in the vicinity of the 'inhibited' nanocrystals (Fig. 3.12).

3.2.9. Special features of nanocrystallisation in Ni–Ti–Cu alloys

The method of production of the nanocrystals of the third type is of some interest for the alloys based on TiNi, characterised by the shape memory effect [107]. The transition to the nanoscale dimensions of the crystals in these alloys, which are difficult to realise by other methods, improves the service characteristics of the alloys and increases their efficiency and durability.

In most cases, special attention was given to the shape memory effect $Ti_{50}Ni_{25}Cu_{25}$ alloy which can be produced in the amorphous state as a result of the presence of copper [108–110].

Figures 3.13 and 3.14 present the results of investigations of the crystallisation kinetics of the $Ti_{50}Ni_{25}Cu_{25}$ amorphous alloys in respectively continuous heating (DTA) and isochronous (30 min) annealing (resistance strain gauges) [111]. In continuous heating crystallisation takes place in the temperature range 465–480°C, and

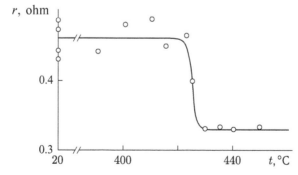

Fig. 3.13. Dependence of electrical resistivity (r) on annealing temperature, annealing time 30 min.

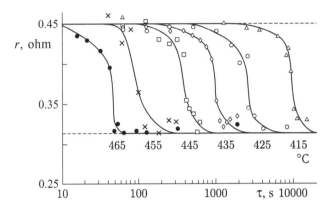

Fig. 3.14. Dependence of electrical resistivity on the isothermal annealing time for different annealing temperatures (the numbers at the curves)

in the isochronous annealing conditions in the range 425–430°C, which is in agreement with the data in [110].

In the isothermal conditions at 425 and 435°C, the kinetics was studied on the same specimens with different holding times. In other cases, a separate specimen was used for every variant of temperature and holding time. Figure 3.14 illustrates the crystallisation kinetics of the amorphous alloy determined on the basis of the electrical resistivity measurements. The data obtained by the method of measuring thermal EMF are on the whole in agreement with the results of measurements of electrical resistivity, but special features are detected at high temperatures (Fig. 3.15). In the initial stages of crystallisation at 435–465°C the curves of the dependence of thermal EMF show a minimum and after passing through this minimum the value of the thermal EMF is restored to the value characteristic

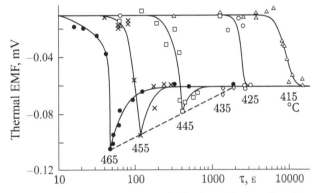

Fig. 3.15. Dependence of thermal EMF on the isothermal annealing time for different annealing temperatures (the numbers at the curves).

of the completely crystallised state. The depth of the minimum decreases with the reduction of annealing temperature and at 425°C, the thermal EMF rapidly reaches its stationary value. The duration of reduction of electrical resistivity to 50% of the initial value (prior to crystallisation) (the half-transformation time) rapidly increases with the reduction of annealing temperature, and in the Arrhenius coordinates the dependence become straight (Fig. 3.16). The effective activation energy of crystallisation is 450 kJ/mol.

The reconstruction of the crystallisation isotherms in the coordinate lg lg $[l/(l-\zeta)] = f(lg\ \tau)$ (ζ is the fraction of the transformed volume, τ is time) gives a family of parallel straight lines (Fig. 3.17), corresponding to the exponent in the Johnson–Mehl–Avrami equation $n \approx 3$.

The data obtained in transmission electron microscopy are in agreement with the course of crystallisation, shown in Fig. 3.13 and

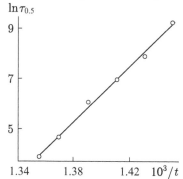

Fig. 3.16. Dependence of half transformation time on annealing temperature.

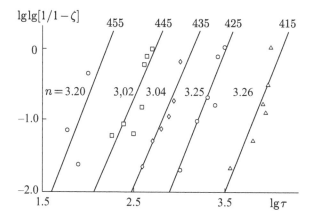

Fig. 3.17. Crystallisation kinetics at different temperatures (numbers at the curves) in the coordinates lg lg $[l/(l-\zeta)]–lg(\tau)$.

3.14. The sharp reduction of electrical resistivity corresponds to the precipitation in the matrix of the phase with the BCC lattice ordered in accordance with the type B2 and having the lattice spacing a = 0.30 nm. Separate single crystal circular inclusions form in the initial stage of crystallisation. Analysis of the images taking into account the angle of inclination of the specimen in the microscope showed that the crystals are almost spherical. The mean diameter of the crystals was approximately 0.5 µm and remained almost completely constant with the variation of temperature in heating and the holding time of the specimen at this temperature.

The mean grain size in the completely crystallised specimens at all investigated an annealing temperatures is similar to the given value of the spherical crystals in the early crystallisation stages: the mean grain size after crystallisation at 415°C was 0.6 µm, and at 465°C 0.45 µm.

The value of the exponent in the Johnson–Mehl–Avrami kinetic equation, published in [111], i.e. $n \approx 3$, together with the spherical form of the growing crystals indicates that crystallisation takes place by the mechanism of growth from a fixed number of centres [33]. The number of these centres in the unit volume N_V can be determined from the mean grain size D: $N_V = D^{1/3}$, which at $D = 0.5$ µm gives $N_V \approx 8 \cdot 10^{12}$ cm^{-3}.

The explanation of the approximate constancy of the grain size at the observed crystallisation kinetics can be explained as follows. The growth of crystals in the polymorphous crystallisation takes place at a constant rate $dr/d\tau$ = const (r is the actual radius of the crystal). Correspondingly, the volume of the crystalline phase increases in proportion to τ^3. In the time period in which crystallisation is recorded experimentally, the size of the crystals varies in a narrow range. For example, when the fraction of the crystallisation increases from 10 to 90%, i.e. nine times, the crystal radius increases on only by $9^{1/3} \approx 2$ times. This indicates that the final grain size after crystallisation is determined by the number of the crystallisation centres. To produce crystals with the size of, for example, 10 nm, the density of the crystallisation area should be 10^{18} cm^{-3}. Since the effective activation energy of the crystallisation process is 450 kJ/ mole, and the kinetics corresponds to the growth of spheres from the fixed number of the centres, the activation energy of the crystals equals one third of the effective activation energy of crystallisation, i.e. approximately 150 kJ/mole.

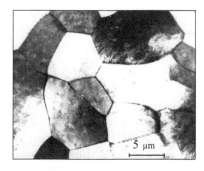

Fig. 3.18. The electron micrograph of the structure of $Ti_{50}Ni_{25}Cu_{25}$ alloy after annealing at 465°C for 1 min.

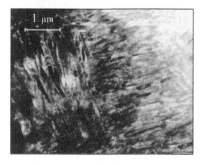

Fig. 3.19. Electron micrograph of the rapidly quenched $Ti_{50}Ni_{25}Cu_{25}$ alloy after isothermal crystallisation in the furnace; annealing 500°C, 1 h.

Fig. 3.20. Electron micrograph of the $Ti_{50}Ni_{25}Cu_{25}$ rapidly quenched alloy after dynamical crystallisation (rapid heating to 500°C, 100 ms).

X-ray diffraction and electron microscopic *in situ* experiments showed that in heating the $Ti_{50}Ni_{50-x}Cu_x$ ternary amorphous alloys with the copper content $x \leqslant 25$ at.% crystallised with the formation of the B2-phase in the entire volume (Fig. 3.18), whereas the alloys with a high copper concentration ($28 \leqslant x \leqslant 40$ at.%) may also undergo the eutectoid transformation with the formation of the two-phase mixture B2(TiNi)+B11(TiCu) in the final stage of crystallisation. Depending on the heating rate and methods, the final temperature in heating, and the chemical composition of the alloys, the mechanisms and dispersion of the products of devitrification can be greatly varied [Chapter 1 Ref.15] [113–116] (Fig. 3.19 and 3.20). The TiCu excess phase has the B11 tetragonal structure with the parameters $a = 0.310$ m, $c = 0.591$ nm, $c/a = 1.9$, the orientation relationships $[001]_{B2}//(011)_{B19}//(001)_{B11}$; $\langle 100 \rangle_{B2}//[100]_{B19}//[100]_{B11}$, the habit plane of the plate-shaped precipitates $\{100\}_{B2}//(011)_{B19}$ [117, 118].

The alloys forming the solid solutions – quasi-binary alloys, alloys with copper with the content lower than 20 at.%, always form in the main polycrystalline state in quenching like other investigated alloys which were however synthesised at lower cooling rates $v_c \leqslant 10^5$ K/s. At $v_c = 10^6$ K/s the electron diffraction data indicate that the alloys solidify with the formation of the sub-microcrystalline B2-structure with the grain size of 0.3–0.9 μm which is approximately 100 times smaller than the grain size of the initial cast alloys. Cooling at the limiting rate of $v_c \cong 10^7$ K/s reduces the mean grain size and increases the scatter of the dimensions in comparison with quenching at 10^6 K/s. The adjacent grains have a random high-angle disorientation but they can also have low-angle or special twinned interphase boundaries. The TiNiCu ternary alloys (Cu > 20 at.%) at $10^5 < v_c < 10^6$ K/s solidified with the formation in the volume of the ribbon of the highly heterogeneous, often bimodal grain structure [119].

Taking into account the data for the *in situ* diffraction experiments, for the heat treatment of amorphous alloys based on titanium nickelide we selected a number of isothermal and stepped conditions for the determination of the optimum conditions of nanocrystallisation of these alloys. The experimental results show that the TiNiCu alloys, which undergo eutectoid crystallisation in annealing at 450–500°C, acquire the nanocomposite structure in which the nanograins (>100 nm) of the main matrix in the volume of the B2 matrix alternate with the nanoinclusions of the excess phases, equilibrium with Ti_2Ni, TiCu or metastable phases based on them. In this case, the optimum heat treatment for obtaining the optimum properties is tempering at 500°C, 5–7 min. The higher or lower temperature or tempering time results in the embrittlement of these alloys. It is far more difficult to realise in melt quenching the process of nanostructured crystallisation in the TiNiCu alloys with a smaller degree of supersaturation (20 < x < 25 at.%) or generally unsaturated alloys (x < 20 at.% Cu). The nanocrystalline structure can form in this alloys only in a relatively narrow temperature–time range where the number of nuclei of the crystalline B2-phase, capable of growth, by the mechanism of volume crystallisation is sufficient and the growth rate of these nuclei is still not high.

As in the process of synthesis from the melt, low-temperature heating in the solid phase state is also accompanied by structural relaxation and, in some multicomponent systems, in particular, those based on B2–TiNi, by the near-range atomic ordering and phase separation. Cooling in melt quenching or subsequent heating, starting at specific critical temperatures, leads to crystallisation. Various structural mechanisms can operate but, in most cases, in actual systems the process starts with primary crystallisation of the metastable or equilibrium phases. Further development of 'bulk' crystallisation in synthesis and subsequent heat treatment of the amorphous alloys take place by the mechanism of volume or 'polymorphous', primary and eutectic (eutectoid) crystallisation, depending on the degree of supersaturation of the alloys. The difference between the mechanisms of volume and primary crystallisation is that after completing the process of crystallisation in the second case, as in eutectoid crystallisation, there may be more extensive secondary reactions of phase transformations, taking place simultaneously or successively in heating or cooling (for example, ageing of supersaturated solid solutions, eutectoidal breakdown).

Thus, the key role in the formation of the amorphous nanostructure states in the alloys based on titanium nickelide by melt quenching is played by the occurrence of considerable deviations from the equilibrium crystallisation conditions and, consequently, suppression of the thermally activated processes of structural and phase transformations in the solid state in them. In the limiting case of heterogeneities in solidification, amorphisation takes place especially in the highly supersaturated concentrated alloys of titanium nickelide. In the case of lower cooling rates or a lower concentration of the alloying elements in relation to the stoichiometric composition $Ti_{50}Ni_{50}$ only partial amorphisation can take place and this is accompanied by an increase of the mean size of the globular phases of the matrix and precipitates (from the nano- to the submicrosize). However, on the other hand, in subsequent heat treatment these alloys remain more stable in a wide range of annealing temperature and time both with respect to the submicrostructure and the properties.

3.3. Martensitic transformation in nanocrystals

3.3.1. The dimensional effect for martensitic transformations

It is clear that the scale factor should have a considerable effect on the shear (martensitic) phase transformations where the important role is played by the accommodative processes and also the nature of the lattice defects and the stress fields at the front of the growing crystal of the new phase. All the structural states for which the effect of the size of the initial phase on the parameters of the martensitic transformation has been investigated, can be divided into three large groups:

1. The polycrystalline materials with small grains;
2. Microcrystalline particles, distributed in the non-transformed solid-phase matrix;
3. Individual free particles in the form of powder.

Each of these structural states has specific dimensional effects, described in the literature. Evidently, the first two groups of the structural states are directly linked with the third group of the nanocrystals, produced by melt quenching.

Already in 1929, Scheil [120] showed that the transformation of austenite to martensite can be inhibited by refining of the austenite grains. For the Fe-based alloys, there are four groups of investigations in which it has been attempted to investigate the impact of the dimensional effect of the initial phase on the relationships governing the martensitic transformation. The first group includes the studies [121–125] in which the size of the initial austenite grains was varied by conventional treatment by the optimum combination of the deformation and thermal effects. It was shown that the reduction of the grain size in the Fe–Ni, Fe–Ni–C, Fe–Mn–C alloys always results in a reduction of the susceptibility to martensitic transformation (reduction of the temperature of the start of transformation M_s) and in a small change of the morphology of the martensite crystals, and the effect is exerted not only by the grain size [121–124] but also by the state of the grain boundaries [125].

The second group of the investigations was carried out in most cases on Cu–Fe alloys [126–130] in which the nanocrystalline (up to 2 nm) particles of γ-iron stable at room temperature formed in the copper matrix and this was followed by either plastic deformation at high strains or by extraction of the particles from the copper matrix by the electrochemical methods. After large deformation, the particles underwent martensitic transformation at any, even very small particle sizes [126, 127]. In addition to this, the particles of γ-iron transformed into α-Fe by the martensitic mechanism during extraction from the copper matrix, whereas the particles, containing a relatively high nickel concentration, transformed into the α-phase only partially or not at all. Subsequent cooling to low temperatures did not result in any additional formation of the martensitic phase [128]. The iron particles in the copper matrix with the size of approximately 200 nm had initially the martensitic structure [129]. At the same time, in the investigation of the particles with the size of up to 20–80 nm, the γ-phase was stable up to the liquid helium temperature, but after preliminary deformation at 77 K the largest particles transformed during cooling to the α-phase, and the critical size was approximately 50 nm [130].

The third group includes the investigations carried out on the ultrafine powders (10–200 nm) of Fe–Ni and Fe–N alloys [131-134]. On the whole, the initial γ-phase was stabilised but this was accompanied by an unexpected result: the existence at room temperature of the α-phase at the nickel concentration at which this phase is not detected in the conventional alloys [131, 134]. Consequently, it has been concluded that the transformation in the nanopowders is made easier at higher temperatures but more difficult at lower temperatures [132]. The calculations also show that there is nevertheless the critical size (approximately 10 nm for pure Fe and approximately 50 nm for Fe–6 at.% N), starting at which the martensitic transformation is 'locked' at any temperatures as a result of the fact that the size of the critical nucleus for the martensitic transformation becomes larger than the size of the nanopowder particles.

The fourth group of the studies analyses the nature of the martensitic transformations in Fe–Ni and Fe–Ni–C alloys produced by melt quenching and having the microcrystalline structure of the initial γ-phase [Chapter 1 Ref. 14] [135–140]. It has been concluded in all the studies that the microcrystalline state of the initial γ-phase, produced by melt quenching, lowers the martensitic point and leads to a number of structural special features of the martensitic phase, especially when the martensitic transformation takes place directly during melt quenching. Another important special feature was discovered in [140]: the martensitic point in the Fe–32% Ni alloy produced by melt quenching is 40–50° lower than in the alloy treated by the deformation–annealing technique at the same grain size of the initial γ-phase (2.5 μm). It is also interesting to note that the amount of martensite, formed in quenching from a temperature of 77 K, was considerably higher in the melt-quenched alloy.

Evidently, all these results indicate that the reduction of the size of the crystals of the initial (high-temperature) phase should be accompanied by the partial or complete suppression of the martensitic transformation. The process has specific features associated with the method of production of the microcrystalline structure of the initial phase. In [141, 142] systematic investigations were conducted into the main stages of the variation of the nature of the martensitic transformation with the reduction of the size of the crystals of the initial (high-temperature) phase and the theoretical interpretation of the dimensional effect with special reference to the martensitic transformation of different nature was proposed.

3.3.2. Martensitic transformation in nanocrystals, separated by the amorphous phase

In [143] it was attempted to produce the nanocrystals of the initial γ-phase distributed in the 'non-transparent' amorphous matrix and subsequently induce the martensitic transformation in the nanocrystals in the process of quenching to very low temperatures. Evidently, the areas of the amorphous phase, separating the individual crystals, permit the localisation of the martensitic transformation in the isolated areas of the initial matrix because, evidently, they 'dampen' the effects of elastic accommodation and create the conditions in which the martensitic transformation takes place independently in the individual crystals.

Nanocrystallisation takes place in the $Fe_{83-x}Ni_xB_{17}$ $(25 \leqslant x \leqslant 33)$ amorphous alloys, and the selection of the composition of the amorphous alloys is determined by the fact that this concentration range is characterised by the transformation of the type of crystal lattice of the primary phase, precipitated in the amorphous matrix (from the α-phase with the BCC structure to the γ-phase with the FCC structure) with the increase of the nickel content [141].

After annealing at 350°C for 1 h the $Fe_{50}Ni_{33}B_{17}$ alloy showed single cases of the martensitic transformation taking place during cooling of the alloy to the boiling point of liquid helium [143]. Transformation took place only in the particles relating to the largest dimensional fraction of the nanoparticles of the initial phase. It may be assumed that the controlling parameter of the susceptibility to martensitic transformation is the maximum size of the separated particles: the transformation takes place in the particles whose size is larger than the critical value. The critical size of the particles, surrounded by the amorphous matrix, for the occurrence of the athermal martensitic transformation in the Fe–(28–29)%Ni alloy is approximately 100–120 nm in cooling of the alloy to the boiling point of liquid helium.

Investigations were carried out into the martensitic transformation in cooling of amorphous alloys containing the nanoparticles of the Ti–Ni–Cu B2-phase which, like the Fe–Ni–B alloys, were in the amorphous state after melt quenching [142]. After annealing in the temperature range 400–450°C which resulted in primary crystallisation, isolated nanocrystalline particles of the B2-phase formed in the amorphous matrix. In the process of subsequent cooling to room temperature in the temperature range 30–100°C the nanoparticles underwent the B2→B19 thermoelastic martensitic transformation [143]. Figure 3.21a shows the typical electron microscopic images of the structure of $Ti_{50}Ni_{25}Cu_{25}$ alloy produced by melt quenching followed by isothermal annealing at 420°C for 1 h. In the process of subsequent cooling to room temperature the isolated particles underwent the martensitic transformation at 40-50°C. It can be seen that the smallest particles, in contrast to the larger ones, did not transform to the B19 martensitic phase. In order to determine reliably the 'locking' size for this type of transformation, the histogram of the size distribution of the isolated nanoparticles was constructed indicating the fraction of the particles of the given size subjected to the B2→B19 martensitic transformation (Fig. 3.21b). It is characteristic that in the range of the variation of the particle size of the initial B2-phase from 10 to 15 nm all the particles larger than 25 nm transformed completely. In the nanoparticles 15–25 nm in size the transformation was not complete, and the fraction of the particles undergoing the transformation decreased with the reduction of the particle size (Fig. 3.21b). Finally, add the nanoparticle size of less than 16 nm, which can be regarded as critical $R*$, the transformation was completed during cooling to room temperature.

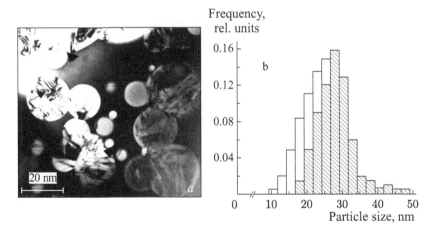

Fig. 3.21. Electron microscopic image (a) and the histogram of the size distribution (b) of the particles of the B2-phase in $Ti_{50}Ni_{25}Cu_{25}$ alloy. Cooling from 400°C to room temperature. The crosshatched area in the histogram indicates the particles undergoing the B2→B12 martensitic transformation.

To explain the special features of the martensitic transformation in this stage of the transformation it is in particular interesting to investigate the stress state and the deformation energy in the formation of the martensite crystal in the closed volume of the initial phase. The authors of [144] analysed the variation of energy during the formation of a martensitic nucleus in the closed volume (in a separate grain of the polycrystal, in an isolated crystal, in a foreign inclusion) in relation to the size of the closed volume under the condition of retention of coherence of the phases.

When solving the given task by the methods of the theory of elasticity it was assumed that the structural components of the alloy can be described by the equations of solid state mechanics. To determine the stress state of the initial phase, it is necessary to solve the so-called second boundary-value problem of the elasticity theory: determine strains and stresses inside the given elastic region, if the displacements at the boundary of the region are given. It is assumed that the martensite crystal has the form of a straight elliptical cylinder [145], and the actual case is characterised by the small ratio of the minor half axis of the ellipse c to the major half axis a [146] (Fig. 3.22). The centres of the ellipse and of the circular region coincide. The stress state in all the planes (xy) is the same. The selected model can be used to investigate the given task as a planar one and solve it by the methods of the theory of functions of the complex variable [147].

The Kolosov–Muskhelishvili complex potentials $\varphi(z)$, $\psi(z)$, are introduced, where $z = x + iy$ is the complex variable, $\bar{z} = x - iy$ is the conjugate quantity; $\bar{\varphi}$, $\bar{\psi}$ are the functions connected with φ, ψ, respectively. The equations determining the stresses in the complex plane have the form

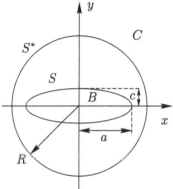

Fig. 3.22. Calculation of the conditions of nucleation of a martensitic crystal in the form of a straight elliptical cylinder in the circular region of the initial phase.

$$\sigma_{xx} + \sigma_{yy} = 2\left[\varphi'(z) + \overline{\varphi}'(z)\right] = 4\,\mathrm{Re}\,\varphi''(z),$$

$$\sigma_{xx} - \sigma_{yy} - 2i\sigma_{xy} = 2\left[\overline{z}\varphi''(z) + \psi'(z)\right], \qquad (3.14)$$

where the prime is the derivative with respect to z. The components of the displacements are expressed by the complex potentials as follows

$$U(z) = u_x + u_y = \frac{1}{2\mu}\left[\chi\varphi(z) - z\overline{\varphi}'(z) - \overline{\psi}\right], \qquad (3.15)$$

where $\chi = 3 - 4\nu$; ν is the Poisson coefficient.

The parameters relating to the inclusion are denoted by the index B, and the parameters of the circular region of the matrix surrounding the inclusion by index A. The boundary-value conditions for the coherent inclusion in the complex representation have the following form:

on the surface of the inclusion S at $z = z_s$ (z_s is the point in the contour)

$$\frac{1}{2\mu_A}[X_A\varphi_A - z\overline{\varphi}'_A - \overline{\psi}_A] - \frac{1}{2\mu_B}[X_B\overline{\varphi}_B' - \psi_B] = U_0(z), \qquad (3.16)$$

$$f_A(z) = f_B(z), \qquad (3.16a)$$

where f is the force. Here $U_0 = u_x^0 + iu_y^0$ are the free displacements of the material at natural distortion u_{ik} of the martensitic transformation $u_i^0 = u_{ik}^0$.

In addition to the boundary-value conditions at the interface of the matrix and the martensite inclusion S, it is also necessary to specify the conditions at the boundary S^* of the closed volume of the matrix (the grain, crystal, inclusion) in which the martensite crystal forms, for example, the equality of the displacements in this volume and the volume surrounding it or the absence of the displacements at the contour S^*:

$$\chi_A\varphi_A = z\overline{\varphi}'_A - \overline{\psi}_A = 0 \text{ at } z = z_{S^*}. \qquad (3.17)$$

Since the elastic deformation in the elliptical inclusion is homogeneous, the elastic displacements in the region B can be represented in the form

$$U_B = N_B z + M_B \overline{z}. \qquad (3.18)$$

Because of (3.16), the following relationship is fulfilled at the boundary S

$$U_A(z_S) = U_0(z_S) - U_B(z_S) = N_A z_S + M_A z_S = \frac{1}{2\mu_A}[\chi\varphi_A - z_S\overline{\varphi}'_A - \overline{\psi}'_A]. \quad (3.19)$$

The determination of the stress state is now reduced to the determination of the functions φ_A, ψ_A, analytical in the region A outside the elliptical contour S and satisfying the equations (3.17), (3.19), and the functions φ_B, ψ_B, analytical inside the ellipse B and satisfying the condition (5) at the boundary S.

We also use the effective method of solving the problems for the outer (in relation to the inclusion) region A – conformal imaging of the region B by the variable z in the region outside the unit circle γ of the complex variable ξ and the determination of the required analytical functions in the plane ξ. This conformal image is given by the functional:

$$z = R(\xi + m\xi^{-1}). \quad (3.20)$$

The parameters R and m are linked with the values of the major and minor half axes of the ellipse a and c by the relationships $R(a + b)/2$, $m = (a - b)/(a + b)$. The ends of the major axis of the ellipse in imaging (3.20) ($z = \pm R(1 + m)$) transfer to the points $\xi = \pm 1$, and the ends of the minor axis $z = \pm iR(1 - m)$ to the points $\xi = \pm i$.

For the potentials in region A we find:

$$\varphi_A = \frac{2\mu_A B_A}{\chi_A}\frac{1}{\xi}, \; \psi_A = \frac{2\mu B_A A_A}{\xi} + \frac{2\mu_A B_A}{\chi_B}\frac{1 + m\xi^2}{\xi(\xi^2 - m)},$$

$$A_A = R(N_A + M_A), \; B_A = R(N_A m + M_A). \quad (3.21)$$

These functions are used to determine the stresses and forces at the contour on the side of the matrix. Determining the forces at the contour on the side of the inclusions by the same method and equating them, we determine the missing coefficients.

The deformation energy of the inclusion is expressed by the work of surface forces acting at the boundary of the elastic region:

$$E_d = \frac{1}{2}\oint f U_0 dS = -\frac{1}{2}\mathrm{Re}\,i\oint U_0 dF = -\frac{1}{2}\oint_S \overline{U}(\xi_s)F(\xi_s)d\gamma. \quad (3.22)$$

At a small difference in the moduli of elasticity of the matrix and the inclusion and a purely shear deformation of the transformation $\varepsilon^*_{12} \equiv s$, we obtain the well-known equation [148] for the deformation energy of the elliptical cylinder in an infinite matrix

$$E_d = \frac{2\mu s^2}{1 - \nu}\left[\frac{4ac}{(a+c)^2}\right]\nu; \; \text{at } a \gg c: E_d \approx A\left(\frac{c}{a}\right)\nu, \; A = \frac{2\mu s^2}{1 - \nu}, \quad (3.23)$$

where $v = \pi ac$ is the volume of the resultant martensite crystal.

The displacements at the contour of the inclusion γ, where $\xi = \xi_s = \exp(i\varphi)$ are determined from the equation

$$U_B(\xi_s) = A_B \xi_s + B_B \xi_s^{-1}. \qquad (3.24)$$

Knowing $U_B(\xi_s)$, we can determine φ_B, ψ_B and also the forces at the contour.

Expanding the displacements U_i and the functions φ_A, ψ_A at the contour the inclusions into Fourier series with respect to ξ_s from the boundary conditions, we determine the non-zero expansion coefficients and from them the forces at both contours. Finally, from (3.22) we obtain the expression for the energy of the inclusion in the finite closed volume with the size $2R$ which in a general case can be determined in the form

$$E_{dR} = E_d + A_1 f(v/V)v, \qquad (3.25)$$

where $A_1 = \alpha\mu s^2/2(1+\chi)$, s is the shear transformation deformation, α is the numerical coefficient ≈ 1, f is the numerical function of the ratio of the volume of the inclusion v to the volume of the matrix $V = \pi R^2$. At $a \approx c$: $f(v/V) \approx (v/V)^2 + O(v/V)^2$, $a \gg c$: $f(v/V) \approx (v/V) + O(v/V)$.

On the basis of the results we can analyse the effects of the dimensional factor for the second stage of the martensitic transformation. The variation of the free energy (per unit length of the crystal), accompanying the formation of the martensite crystal in the finite volume V, is equal to [149]:

$$\Delta F = -\Delta f v + Ss\gamma + A(c/a)v + A_1(v/V)v, \qquad (3.26)$$

here Δf is the difference of the specific free energies of the initial and resultant phases; $S \approx 4a^2$ is the surface area of the inclusion; γ is the specific energy of the boundary of the phases.

The critical dimensions of the inclusions a^*, c^* are determined from the conditions

$$\frac{\partial F}{\partial c} = 0, \quad \frac{\partial F}{\partial a} = 0. \qquad (3.27)$$

With the accuracy to the quadratic terms with respect to (R^*/R), where

$$R^* = \beta a_0^* = \frac{8\beta A\gamma}{\pi(\Delta f)^2}, \quad \beta^2 = \frac{A^1}{A}, \qquad (3.28)$$

we obtain the expression for the energy of the critical nucleus:

$$F^* = \frac{16}{\pi} \frac{A\gamma^2}{\Delta f_{eff}^2}, \quad \Delta f_{eff} = \Delta f \left[1 - \frac{1}{2} \left(\frac{R^*}{R} \right)^2 \right].$$

(3.29)

The parameters a_0^* from the equation (3.28) and $c_0^* = 4\gamma/\pi\Delta f$ are the dimensions of the critical nucleus in the infinite matrix. The equation (3.29) includes the effective difference of the specific free energies of the phases Δf_{eff} taking into account the dimensional effect. At $R \gg R^*$ $\Delta f_{eff} \rightarrow \Delta f$.

In the classic theory of the nucleation [150] the temperature of the athermal martensitic transformation T_M is determined as the temperature at which the rate of nucleation reaches a measurable value ($\Delta f = \Delta f^*$). In [146] it is assumed that in cooling the nucleus grows spontaneously to the crystal of the measurable dimensions as soon as the chemical driving force becomes sufficient for the formation of the required surface energy and the deformation energy ($\Delta F = 0$). The condition of realisation of the martensitic transformation in the stage of intragranular localisation is that the parameter Δf should reach some critical value Δf^*. When the size of the crystal R reaches some critical value R^*, the effective difference of the free energies of the phases starts to depend strongly on R and rapidly decreases with decrease of R (Fig. 3.23).

Substituting into equation (3.29) for Δf_{eff} the value R^* from equation (3.28), we obtain the following condition for the formation of the martensite crystal in the initial small crystal:

$$\Delta f_n \geqslant \Delta f - KR^{-1/2},$$

(3.30)

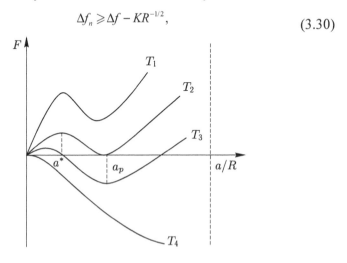

Fig. 3.23. The curve of the dependence of the free energy of the martensite crystal F on its size a/R at the equilibrium ratio (c/a) and at different temperatures.

where $K = k(A\gamma)^{1/2}$, k is a numerical coefficient.

It may be seen that the reduction of the driving force of the transformation as a result of the dimensional effect is inversely proportional to $d^{1/2}$. In the approach proposed in [148] the temperature of the start of transformation T_M in the volume fraction of the martensite phase in quenching to a temperature $T_q < T_M$ decreases in the same manner. We obtain the dependence (3.17), identical to the Hall–Petch equation.

Equation (3.29) can be used to determine the critical size R^{**}, starting at which the formation of the martensite crystals does not take place under any conditions: $\Delta f_{eff} = 0$, if $(R^*/R^{**}) = 2$. Consequently, taking into account (3.28)

$$R^{**} = 8\beta A\gamma\sqrt{2\pi(\Delta f)^2}.$$

(3.31)

For a system of isolated particles of the initial phase, the presence of the critical size of the particle 'locking' the martensitic transformation, R^* indicates that the transformation will take place only in the isolated particles of the initial phase for which $R > R^*$. In other words, the susceptibility to the transformation is determined by the largest nanoparticles, as confirmed by the experimental results obtained previously for the Fe–Ni–B and Ti–Ni–Cu alloys. This situation dramatically differs from the transformation by the relay mechanism in the polycrystal, where the mean grain size of the polycrystalline ensemble is controlling (not the largest size) [143].

Comparison of the results obtained for the Fe–Ni–B and Ti–Ni–Cu alloys shows that, regardless of the obvious differences (the athermal, non-thermoelastic nature of the martensitic transformation in the first case and the thermoelastic transformation in the second case), there are also notable agreements. Firstly, in both cases, the nanoparticles, surrounded by the amorphous matrix, undergo martensitic transformation, starting at the large dimensional fractions. There is a limiting size range for which the following relationship is fulfilled: as the particle size decreases the probability of the particle remaining non-transformed increases. Secondly, there is a critical size of the nanoparticles starting at which the martensitic transformation is completely suppressed. It should be stressed that the specific values of the dimensional parameters, mentioned previously, depend mainly on the chemical composition and the cooling conditions of the alloys. As indicated by the experiments, the resultant values of the 'locking' size in the Fe–Ni alloys are considerably higher than the 'locking' size in the $Ti_{50}Ni_{25}Cu_{25}$ alloy (16 nm). This is associated with a number of reasons: different transformation temperature range, different elastic accommodative conditions. However, this is associated mostly with the different nature of the martensitic transformation: non-thermoelastic (athermal and isothermal) transformation in Fe–Ni and the thermoelastic transformation in the Ti–Ni–Cu alloys. The nature of the accommodative processes is obviously a very important parameter. For example, in the powders where the accommodative processes are evidently easier, the 'locking' size for pure iron is 10 nm [142]. If these results are extrapolated to the alloy with 29% Ni, we obtain the size of approximately 100 nm which is considerably smaller that in the nanocrystal of the same composition. At the same time, the conditions for suppression of the martensitic transformation in the nanocrystal, surrounded by the amorphous matrix, are slightly different than those in the nanocrystal surrounded by the same nanocrystals.

In accordance with the previously described theoretical examination, we can calculate the critical size of the martensite crystal in which twinning is still possible (20–40 nm) or the dislocation deformation mechanism can operate [151]. The estimates show that at small grain sizes the twinned (athermal) martensite has better conditions for its formation: already at $d < 1$ μm one should expect that the formation of lath dislocation martensite will be suppressed and twinned martensite will form instead.

The assumption of the heterogeneous nucleation of martensite is generally accepted at the moment [136]. In most cases, the real materials contain a large number of defects around which the material can be in the form of the nanoscale regions of the new phase. Evidently, the heterogeneous nucleation of martensite crystals is possible in the vicinity of these defects. Each plate grows from a single nucleus nucleus representing the dislocation or other defective configuration, existing in the initial phase above the temperature of the start of martensitic transformation M_s. For the conventional coarse-grained materials, the concentration of martensite nuclei n_i is the parameter independent of the grain size [140]. In the materials with the nanocrystalline structure the situation is different, and the structural state of the material depends strongly on the size of the crystals $n_i = f(d)$.

3.3.3. Martensitic transformation in single-phase nanocrystals

Previously, we investigated the conditions for the occurrence of the martensitic transformation in isolated nanoparticles separated by amorphous interlayers, i.e. in the conditions of incomplete crystallisation for the nanocrystals of the third type.

A completely different pattern of the martensitic transformation is observed when nanocrystallisation has been completed and the initial phase is in the form of a statistical ensemble of the nanograins, contacting together and separated by the grain boundaries. In this case, the relay mechanism of the martensitic transformation operates (the autocatalytic effects) in which the resultant martensite crystals, arrested at the grain boundary, stimulate the formation of martensite crystals in the adjacent grains of the initial phase. The condition of interruption of development of every transformation region for the nanocrystalline materials is of the probabilistic nature and depends on the ratio of the size of the adjacent grains, length and nature of the grain boundaries, and a number of other factors.

Figure 3.24 shows the histogram of the size distribution of the grains of the initial γ-phase in the Fe–32% Ni alloy prior to transformation (a) and of the grains after martensitic transformation after cooling to 77 K (b). The fraction of the transformed volume

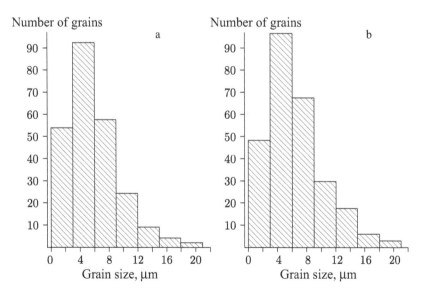

Fig. 3.24. Histogram of size distribution of the grains of the γ-phase in the Fe–32% Ni alloy prior to transformation (a) and in the grains of the γ-phase, after martensitic transformation after cooling to 77 K (b).

is 22%. It can be seen that the form of the histogram, describing
the transformed grains, is identical with the histogram describing
the grains of the initial phase. In addition, the mean size of the
transformed grains (d_m = 6.1 μm) is almost identical with the
mean size of the grains of the initial γ-phase (d_m = 5.8 μm).
If transformation started in the largest grains, the histogram in
Fig. 3.24b would be highly asymmetric and the maximum of
the transformed grains would be obtained for the largest grains
of the initial phase. The similarity of the histograms in Fig. 3.24
evidently shows that the mean grain size of the polycrystalline
ensemble controls the susceptibility of the material to martensitic
transformation. Electron microscopic analysis of the structure after
completion of the transformation confirmed this conclusion [152].
Martensitic transformation has often been detected in smaller grains,
whereas larger austenite grains remained untransformed.

If we examine the dependence of the fraction of the transformed
volume M on d_m of the initial phase for the Fe–Ni alloys, it may
be seen (Fig. 3.25) that for each alloy there is some mean grain
size d^*, corresponding to the complete 'locking' of the martensitic
transformation. In the framework of the 'relay' mechanism of the
martensitic transformation [153, 154], the dependence of M on d_m and

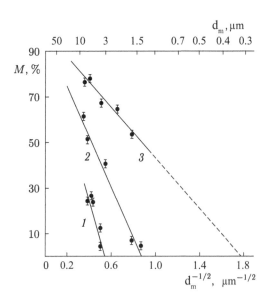

Fig. 3.25. The experimentally determined dependences of the fraction of the
transformed volume M on the mean grain size d_m of the initial γ-phase: 1) the alloy
with 32% Ni, 2) the alloy with 30% Ni, 3) the alloy with 29% Ni.

the expression for $d*$ can be analytically described by the following relationships [155]:

$$M = M_0 - K_M d_m^{-1/2},$$ (3.32)

$$d^* = \left[\frac{K}{T^* - T_3}\right]^2,$$ (3.33)

where $M_0 = K_1(T^* - T_q)$, T^* is the temperature of the thermodynamic equilibrium of the γ- and α-phases (with the accuracy to the coefficient), T_q is the final quenching temperature, K, K_1 and K_M are constants.

The dependences (3.32) and (3.33) describe quite accurately the experimental results (Fig. 3.25). The physical meaning of the critical mean grain size d^*, corresponding to the complete suppression of the transformation, follows from these considerations.

In the grains of the initial phase with the size smaller than the critical size, the only type of martensitic crystals which can form there are those that are not capable of initiating relatively high accommodative stresses required for the formation of the martensite crystals in the adjacent grains. In other words, the 'relay' is broken. The size distribution of the grains in the polycrystalline ensemble and, correspondingly, the probability of finding a grain with a relatively large size in the 'relay' process are of statistical nature and, consequently, the mean grain size of the initial phase is the critical parameter. Strictly speaking, within the framework of the autocatalytic mechanism it is important to consider not the mean but the most frequently encountered size (mode) of the initial crystal.

3.4. Mechanical properties

The problem of the strength and ductility of materials occupies the controlling position in the development of new structural and functional materials of a new generation because reliable service of these materials requires ensuring sufficiently high load carrying capacity and a certain strength margin and resistance to catastrophic failure. Extensive research in the area of the nanostructured materials science is accompanied by special attention given to the examination of the strength of nanomaterials because a large increase of the strength (hardness) and the reduction of the ductility of these objects have been observed quite suddenly. It is believed that the most detailed analysis of the mechanical properties of the nanomaterials has been published in [156]. Also, the tranditional dislocation approach to explaining the main relationships of the mechanical behaviour requires extensive corrections. All these details also relate to the nanocrystals produced by controlled annealing of the amorphous state which in turn was produced mostly by melt quenching.

3.4.1. Variation of the mechanical properties in transition from the amorphous to nanocrystalline state

The measurements of microhardness were taken for both single-component nanocrystals of Se [157] and also for single-phase ($NiZr_2$ [Introduction Ref. 16]) and multiphase (Ni–P [46], Fe–Si–Me–B [158]) systems. The established general relationship suggests that the microhardness usually increases after the formation of the nanocrystals in the amorphous matrix. In other words, the hardness of the nanocrystalline state, with a small number of exceptions, is higher than in the appropriate amorphous state. The generalised data are presented in Table 3.3 [Introduction Ref. 16]. However, when comparing the amorphous and nanocrystalline states, it should be remembered that crystallisation leads to a large (up to 50%) increase of the elastic modulus [19]. This may be the main reason for such a large increase of the hardness, although the physically more justified value $HV(\sigma_T)/E$ can also decrease in crystallisation. In [157] it is stressed that the mechanical behaviour of the completely amorphous metallic materials and the amorphous materials, containing the nanocrystalline precipitates of the second phase, is completely identical.

In both cases, the results show the anomalously high values of the yield limit at room temperature, mechanical instability (deformation softening), low strength and superplasticity or viscous flow at

Table 3.3. Values of hardness (GPa) of several nanocrystals, produced by crystallisation of the amorphous state, in comparison with the amorphous and coarse-crystalline states

Material	Nanostructured state	Amorphous state	Coarse-crystalline state
Ni–P	10.4 (9 nm)	6.5	11.3 (120 nm)
Se	0.98 (8 nm)	0.41	0.34 (25 nm)
Fe–Si–B	11.8 (25 nm)	7.7	6.2 (1 μm)
Fe–Cu–Si–B	9.8 (30 nm)	7.5	7.5 (250 nm)
Fe–Mo–Si–B	10.0 (45 nm)		6.4 (200 nm)
NiZr	6.5 (19 nm)		3.8 (100 nm)

elevated temperatures. Since the small crystalline particles usually do not take part in the plastic flow, the role of these particles in the hardening of the amorphous matrix is in fact reduced to two effects: deceleration of the movement of the shear bands, localised in the amorphous matrix, and the changes in the chemical state of the amorphous matrix (mostly the increase of the concentration of the atoms-metalloids capable of hardening, in the matrix).

More detailed examination shows that the hardening effect depends on the method used to change the grain size; in particular, there is a tendency for some softening of the nanophased materials in the measurement of the hardness of the specimens in which the grain size was varied by subsequent annealing. It was shown in a number of studies that when the individual specimens are subjected to annealing to increase the grain size, initial hardening and subsequent softening may take place. These effects were detected in the nanophased copper and palladium [160], TiAl [161] and NiP [162]. It has been assumed that this is associated with the reduction of the porosity of the compacted nanocrystals, with the transition of the grain boundaries from non-equilibrium to equilibrium configurations, and with local structural relaxation.

It is efficiently to study separately the amorphous alloys of the metal–metalloid and metal–metal type. In the first case, crystallisation is accompanied by the formation of borides, carbides, silicides, and high-strength phases, associated with the presence of the atoms-metalloids in the amorphous matrix. Correspondingly, the strength and hardness of these alloys greatly increase when crystalline phases forms. This effect is especially strong in the presence of borides in the structure. As an example, we consider the variation of hardness in the processes of crystallisation of the $Fe_{70}Cr_{50}B_{15}$ amorphous alloy

Figure 3.26 shows the dependence of microhardness HV on annealing temperature t at a constant holding time. It may be seen that there are two sharp maxima in the range 440–480°C, corresponding to doubling of the hardness.

As shown by electron microscopy experiments, after annealing, corresponding to the first hardness HV maximum, the amorphous matrix is characterised by the high density of the dispersed crystalline particles (in all likelihood, α-Fe or α-Fe–Cr solid solution (Fig. 3.27a). It is interesting to note that the first hardness HV maximum corresponds to the annealing temperature which is lower than the crystallisation temperature of 480°C, determined by calorimetric measurements. Isolated nanocrystals, formed as a result of the

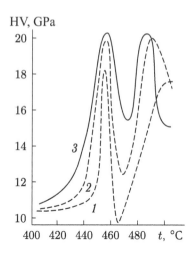

HV, GPa

Fig. 3.26. Dependence of microhardness HV on annealing temperature t_{ann} in the region of crystallisation of the $Fe_{70}Cr_{15}B_{15}$ amorphous alloy: 1) holding time 1 min; 2) 10 min; 3) 100 min.

hardening effect, correspondingly early crystallisation stages. The optimum ratio between the volume fraction, the size and the nature of distribution of the crystalline phase in the amorphous matrix results in a hardness maximum. Evidently, this is also the result of the fact that in the early stage of formation of the α-phase this phase contains a large number of the boron atoms.

The large reduction of the microhardness with the increase of annealing temperature in Fig. 3.26 is associated with the coarsening and increase of the volume fraction of the α-phase and, possibly, with the reduction of the boron atoms to the equilibrium concentration. The subsequently detected second high hardness HV peak is determined by the formation of dispersed borides in the structure. These borides form characteristic colonies of spherulites with the α-phase (Fig. 3.27b).

All the amorphous alloys of the metal–metalloid type become brittle after the thermal effect in the temperature range considerably lower than T_c. Therefore, the previously established anomalies of microhardness in the stage of transition from the amorphous to crystalline state are detected on the background of the almost zero plasticity of these materials. In this connection, tensile strength does not show such high values as microhardness because failure takes place in the elastic the region.

The low plasticity of the nanocrystals is associated with the suppression of the processes of nucleation and movement of the

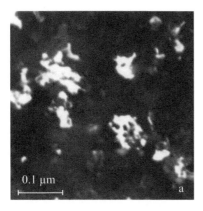

Fig. 3.27. The structure of the $Fe_{70}Cr_{15}B_{15}$ alloy corresponding to the first (a) and second (b) hardness HV maximum in Fig. 3.25: a) the dark field in the reflection of α-Fe; b) the bright field.

dislocations as a result of the small grain size. Nevertheless, a number of multiphase alloys with the nanostructure, for example Al–Mn–Li and Al–Cr–Ce–Cl, produced by crystallisation of the amorphous state, show a combination of high strength and satisfactory plasticity [163].

The unfavourable effects associated with the embrittlement can be overcome to a certain extent by heating the amorphous alloy at a very high rate. On the one hand, this suppresses to a certain extent the thermally activated processes forming the basis of temper brittleness of the amorphous state and, on the other hand, increases the number of crystallisation nuclei, capable of subsequent growth which in the final analysis results in the dispersion of the crystallisation products. Experiments were carried out on $Fe_{70}Cr_{15}B_{15}$ alloy [Chapter 2 Ref. 2]. A strip of the amorphous alloy was heated by direct passage of electric current. The actual heating rate was approximately 10^3 deg/s resulting in considerable refining of the crystallisation products (Fig.

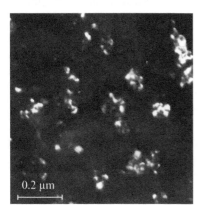

Fig. 3.28. Structure of partially crystallised Fe–Cr–B alloy, heated at a rate of approximately 10^3 deg/s by direct passage of electric current through the specimen. The dark field in the reflection of α-iron.

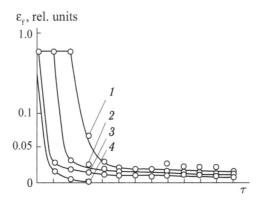

Fig. 3.29. Dependence of plasticity ε_f of $Fe_{70}Cr_{15}B_{15}$ alloy on the duration τ of passage of electric current of different intensity through the specimen: 1) 5.4 A; 2) 5.7; 3) 6.0; 4) 6.3.

3.28) and partial retention of the plasticity typical of the initial amorphous state (Fig. 3.29).

The idea that crystallisation may partially restore plasticity, typical of the amorphous state, was proposed in [164] where the appropriate experiments were carried out on (Fe, Co, Ni)–B amorphous alloys. As in any other amorphous alloys showing temper brittleness, one of these materials – $Fe_{84}B_{16}$ alloy – shows in annealing at 360°C a large decrease of the fracture stress under uniaxial tensile loading and the almost complete disappearance of plasticity (Fig. 3.30). At the same time, the appearance of the first disperse particles of α-iron is accompanied by the renewed growth of σ_p with increase of the annealing time.

The maximum value σ_p of the partially crystallised alloy depends on microstructure, annealing temperature (Fig. 3.30) and also on test temperature. It is important to note that in the alloy with the optimum content of the particles of α-iron, brittle fracture is replaced by microscopically ductile fracture. Here, as in the quenched amorphous state, there are indications of the characteristic 'veiny' pattern on the fracture surface. Comparison of Fig. 3.31 (showing the changes of the structural parameters of the amorphous–nanocrystalline state with the increase of annealing time at 360°C in the same $Fe_{84}B_{16}$ alloy) with the data presented in Fig. 3.30 shows that the maximum values of σ_p are associated with the volume density of the crystals of α-iron with the maximum particle size of 80 mm.

Longer annealing (as indicated by Fig. 3.30) results in a reduction of σ_p and ductility. As suggested in [164], this effect is associated

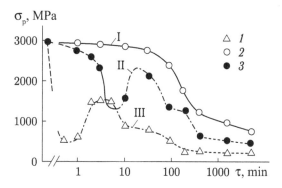

Fig. 3.30. Dependence of fracture stress σ_p for $Fe_{84}B_{16}$ alloy on the annealing time at different temperatures, °C: 1) 380; 2) 360; 3) 290; I – amorphous state; II – partially crystalline state; III – completely crystalline state.

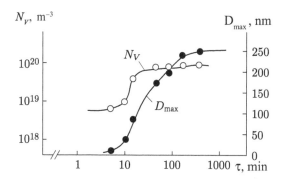

Fig. 3.31. Volume density N_V of the crystals of α-iron and maximum size of the crystals D_{max} depending on annealing time at 360°C.

with the increase of the volume fraction of the particles of the crystalline phase which contains no boron. This is accompanied by the increase of the boron concentration in the amorphous matrix which obviously blocks the migration of the excess free volume. In addition to this, the local stress fields become larger as a result of the variation of the specific volume. The interference of the internal and external tensile stresses may also provide a certain contribution to the susceptibility to brittle fracture.

In [165] investigations were carried out to determine the mechanical properties (cracking resistance and microhardness) of the amorphous alloys $Co_{75.4}Fe_{2.5}Cr_{3.3}Si_{17.8}$ (alloy 1) and $Fe_{58}Ni_{25}B_{17}$ (alloys 2) in the stage of partial transition to the nanocrystalline state under the thermal effect. The microindentation method [166]

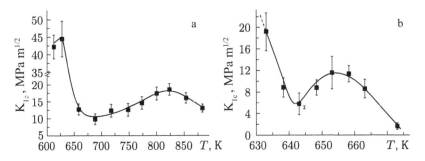

Fig. 3.32. Behaviour of cracking resistance K_{1c} in heat treatment of alloy 1 (a) and alloy 2 (b).

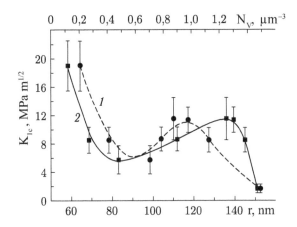

Fig. 3.33. Dependence of the cracking resistance on the size of crystalline particles (1) and their volume density (2).

was used to determine the behaviour of fracture toughness (K_{1c}) in the temperature range previously inaccessible to other methods from the ductile–brittle transition (in the region of existence of the amorphous state) to specific stages of nanocrystallisation. In the experiments carried out on alloy 1, the value of K_{1c} increase on approach to the crystallisation temperature (T_c = 823 K) (Fig. 3.32a). Previously, it was established that this alloy has a region of a large reduction of microhardness in the same temperature range [167]. Thus, in a certain annealing temperature range the alloy 1 showed the plastification effect (a reduction of K_{1c} with a large decrease of HV). The identical but larger plastification effect was also detected in alloy 2 (maximum at the annealing temperature of 653 K) (Fig. 3.32b).

Figure 3.33 shows the graphs of the dependences of the cracking resistance and the structural parameters of the investigated alloy 2.

Electron microscopic studies show a large increase of the size and volume density of the dispersed nanoparticles of the α-phase (BCC) in the region of the detected anomaly of mechanical behaviour. It may be seen that the characteristic size of the particles at which the maximum value of parameter K_{1c} was recorded corresponds to 110-120 nm at the volume density of the crystalline phase of 1.3 μm^{-3}.

The detected plastification phenomenon is caused evidently by the effective retardation of the quasi-brittle cracks formed and propagating in the amorphous matrix as a result of external loading. The possible mechanisms of the retardation of the propagating cracks in different materials have been analysed in detail in [168]. Unfortunately, the behaviour of the cracks in the amorphous and nanocrystalline alloys was not explained in this obviously very interesting publication. In addition, the retardation mechanisms investigated in [168] cannot be used directly for explaining the detected effect. The theoretical and experimental studies were used to propose a new original mechanism of retardation of the propagating quasi-brittle cracks in the amorphous–nanocrystalline state [165]. Figure 3.34 shows the characteristic electron microscopic image of a growing crack in alloy 2 after controlled annealing at 653 K, 1 h. It may be seen that the crack terminated in the area of the nanoparticles which appear to be an insurmountable obstacle in the path of crack propagation. The experimental results show that retardation of the cracks takes place in all likelihood not at the nanocrystal–amorphous matrix interface (both phases have in the present case similar values of the Young modulus) but in the vicinity of the nanoparticles on approach to it. This is associated with the fact that the atmosphere of boron and silicon atoms forms around every particles. These elements are present in large quantities in the amorphous alloy but the solubility in the nanocrystalline phase is low. The Young modulus of a similar atmosphere is considerably higher than in the amorphous matrix where the quasi-brittle crack propagates. Evidently, this should result in deceleration or complete arrest [168]. This conclusion can

Fig. 3.34. Crack deceleration at a particle.

be confirmed by, for example, the fact that the plastification effect in alloy 2 with boron is considerably stronger than in alloy 1 which contains silicon because boron is almost insoluble in the α-Fe–Ni nanocrystals and consequently forms larger atmospheres around the particles.

Thus, the coexistence of the amorphous and crystalline phases is capable of ensuring some increase of plasticity. As soon as the amorphous matrix is completely crystallised, plasticity becomes close to 0. Discussing the reasons for the plastification effect in the stage of precipitation of the α-Fe crystals, the authors of [164] concluded that the interphase boundaries of the amorphous and crystalline phases may become, under specific conditions, effective sources of the free volume required for the efficient course of the plastic deformation processes. It is interesting to note that similar effects can also be realised in a mechanical mixture of a corundum powder and the appropriate lubricant.

Effects of a similar nature were also detected in the annealing of amorphous alloys of the Ni–Ti–B type [169] in which the relaxation embrittlement takes place in the early stages of annealing in the temperature range 260–375°C and the increase of the annealing time resulting in the start of the early stages of crystallisation leads to partial restoration of plasticity. The optimum structure also contains dispersed (20–40 nm) precipitates of the crystalline phase which cause, according to the authors of [169], the more uniform course of the processes of plastic deformation in the amorphous matrix. As in the case of Fe–B alloys, further annealing increases the size of the crystals and leads to the disappearance of the plastification effect.

Recently, studies have been published concerned with the investigation of nanocrystals produced by crystallisation of large metallic glasses [Introduction Ref. 33] [170]. The results of development of new nanostructured and amorphous–nanostructured materials (on the basis of the bulk amorphised alloys) with higher elastic, damping and fatigue properties were published in [170]. The mechanical properties of the bulk alloys ($Zr_{55}Ni_5Cu_{30}Al_{10}$ and $Zr_{53}Ti_5Ni_{10}Cu_{20}Al_{12}$) were published in [Introduction Ref. 33]. Compression tests were carried out at room temperature. In the amorphous–nanocrystalline state this alloy show plasticity values very high for such materials. The results show that both strength and deformation to fracture increase with the increase of the volume fraction of the nanocrystalline structural component.

The variation of the mechanical properties in crystallisation heating of the amorphous alloys of the metal–metal type is slightly different. This will be investigated on the example of Co–Mo–Cr–Zr amorphous alloys [171]. In particular, it should be mentioned that the initial hardness of these alloys is considerably lower than that of the amorphous alloys of the metal–metalloid type with the metallic base of a similar type. Figure 3.35 shows the variation of microhardness in relation to annealing temperature for the $Co_{80}Mo_xCr_{10-x}Zr_{10}$ amorphous alloys and, for comparison, for the amorphous alloy of the metal-metalloid type $Co_{70}Fe_5Si_{15}B_{10}$. It may be seen that in contrast to the alloy of the metal–metalloid type in which the microhardness increases rapidly in the crystallisation stage, the alloys of the metal–metal type are characterised either by a small increase or by complete absence of any significant changes of microhardness. In addition to this, in a number of alloys of this type crystallisation reduces the strength [172].

However, in cases where the difference of the elasticity modulus of the amorphous and crystalline phases is large enough, crystallisation is accompanied, as in the metal–metalloid alloys, by a very significant increase in the strength characteristics. An example is crystallisation of amorphous Cu–Zr alloys, leading to an increase of almost 25% of the value of σ_p in uniaxial tension at room temperature [173].

Almost all the alloys of the metal–metal type show a weak tendency to temper brittleness, so that the temperature threshold of

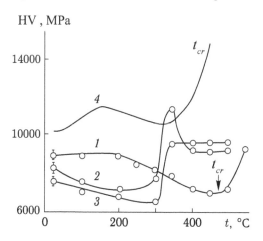

Fig. 3.35. Effect of annealing temperature for 1 h on the microhardness of $Co_{80}Mo_xCr_{10-x}Zr_{10}$ amorphous alloys (1–3) and $Co_{70}Fe_5Si_{15}B_{10}$ alloy (4): 1) $x = 0$; 2) $x = 2$; 3) $x = 10$.

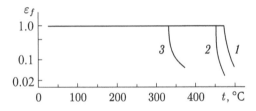

Fig. 3.36. Effect of annealing temperature for 1 h on the plasticity of amorphous alloys $Co_{80}Mo_{10}Zr_{10}$ (1), $Co_{80}Cr_{10}Zr_{10}$ (2) and $Co_{70}Fe_5Si_{15}B_{10}$ (3)

embrittlement t_{br} often coincides with the temperature of crystallisation or sometimes is even higher. This can be seen in the case of the same Co–Mo–Cr–Zr alloys (Fig. 3.36). Their t_{br} value is much higher than the Co–Fe–Si–B alloy. For example, the t_{br} value for the $Co_{80}Mo_{10}Zr_{10}$ alloy is 475°C, which is more than 100°C higher than for the $Co_{70}Fe_5Si_{15}B_{10}$ alloy. Appearance of the first small amounts of the crystalline phase does not lead to a significant change in the plastic properties. However, there are reports that by annealing one can create a ductile alloy with an amorphous–nanocrystalline structure. A number of multiphase nanostructured alloys (e.g. aluminum alloys Al–Mn–Ln and Al–Cr–Ce–Co, obtained by crystallisation from amorphous state, have the combination of high strength and good ductility [174]). It is possible not only dramatically improve the ductility, but also transfer the alloy into the superplastic state [175].

In [176] the authors investigated the mechanism of deformation of the $Pd_{80}Si_{20}$ alloy in micro- and nanocrystalline states obtained from the amorphous state by different heat treatments. The study [176] was published in 1983, when the concepts of nanocrystalline materials were not yet introduced, so the authors wrote about a study of a microcrystalline alloy. After melt quenching, the $Pd_{80}Si_{20}$ alloy was annealed at 563 K for 46 and 63 h. Annealing for 63 h transferred the alloy to the microcrystalline state with a grain size of 0.5 μm and the alloy became brittle. Upon annealing for 46 h the AMC changed to the nanocrystalline state with a grain size of about 5 nm, and such material retain its ductility. The small fraction (not more than 5%) of the amorphous phase was also retained.

The nanocrystals showed a localized plastic flow, completely similar to what occurs in the metal glasses. Bend tests were carried out at 78 K and room temperature to study the structure and morphology of the steps, resulting from the transfer of yield shear bands to the surface of the samples. Microcraks formed at 78 and

300 K on the stretched sample surface, usually on the steps of the output lines on the free surface. The shear deformation in shear bands in nanocrystals to the start of crack propagation is much lower than in amorphous alloys. On the compressed surface of the samples shear bands formed later than on the stretched surface. In nanocrystalline samples deformed at room temperature, the traces of exit of the shear band to the surface, in contrast to the amorphous samples, were wavy, short and often crossed. The steps were smooth. The intersecting traces of the exit bands are the result of secondary shear processes at room temperature. After deformation at 78 K, the nanocrystals showed clear surface steps, similar to those observed in the amorphous alloys from which they were produced.

In [177] the strain relief formed on the sample surface was studied. The relief was observed over the entire surface of nanophase samples and had small relief steps. It was concluded that it reflects the homogeneous nature of plastic deformation of the Fe–Cu–Si alloys in the nanocrystalline state in both the creep conditions and in active tension in the temperature range 573–773 K. However, in the tests with active tensile deformation the strain bands were arranged in two directions at 45° to the tensile axis, and in creep they were small and sinuous and almost perpendicular to the sample axis.

In addition to microhardness, tensile diagrams were plotted for a number of nanocrystals, obtained by crystallisation of the amorphous state, and the values of tensile and yield strength and tensile strength were determined [Introduction Ref. 16] [178, 179]. In [178] the tensile diagram was constructed using wire samples, and the calculated (over the cross serction of the wire samples) tensile strength of the FeCuNbSiB alloy was significantly higher (2.18 GPa) than for the ribbon samples of the appropriate alloy in [179]. The data presented in [179] data show that the alloy in the amorphous state may have the ultimate strength both higher (Fe–Cu–Nb–Si–B) and lower (Fe–Co–Si–B) than the values of the strength of the corresponding nanophase alloy. However, according [179], the ductility always increases in transition of the amorphous alloy to the nanocrystalline state.

3.4.2. Dependence of microhardness and yield strength on nanograin size

Typically, reducing the grain size to nanometer values results in a 4–5-fold increase in the hardness and yield limit. This follows directly from

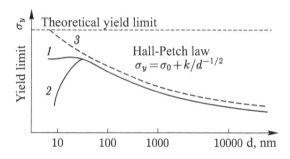

Fig. 3.37. The dependence of the yield stress on the grain size of the material: 1, 2 – experimental curves in the region of violation of the Hall–Petch law – 3 [180].

the well-known and repeatedly experimentally verified dependence of the hardness or strength (in general – deformation stress) on the grain size (Hall–Petch law) [Introduction: 20] (Fig. 3.37):

$$HV(\sigma_{\mathrm{T}}) = H_0(\sigma_0) + k_y d^{-1/2}, \qquad (3.34)$$

where HV is hardness, σ_{T} the yield strength, H_0 is the hardness of the the grain body, σ_0 is the internal stress, preventing the spread of plastic shear in the grain body, k_y is the proportionality factor and d is the average grain size.

Reducing the grain size by two orders of magnitude (from 1 to 10^{-2} μm) we can expect to increase the yield limit and hardness by at least an order of magnitude. If we extrapolate this dependence to the maximum achievable grain size (see Fig. 3.37), we can reach the theoretical shear strength. Nanocrystallisation of the amorphous state can produce materials with extremely small grain sizes. Therefore, the results of microhardness measurements of the third type nanocrystals have established the feasibility of the Hall–Petch law for the maximum achievable grain sizes in the nanometer range.

Nanophase Cu, Pd and Ag [181] with a grain size of 5 to 60 nm, produced by compaction of the ultrafine powders, showed hardness values 2–5 times higher than that of the samples with the usual grain size. Moreover, the hardness of nanocrystalline copper is superior to that of coarse-grained copper [160].

Figure 3.38 provides information on the dependence of the HV hardness on the grain size of the nanocrystals obtained by annealing the amorphous state [Introduction Ref. 16]. We can clearly see a fundamentally different nature of the dependence of HV. The Hall–Petch law of the Fe–Cu–Si–B nanocrystals has the usual character for the nanocrystal sizes down to 25 nm. For the Fe–Mo–Si–B alloy

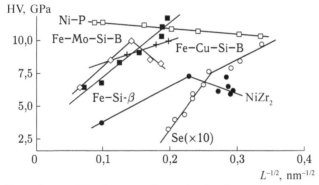

Fig. 3.38. Dependence of hardness on the grain size for nanocrystals, obtained by crystallisation of the amorphous state.

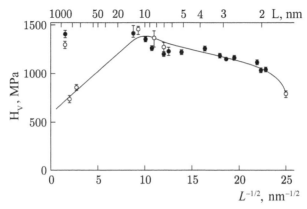

Fig. 3.39. The hardness of the grain size of the Ni–25 at.% W nanocrystal, obtained by electrodeposition and subsequent annealing at different temperatures in vacuum ($\circ$) or air ($\bullet$).

there is a critical grain size (47 nm), below which the dependence becomes anomalous. The characteristic maximum on the dependence of the microhardness on the grain size was also observed for the Ni–W nanocrystals, obtained by electrodeposition (Fig. 3.39) [182]. The anomalous behaviour of ceramic materials with decreasing grain size to $L_{min} \approx 1$ mm was analyzed in detail in [183].

A similar behaviour is also observed for a single-phase nanocrystalline $NiZr_2$ intermetallic compound with a critical grain size of approximately 19 nm [184].

The study of the dependence of the microhardness of the NiP nanocrystal on the grain size is given in [46]. The critical grain size at which the Hall–Petch law starts to be violated was also determined [46]. However, it appears that for NiP the anomalous dependence of the Hall–Petch law is observed for the entire nanometre grain size

Table 3.4. Structural characteristics of investigated Ni–P alloys [185]

Sample	A	B	C	D	E	F	G	H
T, K/ t, min	593/12	599/9	608/4.3	618/4	628/35	643/3	653/3	693/7.7
D, nm	8	13	20	44	52	68	85	107

range from a few nanometres to 120 nm [46]. The dependence of microhardness of the grain size is well described by the relation:

$$HV = 11.8(\text{GPa}) - 4.0(\text{GPa}\cdot\text{nm}^{1/2})d^{-1/2}(\text{nm}^{-1/2}). \qquad (3.35)$$

Evidently, the dependence (3.35) is essentially the inverse Hall–Petch law.

In [185] attention was also given to the behaviour of the fracture stress and ultimate strain to failure depending on the scale of the structure of nanocrystalline Ni–P alloys with a grain size from several to hundreds nanometres. The variation of these properties in dependence on the volume fraction and the specific area of the intergranular boundaries is discussed. Samples with different grain sizes were obtained by changing the heat treatment conditions of the amorphous precursor (Table 3.4). The mean dimensions of the samples were determined by the width of the half maximum amplitude of X-ray diffraction lines, as well as through high-resolution electron microscopy.

As in the case of amorphous alloys, nanocrystals were tested by the bend test using semi-circular specimens produced during nanocrystallisation [1]. The bend test samples were in the form of segments (strips) 2.5 mm wide and 0.03 mm thick, taken from the amorphous ribbon. After annealing under the specified conditions, the samples in the form of half-rings having an outer diameter of 5 mm, were placed between the parallel plates and compression tests were conducted with the closure speed of plates 0.25 mm/s at room temperature. The magnitude of the resultant force F of the bent samples was recorded as a function of the distance between the plates S. In the bend tests of the amorphous alloys, the relationship between the force acting on the plate and the distance between them in the elastic and plastic regions is well known (e.g. [186]). Relative strain to fracture ε_f is defined as:

$$\varepsilon_f = t/(S_f - t), \qquad (3.36)$$

where t is the thickness of the ribbon, S_f is the distance between the plates at the time of fracture. The dependence $F(S)$ does not show any deviations from elastic behaviour, i.e. the stressing of the samples was macroscopically brittle at stresses below the yield limit.

Fracture stress was determined by the maximum force at fracture:

$$\sigma_f = F_f / \omega t \qquad (3.37)$$

where w is the ribbon width. Fracture stress and fracture strain increase almost linearly with decreasing grain size. It was concluded that the observed dependence is associated with the measured increase of the free volume at the interphase boundaries.

Figure 3.40 shows the data for nanocrystals obtained by annealing the amorphous state of $Fe_{73.5}Cu_1Nb_3Si_{13.5}B_9$ [187], $Fe_{81}Si_7B_{12}$ [188] and $Fe_5Co_{70}Si_{15}B_{10}$ [189] (c, d), and also for the nanocrystals obtained by other methods (a, b). The results for microhardness and yield limit for different sizes of the nanograins (nanophases) d indicate that the Hall–Petch ratio is almost always satisfied if $d \geqslant 10$–20 nm. For smaller values of the size of the nanophase the Hall–Petch relationship is valid only for the $Fe_{73.5}Cu_1Nb_3Si_{13.5}B_9$ alloy.

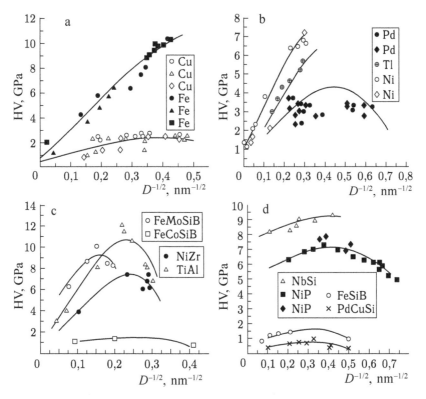

Fig. 3.40. Dependence HV (d) for nanocrystals produced by different methods: a, b – compacting powders, c, d – annealing the amorphous state.

However, the method for producing nanocrystals has a noticeable effect on the character of the $HV(d)$ dependence.

In [190], parallel studies were conducted of the structure and microhardness of the nanocrystals Fe–Mo–Si–B with different grain sizes obtained by crystallisation of the appropriate amorphous alloy. The dependence of microhardness of $d^{-1/2}$ can be described with good accuracy by two straight lines, with the abscissa of the intersection point equal 45 to nm:

$$HV\,(\text{GPa}) = 4.28\,(\text{GPa}) + 36.20\,(\text{GPa}\cdot\text{nm}^{1/2})d^{-1/2}\,(\text{nm}^{-1/2}),$$
$$d > 45\,\text{nm},\tag{3.38}$$

$$HV\,(\text{GPa}) = 13.75\,(\text{GPa}) - 23.77\,(\text{GPa}\cdot\text{nm}^{1/2})d^{-1/2}\,(\text{nm}^{-1/2}),$$
$$d < 45\,\text{nm}.\tag{3.39}$$

Measurements using electron-positron spectroscopy of the volume fraction of the grain boundary free volume and of the subnanopores showed that the dependences of the microhardness and the volume fraction of the nanopores on the grain size are similar. The authors conclude that the presence of the negative branch of the Hall–Petch dependence is due to changes in the structural state of GBs when the grain size varies at $D < 45$ nm.

In [191], the authors presented data which suggest that the change in yield strength or hardness with the grain size is described by the dependence, for which the exponent n is not equal to the usual value $-1/2$, and ranges from -1 to -1.4. Each of these values, obtained from the theoretical estimates, corresponds to its own mechanism of interaction of dislocations with the GBs.

Summing up the set of all the presently known experimental data, one can conclude that the hardness of metals and ceramic materials increases as the size of the grain transfers to the nanophase area. However, the value of the grain size up to which the hardening takes place, depends on a number of factors, and its nature is not yet clear. Usually, the Hall–Petch law holds for a large part of the investigated nanocrystals only to a certain critical grain size, and the opposite effect is observed at the lower values: the hardness (strength) falls with the decrease of the size of nanograins. It is characteristic in this respect that the sample affected by annealing to increase the grain size has higher hardness values than the sample that had exactly the same grain size immediately after preparation [192].

If the nanocrystals are of a sufficient size for uniaxial tensile tests (the longitudinal dimension is much larger than the lateral dimensions, which, in turn, is significantly larger than the grain size), we can determine the limit yield and ultimate strength in uniaxial tension. The received results are so far mainly for the FCC metals. An increase of strength, similar to an increase in hardness [181, 193] is observed. In this case, the nanocrystals exhibit a very low ductility. The degree of increase in tensile strength at a reduction of the grain size to 25 nm was of the same order as that in conventional cold-rolled of the polycrystalline material [193]. The low level of plasticity is clearly linked with the restriction of the nucleation and propagation of plastic shear and, in addition, possibly with the presence of a certain number of cracks in the initial state of the nanocrystals.

In a number of studies attention was given to the high-temperature mechanical properties of nanocrystals. In [194] the authord studied the creep kinetics of Ni–P alloy with an average grain size of 28 nm and, for comparison samples with a grain size of 260 nm in the temperature range from 543 to 593 K. It was found that in the same test conditions the creep rate of nanocrystalline samples is much higher than that of the samples with larger grains. The activation energy of viscous flow for nanocrystals and microcrystalline samples was 0.71 and 1.1 eV, respectively. It was concluded that in the nanocrystalline state the flow rate is controlled by grain boundary diffusion, and in the microcrystalline materials by lattice diffusion.

In [177] the structure and mechanical behaviour of the Pd–Cu-Si alloys in amorphous and nanocrystalline states was studied. It was shown that in crystallisation of these alloys in the creep mode it is easy to produce grains smaller than 10 nm. In this case, the plasticity of these alloys in creep and active tension at elevated temperatures (573 and 773 K) greatly differs.

In [195] it was found that the ductility of the nanocrystalline $Al_{88}Ni_{10}Ce_2$ alloy increased after prior low-temperature annealing in the amorphous state and subsequent tensile strain of 45% in the temperature range 450–465 K. According to [195], low-temperature annealing of the amorphous alloy is accompanied by the precipitation of Al particles stimulating the formation of nanostructures. Deformation at elevated temperatures results in nanocrystallisation and significant plastic deformation.

As is well known, creep and superplasticity phenomena occur at elevated temperatures [179] and they are inherent to the nanocrystals. Their study is also in the focus of researchers (see, e.g. [192, 196, 197–205]). In the experimental study of these phenomena the main difficulties arise in in the preparation of certified samples and in preventing or strict accounting of their recrystallisation. The high-deformation characteristics are often studied by the indentation method [199–202, 206, 207]. The main results of these studied are as follows.

1. The phenomenological relationship between the strain rate and stress is fulfilled for the nanocrystals:

$$\varepsilon = AL^{-n}\sigma^{m}\exp\left(-\frac{Q}{RT}\right),$$ (3.40)

where A, n, m are the numerical coefficients, Q is activation energy. However, the spectrum of the coefficients n and m is quite large [198, 199, 204, 205]. In some cases there is a significant growth of the grains through dynamic recrystallisation, which results in a large reduction of the strain rate. This complicates the study of the superplasticity of metal nanocrystals, and the available data are mainly related to brittle objects. The growth of the grains can be limited by various techniques (porosity, doping of the grain boundaries, the formation of two-phase structures, etc.) [199].

2. The activation energy is usually close in magnitude to that of the grain boundary diffusion. According to [205], $Q = (510\pm33)$ kJ/mol, which is almost identical with the activation energy of grain-boundary diffusion of the Zr cation (506 kJ/mol), and this indicates the dominant role of grain boundary sliding.

3. The temperature dependence of the hardness of nanomaterials has a steeper form compared to the change of hardness for conventional polycrystalline materials [206, 201]. According to estimates [206], the brittle–ductile transition temperature of for TiO_2 is ≈430°C, which is 220°C lower than for conventional polycrystals.

4. The ductility indices, achieved in the experiments [199, 204, 205], are quite impressive: tens and hundreds of percent at moderately high temperatures (650–725°C for Ni_3Al, 700°C for TiO_2, 1150–1250°C for ZrO_2). By switching to the nanocrystalline state the superplasticity temperature was reduced by about 300–400°C in comparison with the to conventional fine-grained material, but the stress level is about one order of magnitude higher than that of the industrial superplastic metallic materials.

3.4.3. *Mechanisms of plastic deformation and the nature of the anomalous dependence of the Hall–Petch relationship in nanocrystals*

Evidently, the role of dislocations in plastic deformation of nanocrystals is negligible, since the dislocation plastic flow is usually accompanied by the formation of a distinct crystallographic texture. In fact, as shown by electron microscopic studies, nanograins can be contain only a very small number of dislocations [208]. In this case, they are often represented by stationary (sessile) configurations. The low dislocation density in the nanocrystals is due to the existence of imaging forces that push the movable dislocations from grains, especially the smaller ones. This happens in the same way as when a point charge is pushed out near the free surface of the conductor. In [209] in the calculation of molecular dynamics it was shown that individual dislocations can from in nanocrystals of α-Fe only if $d < 9$ nm. This effect, of course, negates the role of mobile dislocations in the development of plastic deformation, even if we assume that new sources of mobile dislocations can be activated in the nanocrystals when the external load is applied.

The determining role in the anomalies of the mechanical properties of nanocrystals is played obviously by the grain boundaries (GBs). Early studies of nanocrystals considered GBs as fully disordered layers in contrast to conventional materials, where grain boundaries are generally more ordered. However, observation by direct resolution electron microscopy revealed [Introduction Ref. 6] that the structure of the grain boundaries in nanocrystals is, in principle, similar to the structure of conventional high-angle grain boundaries. There are, however, observations showing that low-temperature annealing can be accompanied by the transformation of metastable configurations of grain boundaries to more equilibrium states [210]. Such grain boundary adjustment may be related either to the actual internal structure of the grain boundaries or to residual stresses existing in the nanocrystals after production and further processing.

An important factor determining the mechanical behaviour of nanocrystals are the internal stresses. They are always found nanosystems, at least because of the large number of closely spaced grain boundaries and triple junctions of the grains. In addition, the internal stresses can arise due to the nature of the method of obtaining nanocrystals.

We know many attempts to explain the effect associated with the deviation, small, very important, and sometimes cardinal, from the Hall–Petch relationship for nanocrystals. Initially, the dependence (3.34) was considered from the positions that GBs are barriers to the movement of dislocations, where the coefficient k_y defines the degree of 'transparency' of the grain boundaries for the dislocations. In the classical Cottrell theory, the dependence (3.34) linked the hardness (yield limit) with flat clusters of dislocations at the grain boundaries, which accumulated shear stresses for the activation of dislocation sources in neighboring grains. In further interpretation of the Hall–Petch dependence the plane dislocation clusters were neglected and the dislocation networks within the grains were considered instead, and the boundaries were regarded as sources of dislocations. In [211] attention was paid to the structure of grain boundaries and it was concluded that special boundaries lead to greater hardening of the polycrystalline ensemble of the grain boundaries than the general boundaries. In [212], the softening with a decrease of the grain size was explained by diffusion creep at the boundaried of nanocrystals in accordance with the Coble mechanism. However, this has not received serious experimental confirmation on the structural level. In some studies, such as [213], it is assumed that below a certain grain size the formation of flat dislocation clusters in crystals is limited which should lead to softening. There is a critical grain size L_{cr} below which flat clusters do not form. Its value can be calculated from the formula [181]

$$L_{cr} = \frac{3Gb}{\pi(1-v)Hv}. \tag{3.41}$$

A similar approach can explain only the inflection in the value of hardness with decreasing grain size, but is not able to describe the inverse Hall–Petch relationship, which, as we have seen, is observed in a number of nanocrystals. To explain this phenomenon, in [214] it is assumed that annealing leads to the relaxation of the intergranular structure of nanocrystals and thus to a lower excess grain boundary energy resulting in this anomaly.

A number of studies were aimed at explaining the observed (in a number of experiments) changes in the magnitude of the exponent at the quantity d in the Hall–Petch relationship. For example, in [191] the authors presented data from which it follows that the change in yield strength or hardness in dependence on the grain size is described by the dependence for which the exponent n is not equal

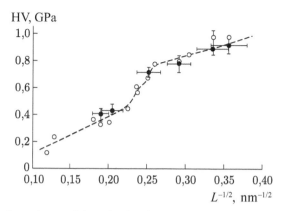

Fig. 3.41. The dependence of the microhardness on the grain size for nanocrystalline Se, obtained by crystallisation from the amorphous state.

to an ordinary value of $-1/2$, and ranges from -1 to -1.4. Each of the values, obtained from the theoretical estimates, are characteristic of the mechanism of interaction of dislocations with grain boundaries. It is also assumed that there is a connection between the type of crystal lattice and the value of n. In [215], measurements of the microhardness of nanocrystalline Se, obtained by crystallisation of the amorphous state, almost unaffected by porosity, impurities and segregation, showed that with a decrease in grain size from 70 to 8 nm there are three distinct stages, corresponding to different values of n in the Hall–Petch relationship (Figure 3.41). Maximum n is observed in the range of d from 15 to 20 nm.

    Optimization of the structure of nanomaterials to make them high strength with satisfactory ductility (i.e. the ability to deform without fracture) is the quintessence of many searches in nanostructured materials [216, 217]. Many nanocrystals are characterised by extremely low elongation in the nanocrystalline state and by the lack of ductility along with high strength [217]. The decrease of the crystallite size for FCC nanocrystals decreases the value of the activation volume v^*:

$$v^* = b\xi l^*$$
(3.42)

and increases the rate sensitivity coefficient of flow stress m

$$m = \left(\frac{\partial \ln \tau}{\partial \ln \dot{\varepsilon}}\right)_T = \frac{k_B T}{\tau v^*},$$
(3.43)

where ξ is the barrier width, l^* is the distance between the forest dislocations, τ is flow stress, $\dot{\varepsilon}$ is the strain rate, k_B is the Boltzmann constant, T is the deformation temperature [156, 218]. For example, reducing the grain size from 100 nm to 20 nm decreases v^* from $\sim 30b^3$ about three times, and the value of m is approximately doubled from ~ 0.02 [219]. For conventional FCC metals $v^* \sim (10^2–10^3)b^3$, and the value of $m = 0.5$ (plastic deformation), or $m = 1$ (diffusion creep).

Eight basic methods of increasing the ductility of nanomaterials based on metals and alloys were described in [216]: 1) the formation of bimodal structures in which the nanocrystalline matrix provides high strength, and the presence of larger inclusions promotes acceptable plasticity [220]; 2) formation of multiphase compositions; 3) the formation of twin structures [221]; 4) preparation of dispersion-strengthened alloys; 5) the use of plasticity effects induced by phase transformation (TRIP) and plasticity effects induced by mechanical twinning (TWIP); 6) dynamic annealing at low temperatures; 7) increased of the coefficient of rate sensitivity of the flow stress up to the values inherent in superplastic alloys ($m \sim 1$), including by repeated mechanical treatment [218, 222]; 8) use of the consolidation methods of nanomaterials ensuring the complete absence of pores and microdiscontinuties in the structure which due to the appearance of stress concentrators negate the positive effects of the nanostructure.

Not all the options in this list are undoubtedly applicable to nanomaterials, but for the methods 1, 3, 7 and 8 there is already experimental evidence of their effectiveness. For example, copper samples with the nanostructured matrix with a grain size of 80–200 nm (75% vol.) with inclusions 1–3 mm in size (25% vol.) showed good indicators of strength ($\sigma_y = 400$ MPa) and ductility (elongation $\delta = 65\%$) [58]. Interestingly, the change in strength depending on the thickness of the twin nanolamelae follows the Hall–Petch relationship [221].

In [223, 224–226] the relationships of the strength and ductility of bimodal structures of nanomaterials (nanocrystalline matrix with microscopic inclusions of larger sizes [220]) were investigated theoretically. The tensile deformation behaviour of these objects, treated as two-phase formations [225, 226], was calculated from the additive considerations in accordance with the rule of mixtures

$$\sigma(\varepsilon)=(1-f)\sigma''(\varepsilon)+f\,\sigma'''(\varepsilon), \qquad (3.44)$$

where f is the volume fraction of microinclusions, σ'' and σ''' are the strength values of unimodal nano- and microstructures. The 'two-phase' approach (grain body + grain interface) is very common in nanostructured materials science and has been analyzed in detail on the example of plastic and elastic properties in [227, 228].

In addition to the above ways of increasing the plasticity of nanocrystals, we should mention the regulation of plasticity by the variation of stacking fault energy (γ_{sf}), which, as is well known, determines the probability of formation of partial dislocations and twins [229].

The starting point for understanding the nature of the plastic deformation of nanocrystals is apparently the study [230], where the authors first proposed athermal grain boundary microsliding as the main mechanism of plastic flow under high shear stresses, which is realised as an alternative to the classical dislocation flow at low temperatures. The model was developed further in [26, 231]. In this model, the relationship between stress and strain (ε) has the form

$$\sigma_y(\varepsilon) = \sigma_S + AG\left(\frac{L}{L'}\right)\frac{\varepsilon}{qhn^2}, \quad \varepsilon < \varepsilon^* = nqh\left(\frac{2s}{L}\right)\gamma^*, \qquad (3.45)$$

$$\sigma_y(\varepsilon) = \sigma_y(\varepsilon^*) + \alpha G\left(\frac{kb\varepsilon}{n\varsigma L}\right)^{1/2}, \quad \varepsilon > \varepsilon^*, \qquad (3.46)$$

where $A = \pi(2 - v) / 4 (1 - v)$, v is Poisson's ratio, G is the shear modulus, σ_S is the shear stress at the flat grain boundary, L' is the average size of the region of grain boundary microsliding, $2s$ is the boundary width, γ^* is the fracture energy, n is the mean orientation factor; q, h, α, k, ς are numerical parameters [26]. In the initial stage of deformation ($\varepsilon < \varepsilon^*$) shear is realised by grain boundary sliding in individual sections of the grain boundaries. Accommodative processes considered here include disclination rotation of grains and the formation of dislocations at the grain boundaries, which then ($\varepsilon > \varepsilon^*$) are emitted from the grain boundaries into the volume.

With decreasing grain size of the nanocrystalline ensemble the classic dislocation flow is smoothly exhausted, giving way to grain boundary microsliding, which becomes easier with decrease of the grain size. From our point of view, this the main reason for the anomalous behaviour of the Hall–Petch relationship. Given the low intensity of grain boundary deformation, this mechanism was subsequently clarified and detailed for the case of bimodal structures [223]. Subsequently, this model has been repeatedly confirmed in both direct and computer simulations [232]. Grain boundary microsliding is most efficient in nanocrystals produced by controlled annealing of the amorphous state, where amorphous grain boundary layers form and where the the 'anti-Hall–Petch' effect is most pronounced [[Introduction Ref. 5].

Similar in nature, the model was considered in [233], in which the plastic shear in nanomaterials with the direct and inverse Hall–Petch effect is proposed to be based on the following relay mechanisms: dislocation (translational) mode at the expense of grain boundary sliding, and the rotational (disclination) mode is implemented through the concerted processes taking place inside the grains and at the boundaries. It has been shown that the transition from the translational to rotational deformation and back may explain the inverse Hall–Petch effect.

However, at the same time no attempts have been made to describe the mechanical behaviour of the nanocrystals by classic dislocation considerations. Apparently, these models can be useful in describing the plastic flow in the 'large' nanocrystals with the grain size is close to 100 nm. In [234] the authors propose a dislocation–kinetic approach based on the accumulation and annihilation of dislocations with the grain boundaries, which play the role as barriers to distribution of dislocations, and their sources and sinks taken into account. After some simplification, the expression for the dependence of the yield limit on strain and grain size has the form

$$\sigma_y = f\alpha G\left(\frac{b}{L}\right)^{1/2}\left[\beta_0 \exp(-f\,k\varepsilon) + \frac{\beta}{k}(1-\exp\,(-f\,k\varepsilon))\right]^{1/2}, \qquad (3.47)$$

where $f \sim 3$ (the Taylor factor for a polycrystal); k is the total factor of dislocation annihilation, depending on L, strain rate and temperature; α is the coefficient of the interaction of dislocations with each other; G is the shear modulus, b the Burgers vector; β_0 is the parameter, depending on the density of grain boundary dislocations; β is a factor determining the accumulation of dislocations and strain. It is easy to see that when $f k\varepsilon \ll 1$ (i.e. with relatively large grains) the yield limit depends on the grain size according to the Hall–Petch relationship and $k_y = f\alpha Gb^{1/2}(\beta_0 + f\beta\varepsilon)^{1/2}$. For small grain sizes ($fk\varepsilon \gg 1$) the dependence $\sigma_y = f(L)$ takes the form inverse to (3.34), and it can be shown that in the downward section of the curve $\sigma_y \sim L$ or $\sigma_y \sim L^{1/2}$, depending on whether single or paired annihilation of dislocations dominates at the boundaries.

Bringing calculations by expressions (3.45)–(3.47) to the numerical results and their comparison requires knowledge of many parameters and requires a number of assumptions. However, according to [235], the calculated values of bending strength according to the type (3.34) $\sigma_y = f(L^{-1/2})$ for Cu and Ni amounted to ~0.25 nm$^{-1/2}$ and ~0.3 nm$^{-1/2}$, which coincides well with the experimental data. Further, this approach was extended to assess the impact on strength of dispersion [235], bimodality [226] and on the analysis of the rate sensitivity of the flow stress [236].

A detailed kinetic analysis of the behaviour of dislocation clusters as barriers at the grain boundaries of the FCC nanometals (range $L = 10-100$ nm) is given in [237]. Considering the rate of changes in the critical shear stress as a thermally activated process, and taking for τ the relation of type (3.34), expressions were obtained for the activation volume and the coefficient k_y.

In a series of studies [227, 238–242] different aspects of the physical mechanics of deformed nanostructures were generalised and analysed: anomalous dependence of the yield strength, the localization of the plastic flow, rotational modes, the role of triple junctions, heterogeneous and homogeneous nucleation of dislocations. For example, based on the energy approach and using nickel and cubic silicon carbide as examples, both the heterogeneous nucleation of dislocation loops at the interfaces and the homogeneous nucleation of full and partial dislocations under stresses (1–7 GPa) in the volume of deformed nanograins was investigated [240, 241]. It was shown that if the size of the emerging loop and the action of the shear stress on the loop are large enough, the formation of a loop can be barrier-free, i.e. athermal. It is also important that the deformation processes in nanostructures of brittle refractory compounds and metals, though different, were considered together. Data of the studies [240–242] give an idea of the possible sources of generation of dislocations for intragranular deformation.

We have already noted (see equation (I.1b)) a substantial increase in the volume fraction of regions occupied by triple junctions of grains in the nanomaterials. One approach to the explanation of the anomalies is associated with the controlling role of triple junctions, as a significant increase in their volume fraction in a polycrystalline system can positively impact on facilitating the processes of plastic deformation [243]. In [162, 244] it was attempted to explain the negative slope in the dependence of strength on the grain size by the fact that the increase in the volume fraction of triple junctions should lead to softening and increasing the bulk plasticity of the polycrystals. The role of triple junctions during plastic deformation of nanomaterials was considered in greater detail [245]. It was shown that plastic deformation of nanocrystals is a result of plastic rotations of grains and their mismatch causes the emergence of partial disclinations in the joints of the intergranular boundaries. The idea

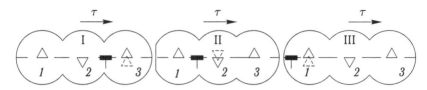

Fig. 3.42. Movement of a disclination under the action of the mechanism of the emission and absorption of dislocations by disclinations, which provides the base of the plastic flow of the nanocrystals.

that the description of the mechanical behaviour of the nanocrystals and polycrystals with ultrafine grains as the evolution of the spatial netwrok of the disclinations was expressed earlier [243]. However, in [245], using the micropolar elasticity theory, the main stages of the plastic shear were strictly described. This was carried out using the relay disclination–dislocation mechanism of deformation, which is the result of processes of emission, absorption and subsequent re-emission of dislocations by wedge disclinations. Plastic deformation begins with the full compensation of negative (positive) components of tensor σ_{12} by the external shear stress τ. The wedge disclination with the Frank vector $\pm\Omega$ can emit both positive and negative edge dislocations, depending on the position of the basal plastic field. Figure 3.42 illustrates the relay mechanism of movement of positive edge dislocations. In stage I the rightmost disclination, formed at a triple junction, emits a positive edge dislocation and its axis moves up. The emitted dislocation moves under the influence of external stress to the negative wedge disclination 2 already shifted from its original position due to previous emission. In stages II and III the movement of dislocations is similar to the above, with the accuracy to the direction of movement of the diclination axes.

The condition of the onset of plastic deformation of nanocrystals taking into account the that the critical grain size L_{cr} at which the deviation from the Hall–Petch dependence takes place, corresponds to critical value Ω_{cr}, can be written as [246]

$$\Omega_{cr} = \Omega x \left(\frac{L_{cr}}{L} \right)^{1/2}.$$

(3.48)

The physical meaning of this relation is that the plastic deformation of the polycrystal with mean grain size L begins, provided that the 'effective' Frank vector corresponds to the critical vector. In other words, the smaller the grain size, the lower the external stress and the value of the Frank vector, correspond to the onset of plastic flow.

Therefore, the special feature of the mechanical behaviour of these systems is a deviation from the Hall–Petch relationship up to the critical grain size of 25 nm. In [247] it was attempted to describe theoretically the dependence of yield stress on the grain size taking account not only the size effect but also the atomic density (degree of relaxation) of the grain boundaries.

It is noteworthy to mention a simple phenomenological model describing the dependence of the deforming stress on the nanocrystal size, based on the concept of the two-phase structure of the NM, proposed by Gleiter [Introduction Ref. 1]. The model suggests that the strength of NM corresponds to the strength of a composite, consisting of two phases: intragranular 'crystalline phase' and the amorphous 'grain boundary phase' [248, 249]. Assuming that the 'crystalline phase' obeys the Hall–Petch relationship, and the 'grain boundary phase' has a constant strength corresponding to the strength of the amorphous condition, the following equation can be written foerthe hardness of the NM:

$$H = v(H_0 + k_y L^{-1/2}) + (1-v)H_{GB}, \tag{3.49}$$

where v is the volume fraction of the 'crystalline phase'; H_0 and k_y are the corresponding constants in the Hall–Petch relation (3.34) and H_{GB} is the hardness of the 'grain boundary phase'.

Assuming for simplicity that the 'crystalline phase' has a cubic shape ($v = (L - s)^2/L^2$, where s is the thickness of the 'grain boundary phase'), and assuming that the amorphous 'grain boundary phase' is described by the relations [19]: $H_{GB} = G_{GB}/6$ and $G_{GB} = 1/G$, where G_{GB} is the shear modulus of the 'grain boundary phase', the following expression for the hardness of NM is obtained [248]:

$$H = \frac{(L-s)^3}{L^3}(H_0 + k_y L^{-1/2}) + \frac{3L^2s - 3Ls^2 + s^3}{L^3} \cdot \frac{G}{12}. \tag{3.50}$$

The authors obtained good agreement between the theoretical values of hardness, following from (3.50), and the experimental data, but could not describe the transition to the region of the anomalous dependence on size L. However, despite the apparent simplicity of the 'two-phase' model, its qualitative agreement with experiment shows the usefulness of a more accurate, physically justified consideration of grain boundary deformation processes involving the liquid-like (amorphous) structure of boundary areas in the nanocrystals.

In [250], the 'combined' version of the mechanism of plastic deformation of nanocrystals was studied: shear in relatively large grains of plastic shear occurs in accordance with the classical dislocation model, assuming that the Hall–Petch relationship is satisfied, and in relatively small-sized grains it takes place by vacancy mechanism of grain boundary sliding.

An important role in identifying the deformation mechanism is played by computer experiments [251–255], which clearly demonstrate the presence of other than Frank–Read sources, and implementation of intragranular plastic shear. In experiments with computer modelling of the processes of plastic flow of nanocrystals [254] calculatation were carried out for Cu nanocrystals, which are equiaxed, free of dislocations crystals with the size from 3.3 to 6.6 nm, separated by narrow lines grain boundaries. The initial and final (after uniaxial strain 10%) configuration of the atoms is shown in Fig. 3.43.

One can see a substantial broadening of the grain boundaries after computer-simulated deformation, indicating a marked contribution of the grain boundary regions to the process of plastic flow. Computer simulations show that nearly half (30–50%) of all the atoms are at the grain boundaries, and the NM after deformation is a two-phase system consisting of equal amounts of the crystal and grain boundary phases. In addition, the process of dislocation glide is sometimes detected in the grains with the formation separate long strips of stacking faults. It was found, firstly, that the dependence of the flow stress and the yield stress on the size of the nanocrystals is governed by the inverse Hall–Petch relationship (Figure 3.44) and, secondly, that the plastic deformation is realised at the grain boundaries in the form of a large number of small shifts, when only a small number of atoms moves relative to each other. Note also that this character of atom displacement is observed in computer

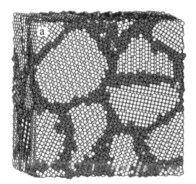

Fig. 3.43. Computer images of *n*-Cu prior (a) and after (b) 10% plastic deformation. The arrow indicates a single stacking fault of deformation origin.

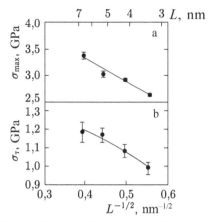

Fig. 3.44. Theoretical dependence of the maximum flow stress (a) and yield stress (b) on the grain size of nanocrystal Cu, obtained by computer simulation.

modelling of plastic deformation in amorphous metallic materials [19]. Based on this review, the authors of [254] concluded that the anomalous dependence of the Hall–Petch relationship is also possible in the absence of porosity and is related to grain boundary sliding, even without thermally activated processes.

It should be noted that other results, obtained by the molecular dynamics method and found to increase the strain rate ε for nanocrystalline Ni with decreasing d in the range of 8 to 3 nm ($\varepsilon \sim 1/d$) and with no change for $d = 10$–12 nm were also obtained [256, 257]. Attention was also paid to the reduction of the yield strength and the elasticity modulus in the size range of 3–5 nm, and the important role of grain boundary sliding and grain rotation during deformation of nanocrystals with a grain size of less than 10 nm was stressed. The possibility of grain rotation in deformation of Au nanofilms was observed *in situ* in electron microscopic studied [258], with the grain size of about 10 nm and with no dislocations detected.

As noted in [[Introduction Ref. 18] [227], all possible deformation mechanisms, contributing to the overall effect, compete with each other and depending on the many features of the nanostructure and the deformation conditions they can play a decisive or a secondary role. Thus, according to the authors of the review [Introduction Ref. 18], the Hall–Petch relationship, based on different dislocation models, is valid only up to a critical size of 10–30 nm and below this value there are fundamental changes in the mechanical behaviour due to a significant contribution of the deformation associated with grain boundaries (sliding, migration, dynamic recrystalliaation). Strict differentiation of the roles each of these mechanisms is required..

3.4.4. Structural mechanism and kinetics of grain boundary microsliding (GBMS)

All of the above consideration in varying degrees emphasize the decisive role nature of grain boundary sliding in determining the specificity of the deformation of nanocrystals. The process of plastic deformation of nanocrystals, regardless of the method by which they were produced, always starts with GBMS, and by itself it is in many respects similar to the process of shear deformation in the amorphous state, since the structure of grain boundaries can be completely or partially correctly described using the model of the amorphous state. The difficulty of the GBMS process obviously leads to brittle behaviour of the nanocrystals.

In [26, 259] attention was given to the initial stage of plastic deformation – microyielding, defined as the stage of deformation at which specific areas of plastic flow form in the elastic matrix by the mechanism of GBMS. Due to the extremely large area of the GBs per unit volume of the nanocrystal, such processes lead to the value of the plastic deformation of a few percent. The dependence of the flow stress on the grain size of the NM and the temperature at this stage of the deformation process was calculated. The stress of resistance to the GBMS was determined through the flow stress of amorphous metal alloys. It has been shown that the critical stress of conversion of the structure of the GBs should be close to the cohesive strength of the boundaries, so a further increase in the tensile stress will result in failure. However, compression or cyclic loading may cause further development of deformation. As the load and the number and size of the GMBS areas increase, they begin to merge, and at some critical volume fraction a percolation transition takes place to the stage of macroplastic flow (Fig. 3.45).

According to current knowledge, any arbitrary GBs can be presented as a limited set of structural elements that coincide with some of the Voronoi–Bernal polyhedrons proposed to describe the structure of liquids and amorphous materials [260]. We can assume that GBMS under the effect of shear stresses takes place due to the restructuring of the structural elements of the boundaries in the formation of microregions of shear transformations of the structure

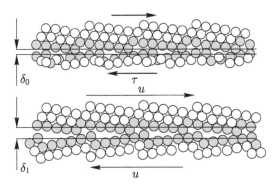

Fig. 3.45. Displacement of adjacent grains with grain boundary microsliding.

of the GBs. The stress of resistance of the general GBs to GBMS will be determined by the critical stress of adjustment of the structural elements of the boundary. Such elementary shift regions are similar to shear transformation that determine the development of heterogeneous plastic deformation in amorphous alloys at low temperatures.

The speed of free slip at the grain boundaries in this case is

$$U = v\chi u_0 \exp\left(-\frac{\Delta G}{kT}\right), \tag{3.51}$$

where v is the frequency of the normal mode of vibration of the shift microregions along the path of activation [261], χ is the volume fraction of centres of formation of shear microregions in the GBs, u_0 is average shift in these regions, ΔG is the free energy of shear transformation microregions with critical parameters.

The analysis of the various mechanisms of the elementary shear processes showed that the energy of formation of shear microregions in the GBs is as follows [259]:

$$\Delta G(\tau_a) = \xi\tau_0\Omega_m\left(1-\frac{\tau_a}{\tau_0}\right)^n, \tag{3.52}$$

where ζ and n are numerical parameters, τ_0 is the stress of athermal stress grain boundary sliding (ideal shear strength of the boundary), Ω_m is the volume of microregions shear transformation.

If the volume fraction of the regions of GBMS changes little during deformation and $L \sim d$ (L is the size of the polycrystalline block with grain size d), then the rate of deformation of nanocrystals in initial stage is:

$$\dot{\varepsilon} = v\chi u_0 \exp\left(-\Delta G \frac{\tau_a - \tau_i}{kT}\right) \cdot \frac{1}{D}, \tag{3.53}$$

where τ_i is the total internal shear stresses in the region of GBMS. For the flow stress (3.53), we obtain the first relation (3.45) with the explicit temperature dependence of the stress of resistance to GBMS:

$$\tau_s(T) = \tau_0 \left[1 - \left(\frac{kT}{\xi\tau_0\Omega_m}\ln\frac{v\chi u_0}{D\dot{\varepsilon}}\right)^{1/n}\right]. \tag{3.54}$$

Dependence $\tau_s(T)$ for the values $n = 1, 2, 4$ is shown in Fig. 3.46 for the following parameter values: $D = 10$ nm, $\varepsilon = 10^{-4}$ s^{-1}, $\Omega_m = 4.3\cdot10^{-28}$ m^3, $v = 10^{12}$ s^{-1}, $u_0 = 0.2$ nm, $\chi = 0.4$.

For $n = 2$ the dependence of the activation energy of microscopic shear stress rearrangement is similar to the dependence of the energy of formation of shear transformations in amorphous alloys in the model [262].

Therefore, the relations (3.53) and (3.54) for $n = 2$ define kinetics of deformation of nanocrystals whose GBs are in the amorphous state. The temperature dependence of the stress of resistance to GBMS and thus of the yield limit of nanocrystals (3.54) must be satisfied to a certain temperature T^*, above which other deformation modes are energetically more favorable.

Depending on the ratio of the structural parameters, the influence of the grain size on the flow stress according to the equation (3.45) will have the form shown in Fig. 3.47. Note that value τ_0 depends on the state of the GBs, therefore, it may also vary with the grain size, but in principle the model of GBMS correctly describes all the experimental dependences $HV(\sigma) = f(d)$.

There are at least four critical sizes of the nanocrystal grains $D^*_1, D^*_2, D^*_3, D^*_4$, (and/or their respective transition regions of grain sizes), determining the conditions of the change of the structural mechanisms of plastic deformation [263] (Fig. 3.48). Depending

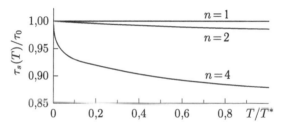

Fig. 3.46. Temperature dependence of the stress of resistance to GBMS, $n = 1, 2, 4$.

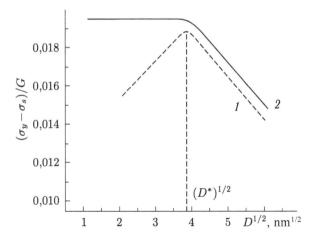

Fig. 3.47. Theoretical dependences of the yield limit of the nanocrystal σ_y on the square root of the grain size d; $1 - L = \text{const}$ (d), $2 - L = \xi_1 d$.

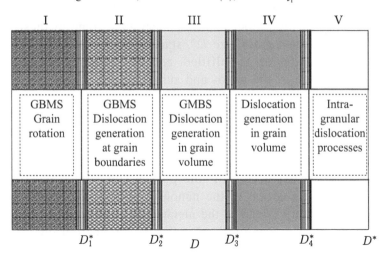

Fig. 3.48. Map of structural mechanisms of plastic deformation of nanomaterials depending on the grain size.

on the grain size and the ratio of stresses of GBMS and of the resistance to movement of dislocations in the lattice, the following options for the development of plastic flow exist.

1. If the grain size of the nanocrystal is less than the minimum critical size D_1^*, the main mechanism of plastic deformation is GBMS and the accommodative process of grain boundary strain – plastic rotation of grains.

2. In the range $D_2^* \geqslant D \geqslant D_1^*$ in GBMS the accommodative processes of dislocations in the grain boundaries have more

favourable energy parameters than plastic rotation of the grains. The acting stress is greater than the stress of formation of dislocations in GBs τ_g, but lower that the stress of exit of the dislocation from the boundary τ_e [264].

3. In the nanograin size range $D_3^* \geqslant D \geqslant D_2^*$, where D_2^* is the second critical grain size below which dislocation clusters cannot form in the nanograins [264], the processes of exit of single or partial lattice dislocations from the GBs to the volume of the grains become possible.

4. At $D > D_3^*$ deformation of nanocrystals (or submicrocrystalline material, depending on the value of D_2^*) will be developed by the generation of trains of dislocations with grain boundaries and the development of the slip bands.

5. When $D > D_4^*$ all the usual intragranular dislocation processes, such as the processes of reproduction and annihilation dislocations [265], can take place.

Depending on the properties of nanocrystals, some of these areas do not form. Critical grain size D^*, starting from which the classical Hall–Petch relationship is fulfilled, can be in one of the intervals (I–IV) of the size of the nano- and submicrograins.

3.4.5. Structural classification of nanocrystals from the position of their deformation behaviour

The concept of 'nanocrystal', introduced by H. Gleiter in 1981, instead of the usual notion of 'the polycrystal with small grains', means that the transition to the nanocrystalline state implements a significant qualitative leap in the mechanical behaviour of the solid. In our view, such a quantum leap is the transition from the classical dislocation modes of plastic deformation of the usual crystals, to the dominant role of grain boundary non-dislocation modes which appear in the nanometer grain size range when image forces 'push' the usual dislocations from the nanosized grains of the polycrystalline ensemble [266]. Thus, the transition to the anomalous Hall–Petch dependence and to other features of plastic deformation and fracture of nanocrystalline materials is, from our point of view, the consequence of a radical change in the structural mechanism of plastic flow.

In the introduction, we examined in detail the relationship between the relative proportion of the crystal matrix, the relative proportion of GBs and the relative proportion of areas of triple junctions, depending on the grain size in an ideal single-phase polycrystalline material. With decreasing grain size there is a smooth increase in the proportion of the grain boundaries to a maximum value of 0.45, and then with a resulting decrease in the crystal size – a gradual decrease. At very small sizes the share taken by the grain boundaries decreases, but the volume fraction occupied by the triple junctions dramatically increases. Consequently, in the nanometer grain size range (less than 15–20 nm), a decisive role in various processes (also in the process of deformation and failure) should be played not as much by the grain boundaries as by the triple (and possibly of higher orders) joints of grain boundaries – the predominant elements of the structure in this size range of nanocrystals.

In accordance with the above, we divide the whole measuring scale of the nanocrystals in the graphs on Fig. I.2 into three areas: the 'large', 'medium' and 'small' nanocrystals (Fig. 3.49). In the 'large' nanocrystals the predominant element of the structure are actually crystals, in the 'medium' – the grain boundaries, and in the 'small' nanocrystals – triple joints. The boundary values of the nanocrystal size, corresponding to transition from one type or

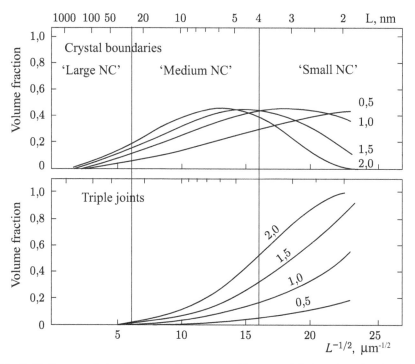

Fig. 3.49. Structural classification of nanocrystalline materials based on the dominant contribution to the structure of the various elements of the structure and defining the plastic flow mode; L is the nominal size of the crystalline phase.

nanocrystal to another, are rather conventional and perhaps slightly 'fuzzy'. Their values depend on a number of structural parameters (such as the width of the grain boundaries and the nature of the size distribution of the nanocrystals).

The main thesis of the proposed classification of nanocrystals in terms of their deformation behaviour is the following [267]: the 'large', 'medium' and 'small' nanocrystals are characterised by the dominance of different mechanisms of plastic deformation determined by the dominant element of the structure (actually crystals, grain boundaries or triple junctions of grains, respectively). We consider briefly these deformation mechanisms.

1. **'Large' nanocrystals.** The dislocation description of the deformation processes is fully justified. Accordingly, for 'large' nanocrystals the Hall–Petch relationship should be satisfied and perhaps there is a slight deviation from it, associated with the constrained generation and multiplication of dislocations or with other reasons, having the dislocation nature [Introduction Ref. 18].

2. **'Medium' nanocrystals.** Plastic shear occurs with the active participation of the grain boundaries. One of the most active real structural mechanisms is the grain boundary microsliding (including the relatively low temperatures). The applicability of GBMS as a possible mechanism of plastic flow of nanocrystals was discussed in detail in sections 3.4.3 and 3.4.4 of this chapter. Both theoretical considerations [26] and computer simulations [254] and structural studies [268] confirmed the reality of the GBMS in nanocrystals. One of the most important consequence of the consequence of the implementation of the GBMS mechanism – the dependence of the flow stress on the size of the nanocrystals – is governed by the relationship, directly opposite to the Hall–Petch dependence: when d decreases σ_y (HV) also decreases.

In the most precise form, the role of grain boundaries in the plastic flow in the 'medium' nanocrystals is seen in the example of nanocrystals formed by controlled crystallisation of the amorphous state, where at certain stages of heat treatment it is possible to form nanocrystals, separated by thin amorphous layers, on which plastic shear takes place. The yield limit of such a material increases linearly with the size of nanocrystals and with reduction of the thickness of the amorphous intergranular interlayers contrary to the Hall-Petch relation. In the limit when the amorphous layer thickness is commensurate with the thickness of the 'normal' grain boundaries we can assume that their amorphous structure is maintained, and we are essentially going to have to consider the GBMS mechanism operating in the amorphous interlayers.

3. **'Small' nanocrystals.** The approach to explaining the mechanical behavioir of the nanocrystals, which is connected to the determining role of the triple junctions, is very fruitful as a significant increase in their volume fraction in the 'small' nanocrystals can have a positive impact on facilitating the processes of plastic deformation [269]. The role of triple junctions during plastic deformation of nanocrystals was examined in greater detail in section 3.4.3 of this chapter. It was noted that the plastic deformation of nanocrystals is a result of plastic rotations of grains and the mismatch of these rotations cases generation of partial disclinations at the junctions of the grain boundaries. A feature of the mechanical behaviour of 'small' nanocrystals is the existence of the dependence according $\sigma_y(d)$, which corresponds to the inverse Hall–Petch relationship, but it may not match the $\sigma_y(d)$ dependence, inherent to the 'medium' nanocrystals.

The proposed separation of nanocrystals into three types ('large', 'medium' and 'small') enables deeper understanding of the cardinal changes in the mechanical behaviour that result from the size effect. In the end, it can be argued that the Hall–Petch relationship, which is directly related to the dislocation nature of plastic flow of polycrystals stops at some stage in the transition to the nanosized parameters of individual crystals accurately reflect the physical and mechanical laws of the plastic flow. As a consequence, it seems more correct to speak not about changing certain parameters of the Hall–Petch relationship in the transition from 'large' to 'medium' nanocrystals, but about the implementation of a fundamentally new relationship $\sigma_y(d)$, based on the non-dislocation mechanisms of plastic deformation in the 'medium' and 'small' nanocrystals.

It is necessary to make two important comments.

1. In the transition from one type of nanocrystals to another, or with the existence in the polycrystalline ensemble of the size distribution, covering different types of nanocrystals, can be implementation of mixed deformation mechanisms that are clearly more complex.

2. From the position of the deformation behaviour, the proposed structural classification can be attributed not to

all nanomaterials, but only to the nanocrystals. In other nanostructured materials, plastic flow processes are subject to others not yet sufficiently studied relationships. For example, the given consideration does not explain the features of the plastic deformation of materials that have passed megaplastic (severe) deformation because this group is not related to nanocrystalline but to nanofragmented materials.

3.4.6. Features of the mechanical behaviour of amorphous–nanocrystalline alloys (ANA)

Since the transition from the amorphous to nanocrystalline state is a phase transition the first kind, nanocrystallisation usually results in the formation of a two-phase structure, which can be interpreted as amorphous–nanocrystalline. The unusual feature of the alloys with the amorphous–nanocrystalline structure is, first of all, that structural (phase) components of such a system are fundamentally different as they have the maximum (crystal) and minimum (amorphous state) degree of atomic ordering. Such a 'unity of contradiction' obviously leads to a number of effects, affecting in particular the mechanical behaviour of these materials. The situation is even more dramatic when the crystalline phase has nanoscale features (less than 100 nm).

There are two basic structural conditions formed in the ANAs [Introduction Ref. 11]: 1) alloys with statistically distributed nanoparticles in an amorphous matrix, and 2) alloys containing predominantly isolated nanocrystals, separated by amorphous layers. The first group of the alloys corresponds to the initial stages of the transition from the amorphous to crystalline state in the process of controlled thermal effects, and the second group – to later (final) stages of the process. Let us consider in detail the features of the mechanical behaviour of both groups.

AHS with randomly (statistically) distributed nanoparticles in an amorphous matrix

There is a well established opinion that the appearance of a crystalline or nanocrystalline phase almost always leads to an increase in the strength characteristics of the amorphous state [270]. It is considered that the increase in strength in nanocrystallisation is due mainly to the appearance of a high-modulus structure of the crystalline phase. This effect is particularly strong in the alloys of the metal–metalloid type in the presence in the structure of borides, silicides and other similar phases [Introduction Ref. 44].

A similar situation is observed in the alloys of the metal–metal type in cases where the difference of the elasticity modulus of the amorphous and crystalline phases is large enough [173]. In [271] the authors reported on a group of newly developed aluminum alloys containing up to 90% Al, produced partially by crystallisation of metallic glasses, which retain their plastic characteristics and have the tensile strength two times higher than the strength of the best commercial aluminum alloys. In [272], X-ray diffraction, DSC and high-resolution electron microscopy were used to study the formation and structure of the Ni–Mo–B nanocrystalline alloys, containing from 27 to 31.5 at.% Mo and 5 or 10 at.% B. It is shown that with the microhardness increses increasing duration of isothermal holding (and simultaneously with the increase of grain size). A dependence similar to the inverse Hall–Petch relationship is observed. It is assumed that the microhardness is largely determined by the amorphous matrix and the increase of microhardness in the formation and growth of nanocrystals is due to a higher yield limit of the amorphous matrix due to the redistribution of chemical components and enrichment of the matrix with boron and molybdenum in the formation of the nanocrystals. Cases in which crystallisation reduced the strength were also reported [171].

In addition to the ratio of the elastic moduli of the nanoparticles and the amorphous matrix, an important role in changing the strength in nanocrystallisation must, of course, be played by the structural parameters of the nanocrystalline phase: the size of the particles, their volume density, volume fraction, the type of crystal lattice structure, the distribution by size and volume of the amorphous matrix, etc.

In [273], a detailed experimental study was conducted of the deformation behavior of ANA with isolated nanocrystals uniformly distributed in the amorphous matrix, which were obtained by controlled annealing of the amorphous state. The objects of study were samples of the three alloys $Fe_{58}Ni_{25}B_{17}$ (alloy 1), $Fe_{50}Ni_{33}B_{17}$ (alloy 2), $Ni_{44}Fe_{29}Co_{15}B_{10}Si_2$ (alloy 3), obtained by melt spinning. The relative intensity of X-ray lines from the amorphous and nanocrystalline phases was used to determine the volume fraction of the nanocrystals, and the degree of broadening diffraction peaks was used to ascertain the average size of the nanocrystals.

Table 3.5 shows the structural features of the initial stage of crystallisation of ANA, accompanied by formation of primary crystals.

Table 3.5. Structural characteristics of ANA [273]

Alloy		Primary nanocrystals				
		Formation temperature, °C	Form	Phase	Lattice type	Lattice spacing, nm
1	$Fe_{58}Ni_{25}B_{17}$	380	Cuboid	α	BCC	0,287
2	$Fe_{50}Ni_{33}B_{17}$	360	Equiaxed	γ	FCC	0,357
3	$Ni_{44}Fe_{29}Co_{15}B_{10}Si_{12}$	340	Equiaxed	γ	FCC	0,357

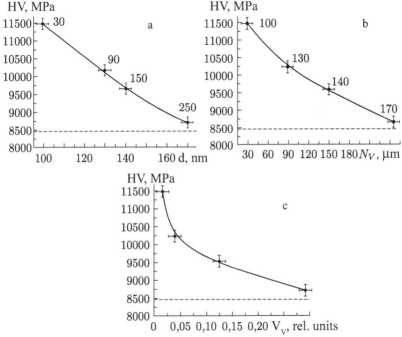

Fig. 3.50. Dependence of the microhardness of alloy 1 of the average size (a), volume density (b) and volume fraction (c) for nanocrystals with a constant annealing temperature of 380°C.

Figure 3.50a shows the dependence of the microhardness HV on the average size of nanoparticles d of the α-phase formed after annealing of alloy 1 at 380°C. Next to each experimental point there is the related bulk density of nanoparticles (μm^{-3}) at which HV was measured. There is a noticeable drop in microhardness with increasing size of the nanoparticles. Similarly, the dependence after after annealing at different temperatures and at constant annealing time is similar. Figure 3.50b shows the dependence of the microhardness of the alloy 1 on another structural parameters of nanoparticles, bulk density N_V, after annealing at 380°C. Each experimental point indicates the average value of the corresponding size of the nanoparticles at which HV was measured. There is a clear decline in microhardness with increasing the bulk density of the nanoparticles. The dependence of the microhardness on the bulk density of nanoparticles for other annealing parameters is similar. Dependence $HV(V_V)$ (Fig. 3.50c) is plotted analytically based on the experimental data presented in Fig. 3.25a, b, given the fact that the volume fraction $V_V = \pi d^3 N_V / 6$ [274].

For values of $V_V \leqslant 0.1$ there is a marked decrease in microhardness, and at higher V_V dependence $HV(V_V)$ reaches saturation. The dependences $HV(V_V)$, corresponding to other temperatures, and the dependence for various fixed annealing times are similar.

In alloy 2 the HV value change is different. As a result of heat treatment HV always increases linearly with respect to the initial amorphous state (8700 MPa) reaching the maximum value (10 300 MPa) after annealing 390°C, 2 h. Since the alloy 2 showed the 'Finemet' effect (see section 3.2) and the size of nanoparticles did not change during annealing (20 nm), the value d was not considered as a parameter that affects the measurement of HV, and only the dependence $HV(N_V)$ was analysed, which is equivalent in this case to the $HV(V_V)$ dependence. The dependence $HV(V_V)$ for alloy 2 after annealing at a constant temperature of 360°C or at a constant annealing time of 0.5 h was close to linear (Fig. 3.51). The values for other annealing parameters were similar. In some cases the dependence $HV(V_V)$ can be described note by linear but by the power law with HV values reaching saturation.

In alloy 3, as in alloy 2, the increase in annealing temperature and time always led to an increase in the values of HV (from 8800 MPa in the initial amorphous state to 12 100 MPa after annealing at 440°C, 2 h). Since the alloy 3 also showed 'the Finemet effect' (constancy of the average size of the nanoparticles (20 nm) at different annealing parameters), we can only analyze the dependence $HV(N_V)$ and the dependence $HV(V_V)$ identical to it. Figure 3.52 presents the $HV(V_V)$ dependence at annealing time (a) and temperature (b). At other temperatures and annealing times the character of the relationship is similar.

As is known, a possible factor in the influence of nanoparticles on the strength of the ANAs is the higher value of the Young's modulus

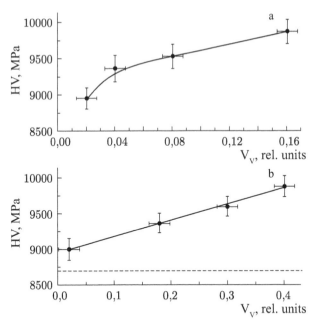

Fig. 3.51. Microhardness of the alloy 2 in dependence on volume fraction of nanocrystals at 360°C (a) and 0.5 h (b).

E [Introduction Ref. 14]. In all investigated ANAs nanoparticles – BCC or FCC substitutional solid solutions based on iron and nickel (see Table 3.5) – form. The boron present in large quantities in the alloys (17 at.% in alloys 1 and 2, and 10 at.% in alloy 3) practically does not dissolve in the nanoparticles (less than 0.1%) [275]. This means that all of the boron atoms remain in the amorphous matrix during nanocrystallisation. In this case, the sharp difference between the values of *E* of the amorphous and crystalline phases, which for the same chemical composition of the alloy can reach 30–50% [Introduction Ref. 14], is significantly levelled. In fact, the values of *E* for the nanocrystals formed in the alloys was in the range of 200–210 GPa [276], and the values of *E* for the amorphous matrix, which preserves a high concentration of boron, is in the range 195– 200 GPa [277].

The yield stress σ_y (or hardness *HV*) of the two-phase ANA increases with the volume fraction of the nanocrystalline particles in accordance with the rule of additive addition of the Young modulus of the structural components in accordance with the dependence of [Chapter 2 Ref. 4]:

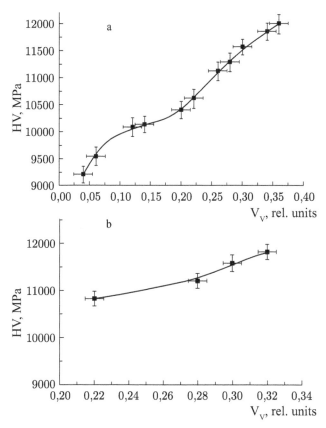

Fig. 3.52. Dependence of microhardness of alloy 3 on the volume fraction of nanocrystals at 1.5 h (a) and at 420°C (b).

$$\sigma_y(HV) = \sigma_y^{am}\left(HV^{am}\left\{1+v_f\left[\left(\frac{E_{nc}}{E_{am}}\right)-1\right]\right\}\right), \qquad (3.55)$$

where σ_y^{am} and HV^{am} are respectively the yield strength and hardness of the amorphous matrix; v_f is the volume fraction of the nanocrystalline phase; E_{am} and E_{nc} is the Young's modulus of the amorphous matrix and the nanocrystalline phase, respectively. For the alloys 1–3 the maximum value of E_{nc}/E_{am} is very close to 1, i.e. 1.076, and therefore the hardening due to the difference of the elastic moduli of the amorphous and crystalline phases can not be considered in this case as the main cause of the changes in the strength of the AMC in crystallisation.

Another possible reason may be the inhibition by nanoparticles of the shear bands which propagate in the amorphous matrix. The process is similar to inhibition of moving dislocations in the crystal,

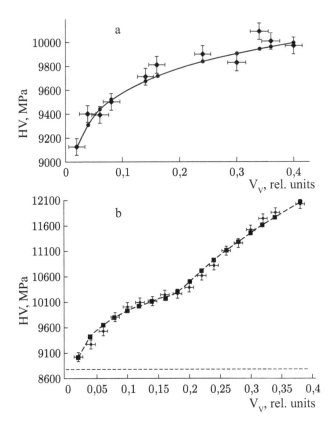

Fig. 3.53. Dependence of the microhardness of alloy 2 (a) and alloy 3 (b) on the volume fraction of nanocrystals for all the states.

containing coherent or incoherent second-phase particles. Such effects have indeed been observed experimentally by TEM [278] and calculated theoretically [279]. Thus, the authors of [278] utilised the unique opportunity to analyze in pure form (without significant impact of the effect of the elastic moduli) the hardening associated with the interaction of shear bands with nanoparticles in plastic deformation, depending on their volume density and size.

Figure 3.53 shows the overall dependence $HV(V_V)$ for all the heat treatment conditions investigated and, therefore, for all the realised two-phase states. In analysis one must take into account two factors. First, comparing the structural states obtained at different annealing temperatures and times, we can define the error due to the fact that the structural state of the amorphous matrix and, possibly, nanocrystalline particles may differ slightly. Second, as for the alloys 2 and 3, there is a stabilising effect of the nanoparticle size previously observed for amorphous alloys of the type 'Finemet' Fe–Si–B–Nb–Cu [74], we can analyse using the graph in Fig. 3.53 the dependence $HV(N_V)$, as $V_V = Kd^3N_V$, where K is a numerical constant.

The graph in Fig. 3.53a for the alloy 2 can be described by the dependence of type $HV = K(V_V)^n$, where $n = 1/3$. The dependence in Fig. 3.53b for alloy 3 is divided into two sections $((V_V)_{cr} = 0.2)$, each of which similar to the dependence obtained for alloy 2.

There is an analogy between deformation of ANA and hardening due to the deceleration of dislocations sliding in the crystal on the incoherent and non-intersected particles. In accordance with the Orowans mechanism, Ashby's theory developed on the basis of this mechanhism, and the modified theory of flat clusters of Orowan loops on the particles, which satisfactorily describe the experimental results, we have [280]:

$$\sigma = \sigma_0 + cG\left(\frac{V_V b}{d}\right)^{1/2} \varepsilon, \tag{3.56}$$

where σ is the flow stress, σ_0 is the flow stress in the crystal containing no particles, c is a constant equal to 0.1–0.6, G is the shear modulus, V_V is the volume fraction of particles, b is the Burgers vector of sliding dislocations, d is the particle size, ε is the degree plastic deformation.

As we can see, there is a marked similarity between the effects of the particle volume fraction and the bulk density of the particles on the hardening in crystals and ANA. The difference lies only in the fact that the exponent n in the crystals is 1/2, and 1/3 in ANA. This analogy is not unexpected, since the shear bands realising plastic shear in the amorphous state, are effective dislocations on the mesolevel and their Burgers vector has not been accurately determined. The degree of shear deformation in such a band is thousands of percent [281]. The lower value of n suggests that the inhibition of shear bands by the particles in ANA is less effective than in crystals with the Orowan mechanism. Apparently, the nanocrystalline particles of about 20 nm in size are partially cut, or is there another, more efficient (than in crystals) mechanism for shear bands to overcome the nanocrystalline phase particles.

The situation with the question of the effect of the size of the nanocrystals on the strengthening of the amorphous matrix is complicated by the fact that in the alloys 2 and 3 the growth of of the nanocrystalline phase during crystallisation is inhibited (Finemet effect) and, therefore, the size effect can not be evaluated. In alloy 1 after the selected heat treatment, there is a change in the average crystal size from 100 to 170 nm, and in principle we could trace the size effect for the value of *HV*. However, such examination will be incorrect, because when the average particle size of the alloy 1 changes their bulk density changes simultaneously. To exclude the impact of parameter N_V, the authors of [278] used the following procedure.

As can be seen from Fig. 3.53, the dependence $HV = f(V_V)$ in the alloys 2 and 3 is the same character: $HV \sim N_V^{1/3}$. It can be assumed that the same relationship is also true for alloy 1. In this case, we can convert all the *HV* values, obtained for alloy 1, to some values corresponding to one and the same constant value $(N_V)_0$. The standard chosen value is $(N_V)_0 = 100$ μm^{-3}, because it is in the middle of the range of the values N_V obtained for the alloy 1 using the heat treatment conditions employed in the study (50–160 μm^{-3}). The adjustment factor was calculated as the average of the coefficients for the alloys 2 and 3 and equalled $K_{corr} = 38.5$ MPa·μm^3. Figure 3.54 shows the depenedence of the corrected values HV_{corr} for alloy 1 on the mean size of the nanocrystals *d*. Solid circles denote the values HV_{corr} corresponding to the initial amorphous state $(d = 0)$ and to the state obtained after the given heat treatments $(d = 100–170$ nm). It can be seen that for values $d > 100$ nm microhardness (strength) decreases with increasing mean size of the nanoparticles. However, comparing the values HV_{corr} for $d = 0$ (amorphous state) and $d > 100$ nm, it can be assumed that the dependence $HV_{corr} = f(d)$ is generally complex and non-monotonic. In order to reproduce this relationship in this form, there are not sufficient values of HV_{corr} in the range $d < 100$ nm.

To obtain these data, additional experiments were carried out in [273]. Samples of alloy 1 were annealed isothermally at 400°C for

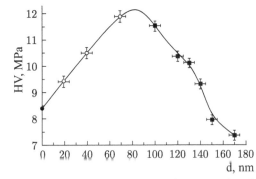

Fig. 3.54. Dependence of microhardness HV_{corr} on the mean size *d* of nanocrystals for alloy 1 for all the investigated states.

very short time periods (3, 5 and 10 min) in order to record the earliest stages of nanocrystallisation. For the obtained values of the mean size of the nanocrystals located in the amorphous matrix, calculations were carried out to determine the corrected values of microhardness at $N_V = 100$ μm^{-3}, which are plotted in the graph of $HV_{corr} = f(d)$, shown in Fig. 3.29 in the form of open circles. In a more detailed form this relationship is a curve with a maximum corresponding to $d = 70$–90 nm.

It is known (see section 3.4.3) that the single nanocrystals exhibit an anomalous dependence of strength on the grain size, which for ordinary polycrystals is generally described by the Hall–Petch relationship [232] (Fig. 3.37). If we compare the dependence 2 in Fig. 3.13 and the dependence $HV_{corr} = f(d)$ in Fig. 3.54, it is striking that there are apparent similarities. It is important to keep in mind that we compare the dependence on the mean size of the nanocrystals for two completely different structural states. The dependences in Fig. 3.37 were obtained for the single-phase polycrystalline (nanocrystalline) state. In the case of ANA (Fig. 3.54) we have a two-phase structure, when the crystalline phase is distributed in the form of statistically arranged equiaxed nanoparticles with the volume fraction not exceeding 0.4. Conventionally, we assume that the dependence, located on the right of the maximum HV in Fig. 3.54 is 'normal' because it smoothly changes into the area of the crystalline particles with the normal size, far from the nanometer range. This relationship is somewhat different from those obtained for crystals containing particles of the second phase [280], but physically it is easy to understand from the following arguments: with the growth of the particle size of the second phase the inhibition of shear bands on the crystalline particles becomes less efficient through more intensive deformation processes in the particles themselves. By reducing their size and with transition to the region of the nanoparticles the slip inside them becomes less active, and the time comes when the nanocrystalline phase behaves as rigid non-intersected second-phase particles in which dislocation and other relaxation processes are completely suppressed. However, when $d < 60$–70 nm abnormal processes that do not fit into the structural model developed above start to take place. Overcoming of the nanoparticles by the shear bands becomes easier and in the limit reaches the stress level at which shear bands propagate in the amorphous matrix containing no nanoparticles ($d = 0$). It should be emphasized that the anomalous behaviour of microhardness (strength) occurs at a time when the size of the nanoparticles of the crystalline phase becomes smaller than the thickness of shear bands (60–70 nm [281]) propagating in the amorphous matrix at plastic deformation.

Before turning to a discussion of the structural mechanisms of plastic deformation of ANA, it is useful to consider briefly the features of electron microscopy studies of shear bands in amorphous alloys, because the amorphous matrix is the basic structural component of the considered ANAs.

Plastic deformation of amorphous alloys can occur differently: homogeneous or inhomogeneous [Introduction Ref. 14]. In homogeneous plastic deformation each element of the solid undergoes plastic changes (Fig. 3.55a), so that the uniformly loaded sample undergoes homogeneous deformation. In inhomogeneous plastic deformation, the plastic flow is localised in thin discrete shear bands, and the rest of the solid remains undeformed (Fig. 3.55b).

The inhomogeneous flow occurs at low temperatures ($T <$ 0.8 T_c) and high applied stresses ($\tau > G/50$). It is usually observed in tension, compression, rolling, drawing and other methods of deformation of ribbon, wire and massive samples of amorphous alloys. The inhomogeneous flow poorly reacts to the rate of active deformation and shows an almost complete lack of strain hardening.

The resulting degree of macroscopic deformation is determined by the number of shear bands formed in the amorphous matrix when the external load is applied. In turn, the tendency to form a large number of shear bands is determined by several factors, and in particular, by the loading scheme. For example, uniaxial tension is accompanied by mechanical instability, the formation of a small number of shear bands to a catastrophic failure, and the total plastic deformation is only about 1–2%. At the same time, the strain in rolling may reach 50–60% [282]. The highest degree of plastic deformation in the

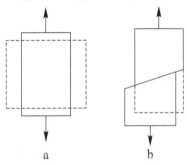

a   b

Fig. 3.55. The scheme illustrating the homogeneous (a) and inhomogeneous (b) plastic deformation of the specimen.

inhomogeneous flow can be achieved in bending ribbon samples. The bending radius in the limit can be commensurate with the thickness, i.e. as high as 30–40 μm [283]. A characteristic feature of the inhomogeneous plastic deformation by which is radically differs from the homogeneous deformation is that for the inhomogeneous flow causes not an increase but rather a decrease of the degree of order in the amorphous matrix.

In the case of inhomogeneous plastic deformation shear steps form on the surface deformed by stretching, compression, bending or rolling of samples. These steps correspond to the exit to the shear bands to the surface. They are usually distributed at an angle of 45–55°C to the axis of the uniaxial tension (or compression) – at the same angle to the rolling direction and also parallel to the axis of bending. The height of the steps on the sample surface reaches 0.1–0.2 μm, and the thickness of the individual shear bands does not exceed 0.05 μm. Hence it is clear that the amorphous alloys have a huge local plasticity in the region of inhomogeneous deformation.

The experiments show that the shear bands are characterised by selective etching [284]; they form in one and the same area of the amorphous matrix with repeated loading and at the same time disappear when the strain of opposite sign is applied to the sample [285].

Especially interesting are experiments *in situ*, recording the character of the nucleation and propagation of the shear bands directly in the deformation process. Structural studies of this kind, made with a scanning electron microscope, show [286] that there are three stages of inhomogeneous deformation: the stage of 'homogeneous' deformation, in which deformation occurs without registration of the images of shear bands, the stage at which the shear deformation proceeds through the formation and propagation of shear bands, and, finally, the final stage, in which the deformation is localised only in some bands to form cracks in the 'head' of the bands or in areas where they intersect. The time of formation and the rate of propagation of shear bands, measured in [286], were respectively 1.5 μs and $5 \cdot 10^{-5}$ m/s. Mutual intersection and branching of individual shear bands was detected, which may explain the small apparent strain hardening observed sometimes on the tensile strain curves.

In [287] the average rate of propagation of the shear bands was 10^{-4}–10^{-3} m/s, and the interaction of slip bands did not occur through the intersection but by avoiding obstacles through formation of secondary shear bands. It is possible that this is due to the scale

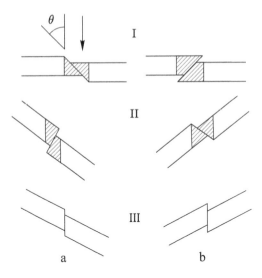

Fig. 3.56. The scheme of the contrast in the study of amorphous alloys by transmission electron microscopy shear bands at different positions of the bands relative to the incident electron beam.

factor, as in [287] *in situ* experiments were carried out by scanning electron microscopy on thin foils of amorphous alloys.

A more realistic opportunity to assess changes in the structure of the shear bands is provided by transmission electron microscopy [281]. As can be seen from Fig. 3.56, the section of the transmission foil corresponding to the shear step has a different effective thickness, compared to the size of the section of the surrounding undeformed matrix. In the case where the angle Θ between the direction of the incident electron beam and the effective shear implemented by the band is less than 180°(I in Fig. 3.56a and II in Fig. 3.56b), the shear band is observed as a bright band on a dark background (Fig. 3.57a); if $\theta > 180°$ (II in Fig. 3.56a and I in Fig. 3.56b) as a dark band on a light background (Fig. 3.57b). The case of the formation of shear bands, shown in Fig. 3.56a, is more likely to be observed in tension and the case shown in Fig. 3.56b in compression. Experimental study of shear in the a light field with the transmitted beam of electrons is associated with difficultiues and should be conducted in the dark field under the influence of the portion of the first diffusion ring. This increases the intensity difference in the shear bands and outside (the contrast from the deformation band becomes clearer).

Dark field images obtained from the study of plastically deformed Fe–Ni–P alloy (Fig. 3.57) [281] allow to identify a

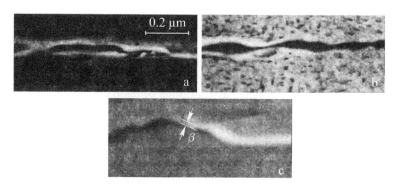

Fig. 3.57. Electron microscopic images of shear bands formed during deformation of $Fe_{65}Ni_{17}P_{18}$ alloy at room temperature: a – θ < 180°; b – θ > 180°; c – $\theta \approx$ 180°.

number of morphological features of shear bands: they are 'non-crystallographic', i.e. easily change the local plane of their orientation and are characterized by branching points (β in Fig. 3.57). Of particular interest is the determination of the thickness of deformation bands. The scheme in Fig. 3.56 is consistent with the considerations regarding the shear bands as a flat formation of zero thickness, which is a deliberate simplification. In the case of the two-dimensional model, the band with the smoothly varying orientation, in transition at some points through position corresponding to angle θ equal to 0 or 180° (III on Fig. 3.56), will be observed on electron microscopic images as a region of light contrast changing to he dark contrast region, or vice versa (Fig. 3.58). The transition of this kind must be accompanied by a narrowing of the local area of the selected contrast, reaching up to the constriction (Fig. 3.58a). The tendency to such behaviour of the contrast is clearly visible in Fig. 3.57. Here some parts of shear bands are observed in the form of very narrow areas because at these points the local orientation of the band is close to the position corresponding to θ = 0 or 180°. If the shear band thickness is not zero but some specific value, the changing nature of the contrast must be accompanied by the formation of a constriction (Fig. 3.58b), the width of which β corresponds to the thickness of the given shear band (or rather, provides an opportunity to assess the maximum value of this quantity).

Figure 3.58 demonstrates the ability to determine the thickness of the shear band to create conditions where $\theta \approx 0$. The accuracy of the measurement of β on electron micrographs is low, and therefore the increase the reliability of estimates of β requires large number of measurements and subsequent statistical analysis of the resultant data.

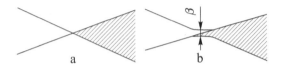

Fig. 3.58. Diagram illustrating the determination of the thickness of shear bands.

It should be noted that any pre-emerging deformation bands (or rather, steps) in the amorphous sample will be destroyed in electropolishing in the preparation of foils for electron microscopic studies, so the absorption (and, obviously, diffraction) contrast from shear bands in this case does not arise. In other words, the shear bands become 'invisible'.

We now turn to the structural studies of the plastic flow pattern in ANA with nanoparticles randomly distributed in the amorphous matrix. In [ntroduction Ref. 10] electron microscopic studies showed for the first time that at a relatively small (less than 0.4–0.5) volume fraction of the nanocrystalline phase plastic deformation is realised by the formation of highly localized shear bands in the amorphous component of ANA. In [288], studies of an amorphous and partially crystallized Zr-based revealed the effect of nanocrystallisation on the geometric parameters of shear bands: reduction of mean length and capacity (height of steps), as well as the increase of the density of shear bands on the surface of the sample tested by scanning electron microscopy. In [279] the authors analysed theoretically possible mechanisms of interaction of nanoparticles with a shear band moving in the amorphous matrix. In [289] a detailed study was made of the deformation behaviour of the ANA produced by controlled annealing of amorphous metal alloys. The object of the investigation was an $Fe_{58}Ni_{25}B_{17}$ amorphous alloy, obtained by quenching from the melt followed by controlled annealing in different conditions. The samples were subjected to microindentation to initiate local shear bands. The structure and phase composition were studied by transmission electron microscopy, and therefore shear bands were created completed foils with a thickness up to 0.1 µm using a diamond pyramid – indenter. As a result, each of the foil contained a large number of shear bands, suitable for electron microscopic study.

Controlled heat treatment (in the optimum conditions) was applied to produce the structural state in the amorphous $Fe_{58}Ni_{25}B_{17}$ allloy [289] in which a nanocrystalline phase and its volume share did not exceed 0.4. Plastic deformation patterns of ANA due to local effects of the indenter on the thin foil, prepared for electron microscopic analysis were studied. Analyzed more than one hundred cases of the interaction of shear bands with nanoparticles.

All the observed interaction acts can be divided into four groups (the value in the brackets is the percentage of implementation of the process):

Group I. The shear band bends round the oncoming nanoparticle. Figure 3.59a shows a typical electron microscope image showing the 'effect of rounding'. Figure 3.60a shows the diagram of the process where the shear band in its path bends around the oncoming nanoparticle changing the trajectory of movement in the amorphous matrix. Movement of the shear bands in rounding resembles the process of double cross slip of a dislocation overcoming rigid barrier.

Group II. The shear band is 'stuck' in the counter nanoparticle. Figure 3.59b shows a typical case of inhibition of the shear band at the particle which stopped its propagation. Figure 3.60b shows a diagram of the 'effect of jamming'. It should be noted that this phenomenon is observed far less frequently than the alternative interaction options listed in groups I and III.

Group III. The shear band propagates through a nanoparticle, as if 'cutting it off'. A typical case is shown in electron micrographs in Fig. 3.59c. This case, obviously, could be realized if the shear band, propagating in the amorphous matrix, could stimulate the dislocation flow within a nanoparticle, which, in turn, would stimulate the formation of shear bands in the amorphous matrix, but on the other side of the nanoparticle. Such a 'plastic relay' is shown schematically in Fig. 3.60c. Note that this mechanism of interaction, along with the mechanism of 'rounding', has been observed most frequently.

Group IV. The shear band, resting on a nanoparticle, causes very large elastic distortions in it which, in turn, initiate a shear band in the amorphous matrix on the other side of the nanoparticles. Typically, the trajectory of movement of the secondary accommodative band coincides with the trajectory of the primary shear band. Typical electron micrographs illustrating the process of 'the elastic relay' are shown in Fig. 3.59d and 3.59d*. The scheme of this mechanism is shown in Fig. 3.60d.

The effects of multiple secondary accommodation have been observed very rarely, when the elastically strained nanocrystal initiated several secondary shear bands in the amorphous matrix. Fig. 3.59e illustrates this relatively rare case: the emission into the amorphous structural component of several new shear bands from the

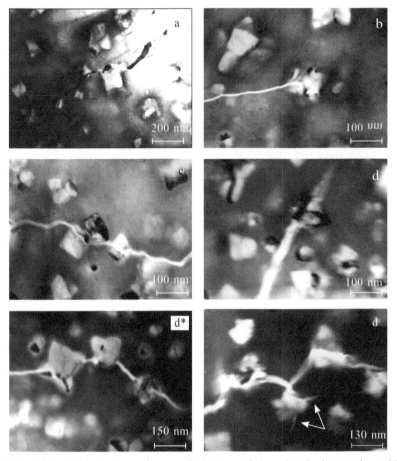

Fig. 3.59. Bright field electron microscopic images of shear bands, interacting with nanocrystals ($T = 653$ K).

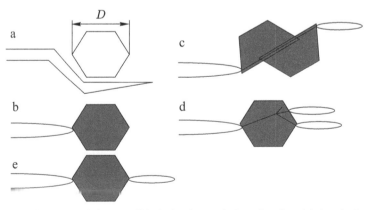

Fig. 3.60. Diagrams of the possible behaviour of shear bands with band phases in the ANC.

boundary of the nanocrystal (indicated by arrows). The corresponding diagram is shown in Fig. 3.60d.

Among the factors influencing the nature of the interaction of propagating shear bands and the nanoparticles located in the amorphous matrix, are the following: the power of shear bands, the propagation rate of shear bands, the relative orientation of the shear bands and nanoparticle size, shape and crystalline structure of the nanoparticles, the difference between the coefficients of thermal expansion of the amorphous matrix and nanoparticles, the ratio of the elastic moduli of the amorphous and crystalline structural components, chemical composition of the formed phases. Which of the above factors is determining in the realisation of a specific variant of interaction of shear bands with nanoparticles is a very important question. Evidently, it is necessary to consider a number of factors in the complex to interpret the results of the study. However, the priority, in our view, is the realisation of the size effect. In the case when the size of the crystalline phases is commensurate with the thickness of the shear band (60–70 nm) processes of bending of the band front between the nanoparticles and, consequently, changes in its trajectory can take place [279]. The variant of cutting the nanoinclusion by the shear band occurs provided that its size exceeds a certain critical size, helping the development of slip bands in the crystal area under the stress concentration. If in the approach of shear bands to the nanocrystals the active slip systems are inappropriately situated (very small Schmidt factor), dislocations can not nucleate in the nanocrystal.

The analysis of the probability of operation of a specific mechanism revealed the cutting mechanism of nanoparticles is the main mechanhism – 52%, while the secondary accommodation is implemented less often – 6% (Figure 3.61). Figure 3.62 shows a tendency that when a certain nanoparticle critical size is reached there is a change of the priority mechanism of their interaction with the shear bands. Thus, the manifestation of the size effect was found in the analysis of the interaction of shear bands with nanocrystals.

Our experimental studies have shown that for the selected alloy the rounding mechanism operates to 85 nm. The motion of the shear band in rounding the counter nanoparticles resembles the process of double cross slip of the dislocation overcoming a hard barrier at the nanoparticles. By analogy with [266] the critical size of the nanocrystals above which the shear band can enter the nanocrystal and 'get stuck' there, implementing the braking mechanism:

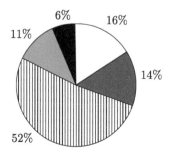

Fig. 3.61. The graph of the probability distribution of operation of the mechanisms of interaction of the shear band with nanocrystals at $T = 643$ K: white – rounding, dark gray – inhibition, shaded – cutting, gray – primary accommodation, black – secondary accommodation.

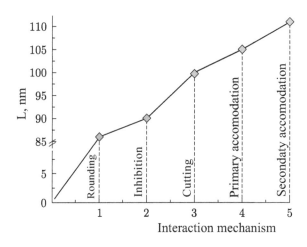

Fig. 3.62. The phenomenon of the size effect in the interaction of the shear band with nanoparticles.

$d_{cr} \approx 90$ nm was determined [290] (Fig. 3.62). For the crystal size greater than 90 nm the cutting mechanism is more likely to operate, when the shear band propagating in an amorphous matrix, can stimulate a dislocation flow within a nanoparticle, which in turn can stimulate the propagation of the accommodative band in the amorphous matrix on the other side of the nanoparticle.

Finally, it is necessary to discuss briefly at the structural level the previously described anomaly of the dependence of the strength (hardness) on the size of the nanoparticles (see Fig. 3.54). When the particle size is less than 70–80 nm strength decreases with a decrease in size (analogue of the Hall–Petch dependence). Attention is drawn to the fact that $\delta = 70$–80 nm is the typical thickness of shear bands propagating in an amorphous matrix [281]. For this reason, it is important to propose another mechanicam of interaction (not yet confirmed by electron microscopic studies) of nanoparticles with shear bands. When $d < \delta$ the mechanism of the 'flow around' is likely to operate in when the nanoparticle appears to plunge into the propagating shear band. As the difference between d and δ becomes greater this process will be facilitated and occur in line with the falling branch on the HV (d) dependence in Fig. 3.54.

We note three important points. First, the pattern of interaction is probabilistic in nature and depends on the size of nanoparticles in addition to a number of other factors (power of the shear band, its propagation rate, the relative orientation of the band and the nanoparticles, its shape, etc.). Second, the finding of the phenomenon of cutting nanoparticles by the shear bands clearly indicates the possibility of the dislocation flow (or its nearest equivalent) in the nanoparticles, which however, is greatly reduced at the nanoparticle size smaller than 80 nm. Third, the results obtained seem to hold some reservations also for the case where the nanoparticles form in the crystal and not in the amorphous matrix.

Amorphous nanocrystalline alloys containing predominantly isolated nanocrystals, separated by amorphous layers
In nanocrystals of this type amorphous layers can be regarded as 'blurred' grain boundaries. If we assume that the individual nanocrystalline regions with the size X are separated by viscous amorphous layers with thickness Δ (Fig. 3.63) and that the deformation of such an amorphous–nanocrystalline system will take place on an amorphous layer (a kind of grain boundary sliding, the mechanism of which is associated with the propagation of plastic shear zones in disordered intercrystalline regions), the yield strength of such a nanocrystalline material, provided that $\Delta \ll X$ and $X = N^{1/3}L$ (N is the mean number of nanocrystals with size L in a polycrystalline block) can be represented in the form [230]:

$$\sigma_T = \left(\frac{\sigma_0}{2}\right)\left(1 + \frac{2}{3}\frac{X}{\Delta}\right). \tag{3.57}$$

This nanocrystal will deform plastically without strain hardening to the strain of the order of $\varepsilon_c \approx \Delta/X$. In a special case which is however very common in practice in which each nanocrystal is separated by an amorphous layer from the adjacent nanocrystals, we have to put in the equation (3.57) $X = L$. As follows from the

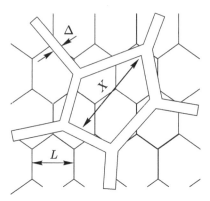

Fig. 3.63. Structural model of a nanocrystal with amorphous intercrystalline layers.

equation, the value of the yield strength for such material increases linearly with the size of the nanocrystals and with a reduction in the thickness of the amorphous intergranular layers in contrast to the Hall–Petch relationship. In support of the above model it can be shown experimentally [Introduction Ref. 10] that the hardness (strength) of the structural state actually increases with a decrease in the thickness of the amorphous layer and increase in the average size of the nanocrystals (Fig. 3.64). In the limit, when the amorphous layer thickness will be commensurate with the thickness of the 'normal' grain boundaries, we can assume that their amorphous structure is maintained [291] and we will have a mechanism of grain boundary microsliding, proceeding in the same way (on amorphous layers). In connection with the above we should mention the interesting results obtained in [292]. The $Fe_{78}Si_9B_{13}$ alloy was investigated which is well-known in the amorphous state after quenching from the melt as an amorphous analogue of the magnetically soft alloy and is

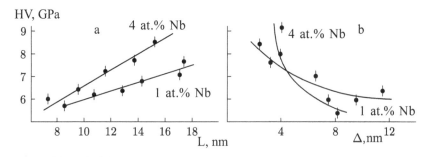

Fig. 3.64. Experimental dependence of microhardness on the mean size of nanocrystals (a) and on the mean thickness of the intercrystalline amorphous layers (b) for the Fe–Si–B–Nb–Cu (Finemet) alloy.

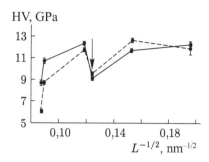

Fig. 3.65. Microhardness of the grain size of the $Fe_{78}Si_9B_{13}$ alloy obtained by annealing the amorphous state; the measurements were performed on contact (solid line) and free (dashed line) surfaces of the ribbon produced by spinning.

used extensively in electrical engineering. The grain size of less than 65 nm was obtained through a series of controlled annealing treatments. As expected, after deformation the alloy did not show any signs of dislocations. It is interesting that in the range $L = 65–70$ nm there was an appreciable decrease in hardness (Fig. 3.65), which according to Mössbauer data and Raman spectroscopy is associated with the decline in the number of nearest neighborhoods and a corresponding reduction of the density of grain boundary regions. This must definitely facilitate the process of grain boundary microsliding and a corresponding decrease in strength. It should, however, be remembered that in the nanocrystalline state this alloy is not is a single-phase alloy and the mechanical properties could have been affected by the redistribution of components between phases (in particular boron atoms).

3.4.7. Failure of nanocrystalline alloys

Studies of features of the propagation of cracks in the nanocrystals are important in the context of finding opportunities to increase the fracture toughness (crack resistance) of brittle materials with a dispersed structure. Experimental data show higher values of fracture toughness of multiphase brittle materials in the nanostructured state [Introduction Refs. 4, 5]. On the other hand, a number of data [293] shows that the plasticity of nanocrystals does not increase. Just beginning to appear are the first theoretical studies of the dimensional effects on the strength of nanocrystals and, in particular, on the propagation of cracks in them [294, 295].

Fractographic investigations of brittle fracture surfaces of nanocrystals revealed the dominant role of the mechanism intercrystalline failure [Introduction Refs. 4, 5] [296]. In [297] the identation method was used to measure the fracture toughness of Fe–Mo–Si–B nanocrystals obtained by controlled annealing of the amorphous state with the grain size of the α-Fe phase from 11 to 35 nm. It was found that the dominant failure mechanism is intercrystalline crack propagation. The mean size of the dimples on the fracture surface for samples with the grain size $d = 11$, 25 and 35 nm was respectively equal to 0.5, 2 and 5 μm. The value of fracture toughness for samples with the grain size increasing from 11 to 35 nm increased from 2.7 to 4.6 MPa·m$^{1/2}$. The failure resistance with grain size variation is independent of plastic deformation [297]. However, the fracture surface of these materials revealed traces of plastic flow [185]. It is assumed that the plastic strain may have a significant impact on the failure conditions of the nanocrystals.

Features of brittle fracture of nanocrystalline materials
The nature of crack propagation in a polycrystal and, therefore, conditions for trans- or intercrystalline failure are determined by the ratio of cohesion γ_0 and grain boundary γ_e fracture energy. Specific energies of trans- and intercrystalline failure have the form

$$\gamma_0 = 2\gamma, \quad \gamma_e = \eta(2\gamma + 2\gamma_s - \gamma_b), \tag{3.58}$$

where γ and γ_b are the specific energy of the free surface and grain boundaries, respectively, γ_s is the energy of cleavage steps, η is the roughness factor the fracture surface.

In conventional polycrystalline materials, the contribution of the junctions of the GBs to fracture energy is negligible and is not included in the analysis of fracture [298]. The volume fraction of material associated with the GBs and triple junctions increases with decreasing grain size. For the nanocrystals the volume fraction of triple junctions becomes comparable to the volume fractions of GBs and intragranular material (see Introduction) and their contribution to the fracture energy must be taken into account. At crack propagation the effective fracture energy in the material is equal to

$$\gamma_0^* = f_0\gamma_0 + f_b\gamma_b + f_j\gamma_j, \tag{3.59}$$

where f_0, f_b, f_j is the fraction of the area of the crack surface related to the internal volume of the grains, boundaries and junctions of the GBs, respectively, depending on the trajectory of the crack, and γ_0, γ_b, γ_j are the contributions of the appropriate structural components to the specific fracture energy of the nanocrystal.

If the plane of crack propagation is perpendicular to the axis of application of external stress σ, and its tip is deflected from its trajectory, then the local stress intensity factors k_1 and k_2 for a kink-shaped crack oriented at angle θ to the main plane [299] are

$$k_1 = \cos^3(\theta/2)K_1, \quad k_2 = \sin(\theta/2)\cos^2(\theta/2)K_1, \quad K_1 = \zeta\sigma\sqrt{L}, \quad (3.60)$$

where ζ is a numerical coefficient, L is the length of the main crack.

The condition of crack propagation along the face of the grain at an angle θ to the main plane [298]

$$k_1^2 + k_2^2 \geqslant \left[\frac{2E\gamma_e}{1-v^2}\right]. \quad (3.61)$$

$E = 2\mu(1 + v)$ is Young's modulus, v is Poisson's ratio, μ is the shear modulus. Provided

$$K_1 \geqslant K_{1c} = \left(\frac{2E\gamma_0^*}{1-v^2}\right)^{1/2} \quad (3.62)$$

the crack will propagate into the volume of the grain and this will lead transcrystalline fracture.

In intercrystalline failure the contribution of the linear tension of the crack surface to the fracture energy becomes significant for nanocrystals. With the crack front bending between grains or bending with a small radius of curvature r the fracture energy is [295]

$$\gamma_e^* = \gamma_e + \frac{T}{r}, \quad (3.63)$$

where T is the linear tension of the crack front, $2r \approx d$ is the grain size. Comparison of (3.61), (3.62) implies the condition of realisation of intercrystalline failure:

$$\left(\frac{\gamma_e^*}{\gamma_0^*}\right) \leqslant \cos^4\left(\frac{\theta_{max}}{2}\right). \quad (3.64)$$

Substitution of the typical values of the parameters in (3.59) and (3.63) gives an assessment of the specific fracture energy for a nanocrystal with a grain size of 10–20 nm due to the inclusion of junctions of the GBs and surface tension of the intercrystalline crack at 15–20%.

The size of a Griffith crack in a homogeneous continuous material [298] is

$$L_G = \frac{4E\gamma_0^*}{\pi\left(1-v^2\right)\sigma_f^2}.$$

At $\gamma_0^* = \mu b/20$, where b is the interatomic distance, $\sigma_f = \mu/50$ we get $L_G \approx 500b \gg d$ (~10 nm), i.e. the size of a Griffith crack can be much larger than the grain size of the nanocrystal. Because of the high density of grain boundaries and their joints, purely transcrystalline fracture can not take place in the nanocrystals. The front of the crack, even with a straight propagation path, periodically passes through the material associated with the interior volume of the grains, GBs and their junctions, so that the fracture energy changes periodically. With intercrystalline fracture of the nanocrystal the fracture energy at a scale much larger than the grain size is also a periodic function of the crack length. The specific fracture energy γ^* (fracture toughness G_c) is a periodic (quasi-periodic) function of the crack propagation path with a period approximately equal to the grain size d. This results in the formation of a number of thermodynamically metastable stable states of the nanocracks [294, 295].

As is known, the nanocrystalline material can be a two-phase material, when one of the phases are the GBs and the other – intragranular areas. The fracture toughness of the nanomaterial is determined by generalization of (3.59). If the volume fractions of the volumes, grain boundaries and their junctions are f_g, f_b and f_j, respectively, fracture toughness can be expressed as

$$G_j = G_b f_b + G_g f_g + G_j f_j, \tag{3.65}$$

where G_b, G_g, G_j are the critical energy release rates for moving cracks in the volume, the boundaries and the grain junctions.

The volume fractions of the grain and grain boundary phases, respectively are:

$$f_g \approx 1 - \frac{\delta}{d}, \quad f_b \approx \frac{\delta}{d}, \tag{3.66}$$

where δ is the thickness of GBs. The dependence of fracture toughness on the grain size has the form

$$G_c = G_g + \delta\left(G_b - G_g\right)d^{-1}. \tag{3.67}$$

At intercrystalline failure we must also take into account the effect of increase the effective length of the crack in $(1 + \psi f_g)^{1/2}$, where the numerical coefficient $\psi \approx 2$ takes into account the geometry of the grains.

With increasing toughness of intercrystalline failure as a result of of the friction of the crack edges K_{II}/K_I times [300] we have

$$G_j^* = G_j \left(1 + \psi f_g\right)^{1/2} \left(1 - f_g^{1/2}\right)^{-1}. \tag{3.68}$$

With the help of (3.67), (3.68) we can also estimate the fracture toughness of heterophase nanocrystals. Combining the dimensions and the properties of the phase components in heterophase materials or nanocomposites, by increasing the fracture energy at increasing path of the intercrystalline or developing (through the interlayer) more brittle phase of the crack we may improve the characteristics of crack resistance.

Failure criterion of nanomaterials with non-equilibrium boundaries and uncompensated grain junctions

The junction of the GBs for which the value of the uncompensated reversal neighbouring grains is Ω, creates a long-range elastic stress field, similar to the stress field of disclinations with power Ω [301]. Additional stress concentration in the junctions of the GBs can be caused uncompensated densities of dislocations in GBs. The equilibrium sizes of cracks formed at the disclination were analyzed in [301–303]. These non-equilibrium grain boundaries and their junctions may have a significant impact not only on the nucleation but also the propagation of the crack in the material.

Consider the condition of passage of a crack with length L through alternating tension and compression zones caused by junction disclinations. The stress intensity factor of the crack in an inhomogeneous stress field [298]:

$$K_{eff} = K_0 + \sqrt{\frac{2}{\pi}} \int_{a-d}^{a} \frac{\sigma(x)dx}{\sqrt{a-x}}, \quad K_0 = \sigma_a \left[\frac{\pi L}{2}\right]. \tag{3.69}$$

If all the junctions of the GBs are uncompensated (Fig. 3.66a), and the strength of the boundaries is so low that intercrystalline failure takes placedestruction, the crack will move through the system of the alternating positive and negative disclinations. The condition of passage of grain boundary (interfacial) cracks through the alternate tension and compression zones from the junction disclinations – excess of the stress intensity factor above the critical value determined from (3.69) by substituting the disclination stress fields [301], is given by the relations:

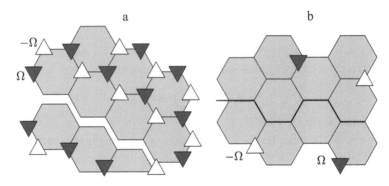

Fig. 3.66. Movement of an intergranular crack a system of uncompensated triple junctions (a) and between uncompensated junctions of the boundaries (b).

$$K_c = K_{cb} + \Delta K, \tag{3.70}$$

$$\Delta K = \zeta \left[\frac{\mu \Omega}{2\pi^{3/2}(1-v)} \right] d^{1/2}, \ \zeta = 6.738, \tag{3.71}$$

where K_{cb} is the critical stress intensity factor for crack propagation along grain boundaries in the material with compensated boundary junctions. When only part of the grain boundary junctions is uncompensated (Fig. 3.66b) it is more easier for the crack to move between them. The condition of passage of the crack between two uncompensated triple junctions, spaced at $Q_n = nd$, is

$$\Delta K = \zeta_1 \sigma^*(Q_n)^{1/2}; \ \sigma^* = 2 \left[\frac{\mu \Omega}{2\pi(1-v)} \right] \left[\ln\left(\frac{2R}{Q_n} \right) - 2 \right]. \tag{3.72}$$

Consider the condition of failure of the nanocrystal with uncompensated triple junctions and non-equilibrium GBs. The formation of the free surface as a result of crack opening in the material with long-range internal stresses leads to relaxation of the latter in crack's neighborhood.

Propagation of the crack tip leads to the relaxation of stress fields of junction disclinations, situated on the line of movement of the front. In addition, the energy of uncompensated joints found not directly in the line of crack's front but at some small distance from it is reduced. The expression for the energy of the disclinations in the centre of the crystal with the size R near its free surface at a distance ξ from it are, respectively, [301]

$$E_{d(0)} = \frac{\mu \Omega^2}{4\pi(1-v)} R^2; \ E_{d(1)} = \frac{\mu \Omega^2}{4\pi(1-v)} \xi^2; \tag{3.73}$$

so that when the crack tip advances by dL the energy of junction disclinations at a distance of no more than L_e from the surface of the crack, i.e. in the area $L_e dL$, significantly decreases. L_e is the screening radius of the dislication field in the disclination ensemble. Analysis of the magnitude of L_e [304] shows that $L_e \approx 3L^*$, where L^* is the average distance between the disclinations in the ensemble. When the crack moves by dL the energy of the disclination system on average decreases by $\Delta E = \varphi \, [\mu\Omega^2/4\pi \, (1-v)]L_e dL$, where φ is a numerical factor of the order of unity. Relaxation of stress fields leads to a reduction of the effective fracture energy by $\Delta\gamma_d$:

$$\Delta\gamma_d = \varphi \left[\frac{\mu\Omega^2}{4\pi(1-v)} \right] L_e .$$

(3.74)

The energy of non-equilibrium GBs [305] is

$$\gamma_B = \gamma_{B0} + \Delta\gamma, \quad \Delta\gamma = \frac{\mu b^2 \rho \ln(R/2b)}{4\pi(1-v)}.$$

(3.75)

Griffith's criterion for a crack with length L in such material will be have the form

$$\sigma_f = \left[\frac{4\mu(\gamma_e - \Delta\gamma_d)}{\pi(1-v)L} \right]^{1/2} + \zeta \left[\frac{\mu\Omega}{2\pi^2(1-v)} \right] \left(\frac{d}{L} \right)^{1/2} .$$

(3.76)

We estimate the characteristic values of the material parameters. The energy of the perfect high-angle grain boundary is about one third the energy of the free surface $\gamma_b \approx \gamma/3$ [302]. The energy of the non-equilibrium boundaries with excess dislocation density $\rho\gamma_{bn} \approx (4/3)\gamma$, then $\gamma_e \approx \mu b/30$. Average value $\langle\Omega\rangle = 0.02–0.04$. When $L = (5–10)d$ we have $\sigma_f \approx \mu/40$.

The effect of plastic deformation on crack nucleation and fracture stress of nanomaterials
The areas of GBMS are modelled by flat inclusions with shear deformation or continual clusters of the effective dislocation loops. In both cases the dependence of the mean shear displacement is

$$u = A(\tau_a - \tau_s)L/\mu.$$

(3.77)

Coefficients A in both cases are close to unity.

At the top of the GBMS areas a stress concentration contributing to the formation of microcracks appear. The most probable mechanisms of formation of microcracks (Fig. 3.67) are similar to Cottrell and Straw mechanisms [298].

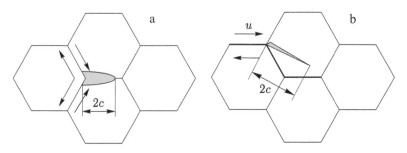

Fig. 3.67. Schematic representation of the mechanisms of formation of nanocracks: a – at the intersection of the areas of GBMS; b – at the top of a single GBMS area.

The energy of the nucleated microcrack with length $2c$ at the top of the intersections GBMS areas (Fig. 3.67, a), in which the value of shear displacement is equal to u, is given by

$$W = \frac{u^2\mu}{4\pi(1-v)}ln\left(\frac{2R}{c}\right) + 4\gamma c - \frac{\pi(1-v)\left(\sigma_a^2+\tau_a^2\right)}{2\mu}c^2 - \theta u\sigma_a c. \qquad (3.78)$$

The first term – the strain energy of dislocation type of the wedge crack, the second – the surface energy, and the third – the elastic energy of the body with a crack, the work of external forces in microcrack formation; θ – numerical parameter depending on orientation of the crack opening plane [298].

The equilibrium length of the crack is determined from the condition $\partial W/\partial c = 0$. For c we obtain a quadratic equation, that is, either there are two stable values of the crack length, or the roots are imaginary, which corresponds to the spontaneous reduction of energy. At the transition point

$$u\left[(\sigma_a^2+\tau_a^2)^{1/2} + \theta\sigma_a\right] = 4\gamma_0. \qquad (3.79)$$

At the nucleation of a crack in a single GBMS region (Fig. 3.67b), expression (3.79) takes a form similar to the Straw's condition of failure: $\sigma = 2\gamma/u$. The expressions (3.45) and (3.46) for the yield strength can be written as

$$\sigma_y = \sigma_0 + k_y D^{-1/2} F(D) \equiv \sigma_0 + D^{-1/2+\chi}, \qquad (3.80)$$

where $\chi = 0$ for $D \geqslant D_1^*$, $\chi \geqslant 1/2$ for $D < D_1^*$. Substituting (3.77), (3.80) into the above equation, for the dependence of fracture stress on the grain size we find

$$\sigma_f \geqslant \frac{2\mu\gamma_0}{k_y} D^{-1/2} F(d). \qquad (3.81)$$

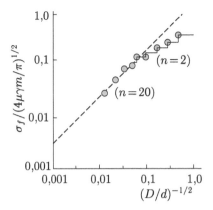

Fig. 3.68. The dependence of the reduced stress of crack formation in the top part of the cluster of dislocations in the nanomaterial on the grain size.

If $d > D^*$ equation (3.81) gives the well-known dependence of the fracture stress on the grain size [298].

The range of nanoscale grain sizes IV (see Fig. 3.48) is characterised by the formation of clusters of dislocations, and microcracks can form at the top of clusters of inhibited at the GBs. The stress of crack formation is calculated using the expression for the dependence on the number n of dislocations on the length L of a cluster of a small number of screw dislocations, locked at a dislocation with the Burgers vector mb [306]:

$$L^{-1/2} = \left(\frac{A}{2\sigma}\right)^{1/2}\left[2(n+m-1)^{1/2} - \Lambda\left(4(n+m-1)^{-1/6}\right)\right], \tag{3.82}$$

where $\Lambda = 1.85575$, $A = \mu b/\pi$.

The resulting expression for the formation of a nanocrack at the top of a blocked cluster has the form

$$\sigma_f = \sqrt{\frac{4\mu m \gamma_0}{\pi}}d^{-1/2}\Phi(D) \equiv Kd^{-1/2}\Phi(d). \tag{3.83}$$

Figure 3.68 shows the classical dependence of the reduced fracture stress on the grain size at $\Phi = 1$ (dashed line), the approximate solution for a small number of dislocations (solid line) and the exact solution (points).

When $d \leqslant D^*$ the form of the functions F and Φ, showing the degree of difference of the dependence of the fracture stress on the grain size for the NM from the classical dependence, is determined by the acting mechanism of plastic deformation of the NM.

Durability and fatigue of nanocrystalline materials
In nanocrystals, the kinetic conditions of cracking in the bulk and
and at the grain boundaries are comparable, which can lead to a
significant dependence of durability on the grain size. Describing
the nanocrystal as a two-phase material, where one of the phases are
the grain boundaries, and the other is the intragranular volume, and
assuming that the processes of formation of microcracks in the bulk
and at the grain boundaries are independent, for the probability of
formation of microcracks in the nanocrystals we can write

$$W = W_g f_g + W_b f_b, \tag{3.84}$$

where $W_g = \tau_g^{-1} = v_0 \exp [-U_g/kT]$, $W_b = \tau_b^{-1} = v_0 \exp [-U_b/kT]$ is the
probability of formation of microcracks in the bulk and at the grain
boundaries, respectively, and f_g and f_b are the volume fractions of the
material bulk and of the grain boundaries, respectively, $v_0 = \tau_0^{-1}$ is the
frequency of atomic vibrations.

As a first approximation $U_g(\sigma) = U(0) - \gamma_g \sigma$, $U_b(\sigma) = U(0) - \gamma_g \sigma$.
$U(0)$ is the part of the activation energy of formation of the
microcrack which is independent of external stress σ. Structural
parameters γ_i also take into account the stress concentration factors.
The durability of the nanocrystal is

$$\tau = (k_1 + k_2) \frac{\tau_g}{f_g + f_b(\tau_g/\tau_b)}, \tag{3.85}$$

where $f_b \approx \delta/d$, $f_g \approx 1 - \delta/d$.

**Influence of grain boundary deformation on the fracture
toughness of nanomaterials**
In the case of quasibrittle fracture of nanocrystals it is important to
know the effect of grain boundary deformation on fracture toughness.
The plastic zone at the crack tip is defined as the area in which the
shear stresses initiated by the crack exceed the stress of resistance
to GBMS.

The model proposed in [307] for the stress field of the crack in
hardening materials with hardening exponent n yields the asymptotic
distribution of stresses at the line extrapolation of the crack (axis x)
within the plastic zone:

$$\sigma_{yy} = S\sigma_y \left[\frac{r_p}{r}\right]^{n/n+1}. \tag{3.86}$$

Here r_p is the distance from the crack tip to the interface between
the elastic and plastic regions along the axis x ($y = 0$); S is the

function of the strain hardening exponent of the nanocrystal. If it is assumed that the main crack propagates by the nucleation of nanocracks at distance $r*$ from its top under the influence of local critical stress σ_c [308], the fracture toughness is

$$K_{1c} = S^{(1+n)/2n} \left(K_{c0} \right) \left[\frac{\sigma_c}{\sigma_y} \right]^{(1-n)/2n} , \qquad (3.87)$$

where K_{c0} is the parameter of the model that is independent of the yield stress and breaking stress [308]. With (3.80) and (3.83) taken into account, for the dependence of the fracture toughness (crack resistance) of the NM on the grain size we have

$$K_{1c} = S^{(1+n)/2n} \left(K_{c0} \right) \left[\frac{KD^{-1/2}\Phi(D)}{\sigma_0 + k_y D^{-1/2} F(D)} \right]^{(1-n)/2n} . \qquad (3.88)$$

For $n = 1/2$, substituting the numerical values of the parameters in (3.36), we obtain that K_{1c} at $d = 10$–15 nm can be about 7–10 MPa·m$^{1/2}$, which is at least two times the fracture toughness of conventional ceramic materials.

An important mechanism of increasing the fracture toughness of materials is to create structures that facilitate bridging at the crack mouth. For ceramic nanocomposites this mechanism was analysed in [309]. Since the nanostructured components of composite materials can not be deformed plastically, it was concluded that the source of increased fracture toughness could be the internal stresses that increase friction when pulling out the nanograins of another phase in bridging.

Analysis of possible fracture mechanisms of nanocrystals showed that the most significant source of increased toughness is grain boundary plastic deformation of nanocrystals [310]. If the stress of resistance to grain boundary microsliding in nanocrystals obtained by a specific technology is lower than the stress of formation and/or propagation of the crack, then such materials undergo plastic flow before failure and show considerable toughness.

Relationships of the formation and behaviour of cracks in nanomaterials are also important for understanding the cyclic and fatigue characteristics of these objects. The study of fatigue strength of nickel samples prepared by pulsed electrochemical deposition with a grain size of 20–40 nm and 300 nm as well as of usual coarse-crystalline Ni ($L = 10$ μm), showed that a reduction of the grain size is accompanied by increased fracture stress in the entire investigated cycle (up to 10^7 cycles at a frequency of

10 Hz), although the difference between $L \sim 30$ nm and $L \sim 300$ nm was not very significant, substantially surpassing the values for conventional Ni [311]. Study of the nature of cracks in specimens after testing revealed the significantly greater length of the cracks in the nanosamples and thus greater rate of growth of cracks in them. A similar pattern was found in dynamic tests of the Al–7.5 wt.% Mg alloy ($L \sim 300$ nm) obtained by low-temperature milling followed by consolidation. Thus, according to these data, the fatigue properties of nanomaterials are reduced due to more active development and mergers of cracks. A detailed explanation of the results of cyclic deformation of copper samples can be found in the monograph [218], in which, in particular, the strong effect of preliminary annealing is emphasized. It is also noted that, compared with coarse-crystalline analogues, nanomaterials have higher saturation stress and the significant Bauschinger effect, which reflects an increase in the density of dislocations in cyclic hardening.

Consider the features of fracture of nanomaterials produced by crystallisation of the amorphous state. The density of the amorphous material is about 1–2% less than that of the corresponding crystals. [1]. Nanocrystallisation is accompanied by the precipitation of the excess free volume at the GBs and IBs of the growing crystals. As a result, there is a high concentration of the free volume at the boundaries.

The excess free volume of the amorphous material is $\Delta V = (V_{am} - V_{cr})/V_{am}$ where V_{am} and V_{cr} is the specific volume of the amorphous and crystalline structural states, respectively. The GB area per unit volume of the nanocrystal is: $S_v = q/d$, where q is the numerical factor determined by the form of the grains ($q \approx 3$). The excess volume for the unit area of the GBs

$$v_s = g\left(\frac{\Delta V}{q}\right)d \equiv id, \qquad (3.89)$$

where g is a factor ($g < 1$) taking into account the shift of part of the excess volume to the free surface of the sample. It can be seen that the smaller total area of the boundaries in the sample, i.e. the larger the grain size, the greater the amount of the excess free volume (and/ or grain boundary vacancies) in the unit area of GBs.

As the GBs have a high concentration of the excess free volume, this causes the nucleation and growth of nanopores. The volume fraction of the nanopores is proportional to the excess free volume formed at the GBs. If the mean size of the nanopores is r_p, and the distance between them is equal to R, then the increment of the surface fraction $p = (r_p/R)^2$ of the nanopores in the GBs with increasing grain size in the first approximation is:

$$dp = \frac{j\,dv_s}{\delta} = jg\left(\frac{\Delta V}{q\delta}\right)dD, \qquad (3.90)$$

where j is a numerical coefficient. In the presence of pores with fraction p on the unit area of the grain boundaries the density of material of the boundaries is:

$$\rho_{gb} = \rho_{gb0}(1-p), \qquad (3.91)$$

where ρ_{gb0} is the density of the material of the grain boundaries in coarse-grained material, $\rho_{gb0} = \chi\rho$; ρ is the density of the single-crystal material, χ is a numerical coefficient approximately equal to 1. Thus, it follows that the density of the material of the grain boundaries decreases linearly with the increase of the grain size in accordance with the experimental data [295].

Increase of the grain size of the nanocrystals of the third type is achieved by increasing short-term annealing temperature. It is reasonable to suggests that the grain boundaries in the studied alloys with the lowest possible grain size d_{min} have an amorphous structure. With the increase in the grain size with increasing annealing temperature processes of structural relaxation and crystallisation take place at the boundaries of the growing crystals.

In the nanocrystals, the decrease in the grain size results not only in an increase in the area of the grain boundaries but also in a reduction of the energy of the GBs and IBs. There is also a tendency to reduce the grain boundary excess free volume with decreasing grain size, which is in agreement with the corresponding change of the mechanical properties – increased fracture stress and strain to failure of the Ni–P alloy [312].

The structural mechanisms that determine the reduction of the fracture stress of the third type nanocrystals with increasing grain size will be discussed. Grain boundary porosity leads to a decrease in the fracture stress of the material, firstly, because of the reduction load-bearing section of the boundaries, causing a reduction in the grain boundary brittle fracture energy, and secondly, because of higher stresses acting on the GBs. The reduction of the load-bearing section of the GBs, caused by the nanopores, lowers the specific energy of fracture of the porous border defined by:

$$\gamma_{ep} = \gamma_e / (1-p), \qquad (3.92)$$

where γ_e is the specific energy of grain boundary fracture.

Additionally, the load is not received by the total area of the boundary with the area A but by the true cross-section area A_s ($A_s < A$). If σ_n is the normal component of the external stress on some grain boundary, then in the presence of porosity with the surface fraction p at the GBs the mean stress acting on the boundary on the basis of the balance of forces at the boundary will be:

$$\sigma_s = \sigma_n / (1-p). \tag{3.93}$$

Consideration of these factors leads to the following expression for the fracture stress of the nanocrystals obtained by nanocrystallisation:

$$\sigma_f = (1-p)^{3/2} \left[\frac{4E\gamma_e}{\pi(1-v^2)X} \right]^{1/2}. \tag{3.94}$$

For the nanocrystals, the size of crack-like defects X, leading to failure, is not associated with the grain size and p is proportional to d. So, from (3.94) it follows that the brittle fracture stress decreases with increasing grain size. If $\sigma_s < \sigma_f$ and fracture is quasibrittle, instead of brittle fracture energy γ_e equation (3.94) will include grain boundary fracture toughness G_c which is almost completely independent on p.

3.5. The magnetic properties

Study of the magnetic properties of the nanocrystals of the third type is stimulated by considerable successes in applications for new high-performance magnetically soft and magnetically hard in the nanocrystalline state [Introduction Ref. 1] [74, 192, 313–315].

In principle, partial or complete crystallisation of amorphous ferromagnetic materials usually leads to the high-coercivity state [316]. Heat treatment above T_c can be result in the precipitation of nanocrystalline phases with high magnetocrystalline anisotropy in the amorphous matrix. For example, coercive force H_c of the $Fe_5Co_{70}Si_{15}B_{10}$ amorphous alloy after annealing at 723 K for 1 h increases to 24 kA/m, due to the release of a nanocrystalline cobalt phase [317]. A high coercive force can also be achieved in partial crystallisation of amorphous alloys that are initially paramagnetic at room temperature [317]. In this case, nanoparticles of the crystalline phase with high Curie temperature T_c form in the amorphous matrix

In [318] it is shown that, depending on the heat treatment, two very different structural states of partially crystallized amorphous

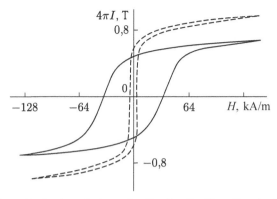

Fig. 3.69. Magnetic hysteresis loop for the $Fe_{67}Cr_{18}B_{15}$ alloy in the amorphous-nanocrystalline (solid lines) and nanocrystalline (dashed curves) states.

$Fe_{85-x}Cr_xB_{15}$ alloys ($x = 10, 12, 15, 18, 20$), characterized by high values of H_c, can be produced. It was found that for each of the alloys there is annealing temperature T^*_{ann}, at which the value of H_c passes through a maximum. Figure 3.69 shows a typical hysteresis loop for the alloy with $x = 18$, annealed under the optimum conditions. The loop completely crystallized alloy with the same composition is shown here for comparison. The maximum value of H_c increases from 16 to 40 kA/m with increase of the chromium content, while the saturation magnetisation I_s decreases from 1 to 0.65 T. The value of T^*_{ann} also decreases from 950 to 870 K with increasing x [318].

The dependence of the magnetic properties of $Fe_{67}Cr_{18}B_{15}$ alloy on annealing temperature is shown in Fig. 3.70. Two peaks of H_c of approximately equal size (35–36 kA/m) appear. The first peak is located very close to the T_c, and the second – about 120° above. The structural state of the amorphous–nanocrystalline alloy corresponding to the second peak is more stable and less sensitive to the heat treatment time. I_s in Fig. 3.70 initially noticeably increases, but then experiences a local minimum corresponding to the second peak of H_c. On the basis on detailed analysis the authors of [318] concluded that the first maximum of H_c is due to the predominance of the process of rotation of the magnetisation vector of single-domain nanoparticles of α-Fe–Cr in the paramagnetic amorphous matrix. The second maximum is due to a delay in displacement of the domain boundaries of the α-Fe–Cr phase precipitates of the $(Fe, Cr)_3B$ phase, where the main role in this process is played by the fluctuations of the magnetisation.

In recent years, a new class of magnetic materials with a mixed amorphous–nanocrystalline structure and higher static and dynamic

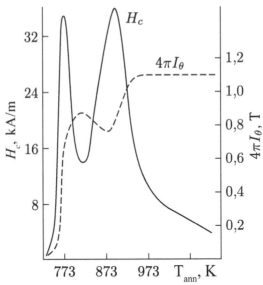

Fig. 3.70. Dependence of the coercive force and saturation magnetisation on annealing temperature for $Fe_{67}Cr_{18}B_{15}$ alloy.

magnetic properties in comparison with those of crystalline and amorphous alloys used for the same applications was developed [74]. For example, the features of magnetically hard materials such as Fe–Nd–B (coercive force, residual induction and maximum magnetic energy) show a clear upward trend in the formation of the nanocrystalline structure. Magnetically hard nanomaterials of this type, obtained by melt quenching, are used on an increasing scale [319].

The paradox lies in the fact that the two-phase amorphous-nanocrystalline state (not the single phase state) is characterised by the uniquely high level of magnetic parameters, intrinsic to ferromagnetic magnetically soft alloys which contradicts, at first glance, the generally accepted views on the nature of ferromagnetism. However, nanocrystalline magnetic materials of the third type with high values of magnetic induction and permeability (initial and maximum) successfully compete with well-known alloys based on Fe–Si and Fe–Co, both amorphous and crystalline. Figure 3.11 shows clearly the advantages of nanocrystalline alloys produced by controlled annealing the amorphous state, as compared to traditional magneticalloy soft materials, including amorphous ones. These advantages become apparent to an even greater extent in the high-frequency magnetic reversal range. We consider the physical basis of this unusual phenomenon.

3.5.1. Theory of magnetism in nanocrystals with strong intergrain interaction

The most important fact in understanding the optimal parameters of the structure of the nanocrystalline magnetically soft state is the fact that the measure of the magnetic hardness – coercive force (H_c) – is inversely proportional to the grain size in the range of 0.1–1 mm in which d exceeds the thickness of the domain (Bloch) wall δ_w; $d \gg \delta_w$. In such cases, the grain boundaries act as obstacles to the motion of domain walls and, therefore, fine-grained materials are generally magnetically harder than coarse materials. Recent advances in understanding the nature of the coercive force have led to the conclusion that a very small grain size, $d \ll 100$ nm, leads to a sharp reduction in H_c [320–325] (Fig. 3.71). This is due to the fact that d is significantly smaller than δ_w ($d \ll \delta_w$). In this case, the wall incorporates several grains, so that the fluctuations of the magnetic anisotropy on the scale of a single grain do not lead to the inhibition of the entire domain wall. This important concept suggests that nanocrystalline alloys have a significant potential as a magnetically soft material because its properties require that the nanocrystalline grains are in the magnetic sense a single unit. Similar ideas were expressed for the so-called 'exchange-spring' magnetically hard materials [326–328].

In principle, the reduced coordination number of atoms at the surface should affect the Curie point of ferromagnetic nanocrystals where the percentage of grain boundaries is high. However, in most cases described in the literature, the value of the Curie point (T_C) does not deviate much from the values that are characteristic of bulk

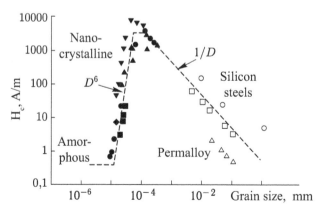

Fig. 3.71. Diagram illustrating the dependence of the coercive force on the grain size of magnetic materials [320]. Data from different experiments are given.

materials. For example, T_C value for Ni is 360°C in both coarse- and nanocrystalline states [329]. In [330] the value obtained T_C = 1366 K for the Co nanoparticles and T_C = 1388 K for the bulk state.

The amorphous–nanocrystalline materials are in the two-phase state and have two Curie points (nanocrystalline and amorphous phases). Both of these are important parameters in the description of the magnetic properties. The amorphous phase in which nanograins form during crystallisation, is usually enriched with non-magnetic atoms and, therefore, has lower parameters of magnetic ordering and a lower Curie temperature. This is also the case for a mixture of two crystalline phases, such as α-Fe as the main phase and the second phase in the form of carbides with a lower T_C.

When discussing the benefits of nanocrystalline alloys as magnetically soft materials, we should first consider the properties such as coercivity and permeability. Reduction of the coercive force and the resulting increase in permeability are welcomed developments, which allow to give preference to amorphous and nanocrystalline alloys.

When considering the magnetic anisotropy in magnetically soft nanocrystals, important characteristics are the length of the magnetic exchange interaction (exchange length) and its relationship with the width of the domain wall and the size of the monodomain [331]. These parameters can be determined using the following equations [326]:

$$\delta_w = \pi \sqrt{\frac{A}{K}} \text{ and } L_{ex} = \sqrt{\frac{A}{4\pi M_s^2}}, \tag{3.95}$$

where δ_w is the domain wall thickness, L_{ex} is the the exchange length, A is the rigidity of the exchange interaction, K is the magnetic anisotropy constant, M_s is saturation magnetisation.

The model of the random distribution of anisotropy, proposed by G. Herzer [321–325, 332, 333], was a prerequisite for the explanation of the magnetically soft properties of ferromagnetic nanocrystals. As part of this model, consideration of the effective anisotropy of the nanocrystals was based on the concept of the random distribution of anisotropy in the amorphous alloys. In particular, the concept of the characteristic volume whose linear dimensions correspond to the characteristic exchange length $L_{ex} \approx (A/K)^{1/2}$ (Fig. 3.72) was introduced. N grains with randomly distributed axes of easy magnetisation in the volume L_{ex}^3, having the exchange interaction. Since the axes of easy magnetisation are distributed randomly, one

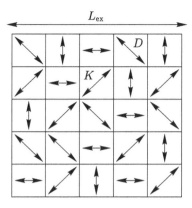

Fig. 3.72. The model of chaotic anisotropy. The arrows indicate the randomly oriented axes of easy magnetisation of magneto-crystalline anisotropy [324].

can perform a statistical average over all N grains and the effective anisotropy is equal to $K_{eff} = K/(N)^{1/2}$, where is the anisotropy constant for any of the grains. The number of grains that are in the exchange interaction is equal to: $N(L_{ex}/d)^3$, where d is the mean diameter of individual grains. Transforming the expression for K_{eff}, we get:

$$K_{eff} \cong Kd^{3/2} \approx \left(\frac{K_{eff}}{A} \right) \approx \left(\frac{K^4 d^6}{A^3} \right). \tag{3.96}$$

Since the coercive force is proportional to the effective anisotropy, this analysis leads to the conclusion that the effective anisotropy and, therefore, the coercive force should grow as the sixth degree of the grain size:

$$H_c \sim d^6. \tag{3.97}$$

An important condition of this relationship is that the nanocrystalline grains should necessarily have the exchange interaction. It does not hold for non-interacting particles which are characterised by the interaction parameter commensurate with the diameter of each particle and sensitive to the superparamagnetic susceptibility.

Another dependence of the coercive force on the grain size was proposed in the literature for low-dimensional systems and was later confirmed experimentally. In [334] the contribution of domain walls to coercive force, H_w, was calculated It has been shown that it depends on the parameter of the fluctuations and equiaxed magnetic anisotropy. The authors concluded that the coercive force H_c should, in principle, be determined by one of three different parameters:

1) the field of reverse domain nucleation, H_N;
2) the field in which the domain walls start to grow, H_G;
3) the field in which the domain walls become mobile, H_W.

The coercive force is determined by the highest of these three values.

The initial model of the randomly distributed anisotropy was addressed to the homogeneous amorphous or nanocrystalline phases. The next step was the extension of this model to the two-phase amorphous–nanocrystalline systems characteristic of magnetically soft nanocrystals of the third type (for example, for Finemet alloy) [323, 333].

The effective anisotropy for multiphase materials is

$$d_{eff} = \left(\sum_i \frac{X_i d_i^3 K_i^2}{A^{3/2}} \right)^2,$$

(3.98)

where the sum corresponds to i phases in the material. In Finemet-type alloys (see section 3.5.3) $K_{am} \ll K_{BCC}$ (K_{am} and K_{BCC} are the magnetic anisotropy constants of the amorphous and nanocrystalline BCC phases, respectively), and a simple two-phase nanocrystalline material has the effective magnetic anisotropy

$$K_{eff} \approx \left(1 - X_{am}\right)^2 \frac{K_i^4 d^6}{A^3}$$

(3.99)

assuming that the volume fraction of the amorphous phase (X_{am}) is small. This simple expression predicts the effect of 'dilution' associated with increase of the relative amount of the amorphous phase.

In the literature, there was also a slightly different power-law dependence d^n for H_c at $n < 6$ [335–337]. Unlike the Finemet alloys, the coercive force in the alloys Fe–Zr–B–(Cu) [338] and Fe–P–C–Ga–Si–Cu [339] is subject to a simpler law of d^3. A similar dependence was explained in [337, 340, 341] as a special case of the Herzer model the additional effect of the long-range equiaxed anisotropy (K_u), i.e. with the value of the exchange interaction parameter much greater than L_{ex}. Under these conditions, the effective anisotropy can be represented as follows:

$$K_{eff}^{gen} = \sqrt{\left(K_u^2 + \left(K_{eff}^{nc} \right)^2 \right)},$$

(3.100)

where K_{eff}^{gen}, K_{eff}^{nc} is the effective anisotropy, respectively general and of the nanocrystal, in the absence of additional equiaxed anisotropy. Substituting into (3.100) the expression for the effective magnetic

magnetic anisotropy from the model of the randomly distributed anisotropy, taken out of equation (3.96), we obtain

$$K_{eff}^{gen} = \left(K_u^2 + \frac{K_j^2 d^3 \left(K_{eff}^{gen} \right)^{3/2}}{A^{3/2}} \right)^{1/2},$$
(3.101)

Equation (3.101), unfortunately, can not be solved analytically, but in the extreme case, when $K_u \gg K_{eff}^{nc}$, it simplifies to

$$K_{eff}^{gen} = K_u + \frac{1}{2} \left(\frac{\sqrt{K_u} K_i^2 d^3}{A^{3/2}} \right) = a + b d^3,$$
(3.102)

as observed experimentally.

The macroscopic characteristics of equiaxed magnetic anisotropy may be due to induced anisotropy caused by the domain structure formed during annealing or due to the magnetoelastic interaction. In [338] the authors proposed a modification of the model of randomly distributed anisotropy for low-dimensional systems, where the coercive force is expressed as

$$H_c \approx K \left(\frac{d}{\sqrt{A/K}} \right)^{2n/(4-n)},$$
(3.103)

where A is the rigidity of the exchange interaction and n is the dimension of the exchange interaction region.

The two-phase model of effective anisotropy deals with the case when the rigidities of the exchange interaction of the two phases (amorphous (am) and nanocrystalline (nc)) are comparable values.

In [342], the two-phase model of the random distribution of anisotropy was applied to more realistic case where $A_{am} < A_{nc}$. In this model the effective anisotropy is:

$$K_{eff} \approx \frac{1}{\phi^6} (1 - X_{am})^4 K_1^4 d^6 \left(\frac{1}{A_{nc}^{1/2}} + \frac{(1 - X_{am})^{-1/3} - 1}{A_{am}^{1/2}} \right)^6,$$
(3.104)

where ϕ is a coefficient reflecting the symmetry of K_{eff} and the rotation angle of spin along L_{ex}. Note that in the classical model of $\phi \approx 1$.

The nature of saturation magnetostriction for nanocrystalline alloys is discussed in [332]. It proposes to consider that λ_s is determined by the balance between the contributions of nanocrystals and the amorphous matrix. When $\lambda_{nc} < 0$ and $\lambda_{am} > 0$, a very small value of λ can be obtained for the amorphous–nanocrystalline composite. A simple two-phase model for the induction and saturation magnetostriction, which follows from the simple rule of mixtures [321, 322], was proposed.

As we said, the Herzer model [323–325, 332, 333] described quite accurately the data on the coercive force for many ferromagnetic nanocrystals. However, to comply with this theory, two essential conditions must be fulfilled:

1. The grain size should be smaller than the characteristic parameter of the exchange interaction.
2. The grains must retain the ferromagnetic interaction.

In alloys with a single nanocrystalline phase with the condition 1 satisfied the model works at temperatures below the Curie point. But this is not required in the case of multiphase systems which we encounter in the case of magnetic alloys produced by crystallisation from the amorphous state.

For two-phase microstructures with ferromagnetic amorphous layers (AL) and ferromagnetic nanocrystals (NC) of unique nature the mechanism of NC–AL–NC exchange interaction is crucial for the formation of the magnetic properties of these materials. This interaction depends on the size of the nanocrystals and, more importantly, on the chemical composition, size and volume fraction of the AL. The highest properties apparently correspond to the state when both criteria are met at a temperature below the Curie point of the amorphous phase, which, as a rule, in turn, is below the Curie point of the nanocrystals.

Deviations from of the lower H_c value from the values of the randomly distributed anisotropy proposed by the model were first measured at temperatures close to or higher than the Curie point of the amorphous phase [321]. The decrease in the intensity of interaction between the ferromagnetic NC particles through the AL is directly correlated with an increase in H_c. This discovery became the basis for many studies which determined the limiting parameters of the alloys containing a sufficient number of very small nanocrystals or isolated nanoparticles in the amorphous matrix.

In [343, 344] the authors determined the temperature-dependent magnetic susceptibility in the partially and fully nanocrystallised alloys based on Fe–Si–B. It was concluded that for small nanocrystals with a sufficient volume fraction of the AP the exchange interaction between NC can be minimized or even completely eliminated and the superparamagnetic behaviour of the material can be observed. In [345] It was shown that the magnetic interaction increases with increasing of the volume fraction of the NC. These interactions tend to suppress superparamagnetism. A peak of $H_c(T)$ near the Curie point of the amorphous phase was also detected. The ascending branch of the curves is associated the suppression of the exchange interaction between the particles. High above T_C of the amorphous phase the value of H_c decreases in accordance with predictions of the theory of superparamagnetism.

The phenomenological parameter γ_{am}, which defines the exchange interaction in the amorphous NC phase, was introduced in [346, 347]. A model, which predicts a peak of the dependence $H_c(T)$ at the Curie temperature of the amorphous phase, was also proposed there. It has been shown [348] that at the value of the parameter $\gamma_{am} >$ 0.85 the material behaves as an ensemble of single domain particles, which are superparamagnetic near T_C for the AL. Highly diluted NCs (volume fraction = 0.15–0.25), located in the amorphous matrix, act as inclusions and create the effect of magnetic hardness [349, 350].

Further improvement of the magnetic characteristics of the ferromagnetic two-phase materials can be achieved by increasing the spontaneous magnetisation of the amorphous phase [337, 340]. This is due to the dominant role of the rigidity of the exchange interaction of the amorphous phase A_{am} in the temperature range below T_C.

3.5.2. Magnetic properties of 'Finemet' alloys

As we have already noted, the systems with the nanocrystalline phase have been regarded primarily as magnetically hard materials (for example, to produce high-quality Nd–Fe–B, Pr–Fe–B, Sm–Co and other alloys). To obtain good magnetically soft properties (low coercivity and high permeability) it is necessary to produce as large grains as possible. A typical example: electrical steel (Fe–3% Si), where the maximum properties are obtained as a result of secondary recrystallisation [351]. The situation changed dramatically after Japanese researchers accidentally discovered unique magnetically soft parameters in an amorphous–nanocrystalline alloys based on Fe–Si–B with small amounts of copper and niobium, which was subsequently named 'Finemet' [76]. Initially, the chemical composition of the alloy was $Fe_{73.5}Si_{13.5}B_9Nb_3Cu_1$, but layer it underwent some changes.

Detailed studies were made of a wide variety of Fe–Si–B alloys with small additions of refractory elements (Nb, W, Ta, Zr, Hf, Ti and Mo) and copper. The composition of these alloys was located in the immediate vicinity of the classic Finemet alloy [25]. In the initial state (after melt quenching), they are amorphous, and the optimal level of the properties was achieved after partial crystallisation which resulted in the precipitation of the nanocrystals ordered Fe–Si phase in the amorphous matrix. It is important to note that the nanocrystalline phase, reaching a size of about 10–20 nm, did not grow any further, which appeared to be associated with hindered diffusion and atmospheres that create the atoms of boron, niobium and copper in an amorphous matrix around growing nanocrystals [352] as well as with the inhibition of growth of nanocrystals by ultrafine precipitates of the metastable boride phase at the interface [353]. It is also shown [353] that in an amorphous matrix at the stage preceding nanocrystallisation there are Cu clusters, which stimulate the release on them, as on a substrate, of nanocrystals the FeSi phase, ordered by type DO_3 [354]. More detailed features of structure formation in Finemet-type alloys are set forth in section 3.2.8.

In the classic Finemet alloy the magnetic domain structure in nanocrystals is missing which, combined with the mutual compensation of the magnetostrictive effect in nanocrystals and in the amorphous matrix, leads to the formation of very low coercivity (5–10 A/m), high initial permeability at normal (100000) and high (10000) frequencies and small magnetic reversal losses (200 kW/m³). The additional positive effect on the properties is also exerted by treatment in a magnetic field [355] and by annealing in a nitrogen-containing atmosphere [356].

As follows from the previous section, the formation of high magnetic properties of Finemet amorphous–nanocrystalline alloys is controlled by the magnetic interaction between the nanocrystals. The intensity of this interaction is reduced or suppressed above the Curie point of the amorphous phase. Since the non-magnetic alloy components may be mainly in the amorphous phase and, therefore, reduce its Curie point, it is necessary to closely monitor the composition of the alloy.

In the magnetically soft materials, their composition and structure should ensure the maximum reduction in the magnetocrystalline anisotropy which is determined mainly by the induced anisotropy, associated with the magnetostrictive strain. The best materials (including Finemet) have the lowest values of magnetostriction. Crystallisation annealing of amorphous melt-quenched Finemet is accompanied by a decrease in saturation magnetostriction λ_s (from 20 to $3 \cdot 10^{-6}$), which can be attributed to a small positive value of λ for the amorphous phase, a small negative value for the nanocrystalline α-FeSi phase and compensation of areas +λ with areas −λ. The maximum value of the constant of magnetic anisotropy K is observed after annealing at 450°C, corresponding to the onset of crystallisation [357]. At higher temperatures there is a decrease of K with a decrease of the distance between the NC-phase and with increase of the interparticle exchange interaction.

Saturation magnetisation of Finemet alloy is determined by reversible rotation of the magnetisation vector in accordance with the law

$$M(H) = M_s \left[1 - \frac{a_1}{H} - \frac{a_2}{H^2} \right] + bH^{1/2}, \qquad (3.105)$$

where the term a_2/H^2 describes the contribution, directly resulting from the model of randomly distributed anisotropy axes. It is associated with FeSi nanocrystals. Coefficient a_2 reflects the theoretically predicted effective magnetic anisotropy of the material, while in amorphous alloys it is due to the local stress and magnetoelastic interaction.

According to the authors of [358], the Finemet compositions proposed in [77] are 'overalloyed' with metalloids–amorphisers and this reduces the magnetic properties and degrades the quality of the ribbons. In this context, in [358] it was attempted to improve the alloys by changing their chemical composition, as well as by optimising heat treatment. A Russian equivalent of the Finemet alloy, under the grade 5BDSR, was developed as a result.

Figure 3.73 shows the initial permeability and coercive force H_c, measured in the quasi-static magnetisation reversal mode, in dependence on the content of niobium in the $Fe_{77-x}Nb_xCu_1Si_{16}B_6$. The maximum μ_0 and minimum H_c on the curves indicate that to improve the magnetically soft properties should ensure a certain ratio of the components in the phases. In addition, as the concentration of niobium increases, the width of the temperature range of the existence of the desired amorphous–nanocrystalline structural state increases. This facilitates the selection of the annealing mode to obtain the desired structure and allows higher stress relieving temperatures to be applied.

Alloying of copper, as expected, contributes to the high density of crystallisation sites and reduces the size of critical nuclei, which provides the required nanocrystalline structure [352]. Alloying with copper embrittles the amorphous ribbon (impairs its production technology), so the concentration of copper in the alloy should be minimized. The dependence of μ_0 for alloys with and without copper

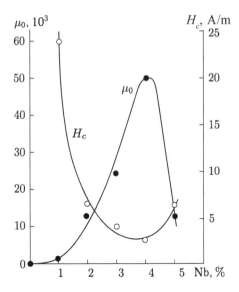

Fig. 3.73. Dependence of the magnetic properties of $Fe_{77-x}Nb_xCu_1Si_{16}B_6$ alloys on the Nb concentration.

and copper on annealing temperature is shown in Fig. 3.74. It can be seen that the level of the properties in the alloy containing copper, increases by about an order of magnitude due to the formation of the required two-phase amorphous–nanocrystalline structure (80% of the FeSi nanocrystalline phase with the size of about 20 nm and 20% of the amorphous phase).

In the study of $Fe_{73-x}Cu_1Nb_3Si_{13}B_x$ alloys with different B contents it was found that a high level of the properties is achieved in a fairly wide range of concentrations, but the highest values of μ_0

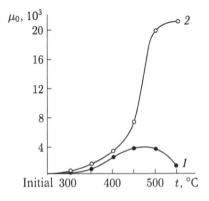

Fig. 3.74. The dependence of the initial permeability μ_0 of $Fe_{75}Nb_3Si_{16}B_6$ (1) and $Fe_{74}Nb_3Cu_1Si_{16}B_6$ (2) alloys on annealing temperature, holding time 1 h.

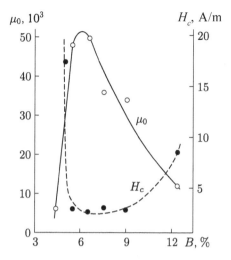

Fig. 3.75. Dependence of the magnetic properties on the boron concentration in Fe$_{73-x}$Cu$_1$Nb$_3$Si$_{13}$B$_x$ alloy. Measurements of μ_0 at H_c = 0.08 A/m.

can be obtained only in a narrow range of compositions (Fig. 3.75). According to the electron microscopy results, at a boron content higher than 9 at.% after heat treatment at 550°C for 1 h, in addition to the FeSi phase there were also undesirable boride precipitates [359].

The dependence of the properties of the (Fe$_{89-x}$Cu$_1$Nb$_3$Si$_x$B$_7$) nanocrystalline alloys on the Si content after various heat treatments is shown in Fig. 3.76 [358]. In alloys with a high Si content the values of H_c can be lower. The right corner of Fig. 3.76 shows the dependence of μ_0 after annealing at 550°C. The growth of this characteristic with increasing Si concentration is due to the fact that close to 16–17% Si the saturation magnetostriction its sign (i.e. $\lambda_s \approx 0$) [332]. However, with increasing silicon concentration the workability of the alloy is impaired. It is very difficult to produce ductile ribbons from high-silicon alloys.

Figure 3.77 shows the dependence of saturation magnetostriction λ_s on annealing temperature for two compositions Fe$_{74}$Cu$_1$Nb$_3$Si$_{13}$B$_9$ (1) and Fe$_{74}$Cu$_1$Nb$_3$Si$_{16}$B$_6$ (2) described in [77]. It is evident that as a result of annealing λ_s decreases rapidly near the temperature of the onset of nanocrystallisation and passes through a minimum for the composition (1), and for composition (2) it transfers to the negative region and again changes its sign with increasing temperature. The observed changes of λ_s demonstrate the possibility of producing a

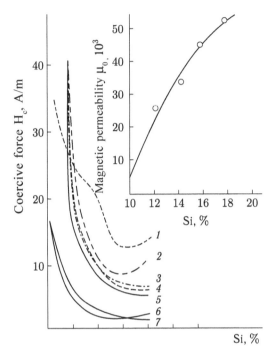

Fig. 3.76. Dependence of the magnetic properties of $Fe_{89-x}Nb_3Cu_1Si_xB_7$ alloy on the concentration of silicon ($x = 9, 12, 14, 15.5$ and 17.5%) after annealing in a magnetic field of 800 A / m from 400°C (1) up to 575°C (7).

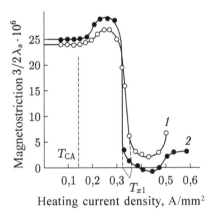

Fig. 3.77. The dependence of the saturation magnetostriction λ_s of $Fe_{74}Nb_3Cu_1Si_{13}B_9$ (1) and $Fe_{74}Nb_3Cu_1Si_{16}B_6$ (2) alloys on current density (the heating temperature) during annealing.

composition with $\lambda_s \approx 0$, less dependent on the annealing mode than alloy 2 and with a higher level properties than alloy 1.

The results of the study in [358] were used to determine the composition of the alloy, marketed under the brand name 5BDSR. The material in the form of ribbon with a thickness of 20–30 μm and a width of 40 mm is used for the manufacture of magnetic high-frequency switching transformers, inductors, magnetic amplifiers, current sensors etc. The normalised properties after heat treatment in the standard mode (without applying a magnetic field) are characterized by the following values.

Saturation flux density B_s, T	Not less than 1.2
Initial magnetic permeability	Not less than 30 000
Maximum permeability μ_{max}	Not less than 100,000
Coercive force H_c, A/m	Not more than 1.6

The alloy has the following physical properties

Density,	7.6 g/cm^3
Resistivity, μO·m	1.35
Vickers hardness HV	600
Curie temperature,°C:	
in the amorphous state	350
in the nanocrystalline state	550
Saturation magnetostriction constant λ_s	less than 10^{-6}

Thermomagnetic treatment in a longitudinal or transverse field in magneticcircuits made of alloy 5BDSR may lead to the formation of respectively rectangular or linear hysteresis loops (Fig. 3.78). In other words, the heat treatment can correct the magnetic parameters of the alloy according to its use in a particular product. The frequency dependence of the initial magnetic permeability $\mu^{\sim}$ ($H_c = 0.08$ A/m) after standard heat processing is shown in Fig. 3.79.

Comparing the properties of the 5BDSR alloy with the properties of the known high-cobalt coarse-crystalline alloys, we can see that the main parameters of this are not inferior to these materials, and its induction saturation is ~1.5 times higher. Obvious advantages of the 5BDSR alloy are a relatively low cost, as well as higher temperature stability of the properties.

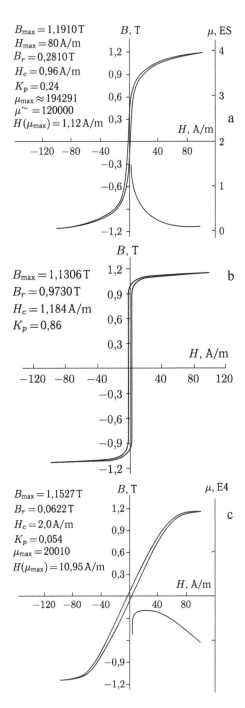

$B_{max} = 1,1910\,T$
$H_{max} = 80\,A/m$
$B_r = 0,2810\,T$
$H_c = 0,96\,A/m$
$K_p = 0,24$
$\mu_{max} \approx 194291$
$\mu^\sim = 120000$
$H(\mu_{max}) = 1,12\,A/m$

$B_{max} = 1,1306\,T$
$B_r = 0,9730\,T$
$H_c = 1,184\,A/m$
$K_p = 0,86$

$B_{max} = 1,1527\,T$
$B_r = 0,0622\,T$
$H_c = 2,0\,A/m$
$K_p = 0,054$
$\mu_{max} = 20010$
$H(\mu_{max}) = 10,95\,A/m$

Fig. 3.78. Hysteresis loop of the 5BDSR alloy after annealing: a) without the field; b) in a longitudinal magnetic field; c) in a transverse magnetic field.

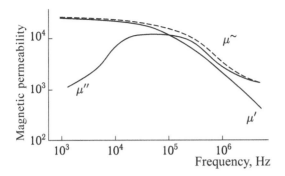

Fig. 3.79. The frequency dependence of the effective magnetic permeability of 5BDSR alloy in the field of 0.08 A/m (μ' and μ'' are the real and imaginary parts, respectively, $\mu^\sim$ is dynamic magnetic permeability).

3.5.3. Magnetic properties of Nanoperm and Thermoperm alloys

Magnetically soft alloys are often subject to the following additional requirements [360]:

1) high combined values of magnetic induction and permeability in constant and variable fields;
2) the ability to maintain high magnetic properties at very high reversal frequencies at elevated temperatures, as well as the presence of many other important properties of non-magnetic nature, such as mechanical, corrosion, and others.

Unfortunately, alloys such as Finemet does not satisfy both of these requirements, as they have insufficiently high saturation magnetisation and can not be used at elevated temperatures. The first is related to the presence of a large number of metalloids in the alloy (more than 20 at.%), which have to be introduced in order to obtain an alloy in the amorphous state after melt quenched. For the same reason, the Finemet alloys are characterised relatively low Curie points of the amorphous matrix (350°C), and this, as we showed above, excludes the formation of high magnetic properties at temperatures above 250–300°C.

For this reason, attempts have been made on other systems to obtain alloys with the Finemet effect with improved magnetic characteristics. In particular, Japanese researchers have developed magnetically soft alloys Fe–M–B–Cu (M = Zr, Nb, Hf), which are called Nanoperm [361]. The composition of these nanocrystalline alloys has been optimised so as to achieve a low magnetostriction coefficient and, consequently, high permeability. They can be transformed by melt quenching to the amorphous state at a much lower concentration of non-magnetic elements and, therefore, it is possible to significantly increase their saturation magnetisation while maintaining high permeability. In subsequent annealing, alloys of the Nanoperm type are characterised by the formation of α-Fe nanocrystals with the structural parameters similar to those of the nanocrystals, formed in Finemet.

Figure 3.80 shows the temperature dependence of the magnetisation $M(T)$ of the tempered and annealed $Fe_{86}Zr_7Cu_1B_6$ alloy samples compared with the saturation magnetisation of pure α-Fe [362]. Dependence $M(T)$ is particularly useful for identifying changes in the structure associated with the crystallisation process. The Curie point of the amorphous phase is (333 ± 5) K. After crystallisation the composite view of the $M(T)$ curve represents the contribution from the α-Fe nanocrystals the intergcrystalline amorphous phase with a significantly lower Curie point.

The magnetic moment of the alloy is slightly increased during crystallisation due to the displacement of the nanocrystals in the amorphous matrix of the atoms of B and Zr, reducing the magnetic moment. The Curie point of the amorphous phase in crystallisation remains virtually unchanged at only 340 K.

In [363] measurements were taken of the temperature dependence of magnetisation for the $Fe_{93-x}Zr_7B_x$ and $Fe_{93-x}Zr_7B_xCu_2$ alloys ($x = 4$–14). The temperature dependence, shown in Fig. 3.81, is identical to that observed in Fe–Ni crystalline Invar alloys. It was shown [364, 365] that the increase in the magnetically soft properties

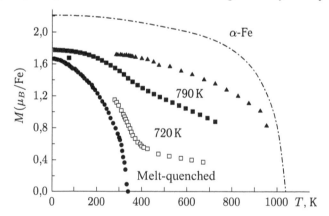

Fig. 3.80. Dependence $M(T)$ for melt-quenched and then annealed, at appropriate temperatures, $Fe_{86}Zr_7Cu_1B_6$ alloy, as well as for pure α-Fe in the nanocrystalline state [362].

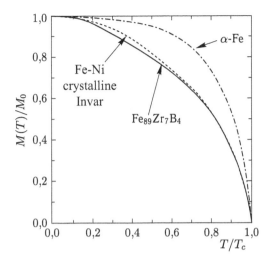

Fig. 3.81. Dependence $M(T)$ for the amorphous–nanocrystalline alloy $Fe_{89}Zr_7B_4$, for crystalline Fe–Ni Invar and α-Fe [254].

in Nanoperm alloys takes place due to two reasons: low effective anisotropy at the onset of crystallisation and decrease of saturation magnetostriction. Cu additions increase the intergranular exchange interaction due to the greater number of α-Fe nanocrystals in an interacting unit volume. This obviously reflects the important role of Cu clusters in the nucleation of α-Fe particles.

Other amorphous–nanocrystalline alloys – analogues of Finemet – are (Fe, Co)–M–B–Cu alloys (M = Nb, Hf or Zr), named Thermoperm [366]. They have high induction (1.6–2.1 T) in combination with high permeability and high Curie point. In these alloys [338, 367] the amorphous matrix is characterised by the formation of nanocrystalline phases based on BCC superstructures B2 (α-FeSi and α'-FeCo) with much improved high-temperature magnetic properties compared with the Finemet and Nanoperm alloys. Thermoperm-type alloys have been developed for use as materiala with low permeability but high induction at high temperatures.

Figure 3.82 shows the frequency dependence of the real and imaginary components of the permeability µ' and µ", respectively. The value of µ' reflects the density of the loss due to eddy currents and hysteresis. Maximum permeability of the material is 1 800. The dependence µ"(T) has a maximum at a frequency of ~20 kHz. The frequency peak appears to be associated with higher electrical resistance in nanocrystalline materials, and the losses in the variable field reflect the behaviour of the domain wall in a viscous

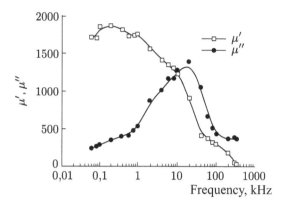

Fig. 3.82. The frequency dependence of the real and imaginary components of permeability of Thermoperm alloy; annealing at 650°C, 1 h [368].

continuum. A higher value of ρ (50 μohm·cm at 300 K) shifts the high permeability to higher frequencies where the eddy current losses, including the movement of domain walls, dominate.

In [366] the $(Fe_{1-x}Co_x)_{88}Hs_7B_4Cu_1$ alloys regarded as potential candidates for magnetically soft materials for outside generators were studied. The materials are used in high-temperature atmospheres (500–600°C), and they should have an induction at 2 T and over at 500°C, and also possess thermal stability at 600°C for 5000 h. In addition, the losses in magnetisation reversal should be less than 480 W/kg at 5 kHz and 500°C. The characteristic exchange length for the equiatomic ordered FeCo alloy is 46 nm, assuming that $A = 1.7\cdot10^{-11}$ J/m and $K = 8$ kJ/m^3. If high values of K are required (for an alloy with 30% Co $K \sim 25$ kJ/m^3), but all other parameters should be the same as in the equiatomic alloy, this parameter will be 26 nm. Analysis by Herzer's theory suggests that a material with a particle size of ~30 nm should have $H_c \approx 100$ A/m. This corresponds to the value $H_c \approx 10^{-2}$ A/m, obtained in the experiment for the size of the nanocrystals $d \approx 10$ nm, and shows that further refinement of the nanocrystalline phase should lead to better magnetic properties.

3.6. The shape memory effect

Recently, the nanocrystals of type III have been widely used as materials with the shape memory effect (SME). The essence of this effect is as follows (Fig. 3.83): the sample of the SME material recovers its initial shape when heated, if it was deformed at low temperatures. The SME was first observed on the TiNi equiatomic

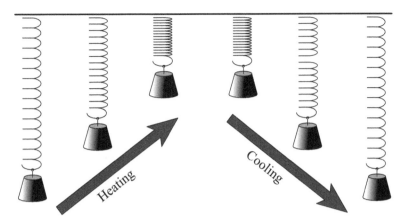

Fig. 3.83. The phenomenon of the shape memory effect.

alloy in [369], and was later found in a wide range of alloys, belonging to different systems, particularly in alloys CuAl, CuMn, AuCd, CuAlNi and many others [370].

The SME is based on a thermoelastic martensitic transformation, discovered in 1949 by G.V. Kurdyumov and L.G. Handros alloys CuAlNi and CuSn alloys [371]. It is known [372] that depending on the method of relaxation of internal stresses in the martensitic transformation, all diffusionless phase transitions can be divided into two large classes. In the first, the most numerous, internal stresses at a certain stage of growth of the martensite crystal exceed the yield strength of the matrix or the growing phase and this disrupts the coherence and the martensite crystal almost ceases its growth, and the relaxation of the accumulated stress takes place by plastic deformation. Further transformation develops by the nucleation and growth of new plates. The accommodative plastic deformation and martensite crystals usually disappear when in heating due to the emergence of new boundaries in the volume of the martensite phase, and not through reverse motion of the initial interface. This transformation requires a significant driving force and, as a consequence, a significant thermal hysteresis of hundreds of degrees. This is the so-called 'explosive' martensitic transformation which has been the object of study for a long time. Transformations of the second class are characterised by the fact that the change in the shape and volume is compensated not plastic but by elastic deformation and the growth of the martensitic phase crystals is terminated before the yield point is reached. This sets the thermoelastic equilibrium between the crystal and the initial martensite matrix, which was first predicted by G.V. Kurdyumov and then successfully implemented experimentally.

For successful implementation of the SME the alloy in which there is a thermoelastic martensitic transformation must be deformed at a low temperature so that it fully or partially undergoes the thermoelastic transformation, and then it should be heated to a temperature above the temperature of completion of the reverse phase transition. In practice, the SME is implemented by the variant of deformation of the sample below the temperature of the onset of martensitic transformation either in or cooling (M_s), or by deformation (M_d). Application of a heavy load at these temperatures causes the ordering of local martensitic shifts with different orientation, accompanied by an increase in the number and volume of the domains with martensite deformation in the same direction as that of the applied stress, due to the high mobility of the interphase and interdomain boundaries.

Later the reversible shape memory effect (RSME) was observed in the TiNi alloy. In contrast to the SME, the RSME is expressed in the spontaneous deformation of the sample at in cooling–heating cycle and can be repeated many times (for example, several million cycles for TiNi [373]) with no significant changes in the structure and material properties. The role of external stresses, ordering martensitic deformation, is played by internal stresses that can be created, for example, by prelimineary directional plastic deformation [374, 375]. In this case, there are local places that are sufficiently strong oriented in the microstress space, forcing the sample to spontaneously deform in the range of direct transformation, taking the low-temperature form. When heated, the sample is returned to the original state and the process is repeated. The magnitude and direction of reversible deformation can vary widely depending on the temperature and the degree of preliminary plastic deformation and subsequent annealing. This allows one to teach the material to 'memorise' two arbitrary shapes in a general case. Despite the fact that in practice it is not possible to reach the limiting deformation from one form to another, equal in size to ε_m, reversible deformation can be significant. Formation for the titanium nickelide it is 4–5% at the maximum possible value of 11%. It is obvious that RSME greatly extends the use of materials that undergo thermoelastic martensitic transformation in constructions and devices in with multiple cyclic effects [376].

SME materials are nanocrystals of the third type on the basis of the Ni–Ti–Cu system [377]. They are produced in the amorphous state by melt quenching and then annealed to produce by crystallisation the high-temperature nanocrystal metallic phase B2. The regularities of nanocrystallisation of the Ti–Ni and Ti–Ni–Cu alloys were reviewed in detail in section 3.2.9 of this chapter. Here we will focus on the functional properties of the nanocrystalline materials exhibiting SME.

The thermoelastic martensitic transformation during cooling of alloys of the $Ti_{50}Ni_{50-x}Cu_x$ system is a one-step phase transition B2 → B19′ if $x < 6$, and the transformation temperature decreases (down to room temperature) with increase of the concentration of copper in the alloy. Starting at ~6 at.% Cu, a cascade of B2→B19→B19′ transition takes place in which the temperatures of the second martensitic transition smoothly decrease into the cryogenic temperature range. At a concentration of more than ~17 at.% Cu only the B2→B19 transition occurs above room temperature. In alloys based on TiNi–TiCu with nano- and submicrocrystalline structures, the critical temperatures of martensitic transformations are somewhat reduced, but in general the structural types and lattice parameters of the martensitic phases and also the sequence of transformations do not change.

Electron microscopy studies revealed that the martensitic transformation in the B2-phase nanocrystals is implemented by the 'single crystal–single crystal' mechanism for all martensitic phases B19, B19', and only in the larger grains (micron-size) there are formed martensite packets, but no more than one or two packets in each grain [378].

All alloys based on TiNi, obtained by controlled crystallisation of the amorphous state, exhibit a number of characteristic features of the mechanical behaviour. The tensile strength σ_B of nanocrystalline TiNiCu alloys can reach 1600 MPa with the elongation δ to 6–8%. σ_B and microhardnee HV can twice as high as the same characteristics of the bulk alloys of the same chemical composition, which is due to their nanocrystalline state.

It was found that the alloys have narrow-hysteresi effects of single and spontaneous reversible shape memory. The bilateral reversible shape memory effect is due to the crystallographic and microstructural texture of the rapidly quenched ribbon. Thermal cycles in the range –289÷200°C practically cause no changes in these alloys in the critical temperatures of direct and reverse martensitic transformations, indicating their high microstructure reversibility and thermal stability in the austenitic state. In the condition of 'constraint' under loading the reverse martensitic transition in heating leads to the accumulation of reactive stresses and restoration of the shape, accompanied by the completion of the operation.

$Ti_{50}Ni_{25}Cu_{25}$ alloy attracts attention because of the possible occurrence of one-step martensitic transformation B2↔B19 with a narrow temperature hysteresis, and also a high level of thermomechanical parameters during the manifestation of the shape memory effect [379]. Because of this, the alloy is used widely in functional elements of high-speed miniaturised sensors and actuators [380–382]. The almost linear relationship with the electrical resistance of the element with the SME and the degree of its deformation during thermal cycling under constant load with heating by passing an electric current was found [383]. Next, it is critical to clarify the effect on this ratio and other SME properties the applied stress and the number of applied thermal cycles. In [384] $Ti_{50}Ni_{25}Cu_{25}$ alloy was subjected to repeated thermal cycling (up to 8000 cycles) in the presence of external loads of 50 and 100 MPa.

Thermal cycling of samples under a constant load was performed by passing an electric current. The electric current was of the stepped sawtooth shape, and after each stepped change in the current intensity by 2.5 mA for 300 ms measurements were taken of the electrical resistance NER and the strain ε of the sample as a function of current density J. The selected parameters ensured thermal balance with the environment. Mechanical load was applied to the sample in the original austenite phase, i.e. at maximum current J_{max}. Each austenite → martensite → austenite cycle (sequence $J_{max} \rightarrow J_{min} \rightarrow J_{max}$) was carried out at the current changing in the range from 0.15 A to 0.65 A. Strain ε, the normalized electrical resistance NER, reversible strain ε_{rev} and plastic strain ε_p are defined as follows:

$$\varepsilon = \frac{\varepsilon(J) - \varepsilon(J_{max})}{\varepsilon(J_{max})}, \quad NER = \frac{ER(J) - ER(J_{max})}{ER(J_{max})},$$

$$\varepsilon_r = \varepsilon(J_{min}) - \varepsilon(J_{max}), \quad \varepsilon_p(\text{cycle } n) = \varepsilon_n(J_{max}) - \varepsilon_1(J_{max}).$$

Figure 3.84 shows the typical dependences ε(J) and NER(J) at loads of 5, 50 and 100 MPa after two thermal cycles with the hysteresis, characteristic of these alloys. At the same time, the dependence NER(ε) is very close to linear and has a much narrower hysteresis, which at a small number of thermal cycles narrows with increasing applied stress (Fig. 3.85). At 50 MPa the hysteresis almost disappears when the number of cycles is increased to 5230. This may be due

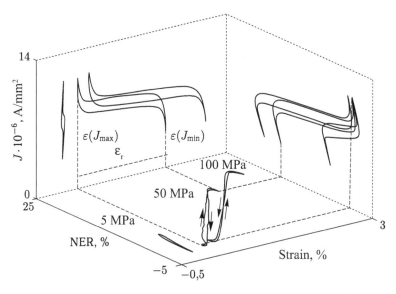

Fig. 3.84. Strain ε and normalized electrical resistance NER of rapidly quenched Ti$_{50}$Ni$_{25}$Cu$_{25}$ as a function of current density, and the dependence NER(ε), during thermal cycling under a load of 5, 50 and 100 MPa.

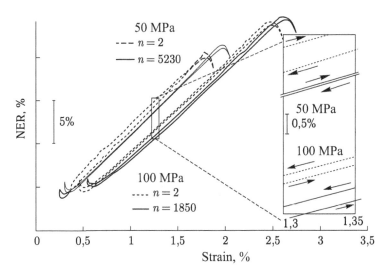

Fig. 3.85. Relationship of strain and normalized resistivity of rapidly quenched Ti$_{50}$Ni$_{25}$Cu$_{25}$ alloy with repeated thermal cycling under a load 50 and 100 MPa; *n* is the the number of cycles.

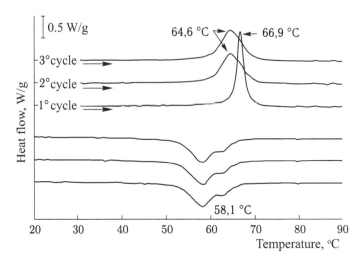

Fig. 3.86. DSC curves of rapidly quenched Ti$_{50}$Ni$_{25}$Cu$_{25}$ after 5230 thermal cycles at a constant load of 50 MPa.

to the fact that during the first thermal cycles the growth oriented martensite variants and their reorientation are not fully completed and this takes place only after a certain number of cycles. The characteristic curves obtained by differential scanning calorimetry (DSC) after repeated thermocycling of the alloy under a constant load are shown in Fig. 3.86. After 5230 cycles under a load of 50 MPa, the direct martensitic transformation (MT) (in cooling) shows two temperature peaks while the reverse MT (in heating) shows only one peak. As shown by further studies, the appearance of a second peak at direct transformation is due to the presence of a thin surface crystal layer on the free surface of the ribbon in the initial state after quenching.

In [384] changes were studied in the strain sample at the maximum ε (J_{max}) and minimum ε (J_{min}) values of the current density, as well as of the reversible strain ε_r and platic strain ε_p, depending on the number of thermal cycles performed. The samples fractured at 100 MPa in the range of 3000 to 4000 thermal cycles, and at 50 MPa there was no failure samples even after 8000 thermal cycles.

ε_r increases slightly during the first 1500–2000 cycles, and then remains almost constant, about 1.5 and 2% for 50 and 100 MPa, respectively. ε_p increases at 100 MPa far more rapidly, reaching 0.1% compared with 0.05% at 50 MPa.

To obtain the required temperaturea of the start and end of thermoelastic martensitic transformation and, therefore, to obtain the

SME in the required temperature range copper in ternary Ti–Ni–Cu alloys is replaced by other elements. Several studies have shown [385, 386] that the alloys Ti–Ni–Hf and Ti–Ni–Zr show the SME at higher temperatures (100°C) and that are characterised by high level of the SME characteristics compared with other alloys. This allows us to consider them as the most promising SME materials for use at high temperatures. The practical need for such materials is acute and growing rapidly. First of all, these materials are needed to create the thermal actuators (actuators) and sensors for the automotive, nuclear, and oil and gas extraction industries, automation and robotics. Many of these applications require high performance of SME components, so that ribbon materials should be developed. However, the addition of Zr and Hf causes a noticeable hardening of the alloy and reduces its machinability so that it is extremely difficult to produce thin profiles. In this respect, the technology of rapid melt quenching is most effective in overcoming this problem. Methods of preparation and some properties of the rapidly quenched Ti–Ni–Hf alloys are presented in [387, 388]. In [389] the microstructure and the effect of thermal cycling under the influence of constant load on the properties of SME in rapidly quenched Ti–Ni–Hf alloys was studied.

X-ray analysis showed that at room temperature the alloys with 10 and 15% Hf after melt quenching contain predominantly monoclinic B19′-martensite and a small amount of the austenitic B2-phase, while the alloy with 20% Hf is in the amorphous state. Electron microscopic studies confirmed the amorphous state of the alloy with 20 at.% Hf. Some amount of the amorphous phase was also found in the alloy with 15 at.% Hf. In the alloy with 10 at.% Hf there are four types of typical morphology of martensite: needle, wedge and packet martensite. The fourth morphological type of structure corresponds to an irregular structure in which martensite coexists with retained austenite. The crystallographic orientation relationship between martensite and austenite grains is defined as $[100]'_{B19}//[100]_{B2}$; $(011)'_{B19}//(001)_{B2}$, i.e. similar to bulk material [390].

The above-mentioned structures were also found in the alloy with 5 at.% Hf, but the remarkable feature of the morphology of this alloy is the presence of spherical particles with a diameter from several tens of nanometers to several tens of microns. Surrounded by these particles are martensite and austenite and also the amorphous phase. The particles contain packet martensite and austenite. Another interesting fact is the appearance of nanoparticles in the structure of the alloy (Fig. 3.87a). The corresponding microdiffraction pattern

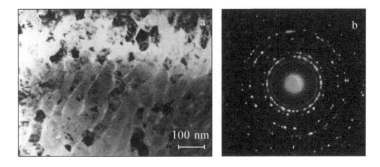

Fig. 3.87. Nanosized particles observed in the alloy with 15% Hf (a) and the appropriate electron diffraction pattern (b).

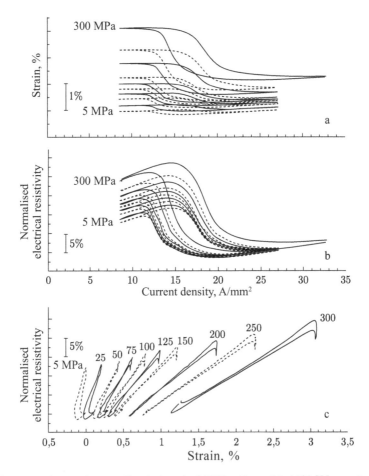

Fig. 3.88. Strain ε and normalized electrical NER alloy with 10% Hf as a function of current density, and the dependence of NER(ε) during thermal cycling under load from 5 to 300 MPa.

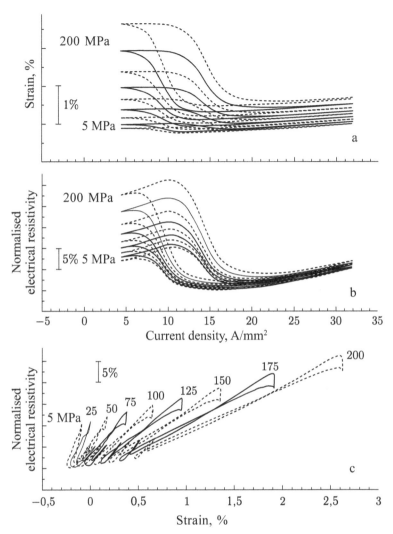

Fig. 3.89. Deformation ε and normalized electrical NER alloy with 20% Hf as a function of current density, and the dependence of NER (ε), during thermal cycling under load from 5 to 200 MPa.

(Fig. 3.87b) indicates that these nanocrystals do not have a dedicated crystallographic orientation. The alloy with 20% Hf also contains spherical particles, but they are embedded in the amorphous matrix and with no martensite or austenite in their vicinity.

Thermal cycling of the alloy by heating by passing electric current was carried out at different applied mechanical stress in the range of 5 to 300 MPa. Figures 3.88 and 3.89 show the typical dependence of the normalized strain ε and electrical NER on current density for alloys with 10 and 20% Hf, respectively. In both cases the single-stage nature of the martensitic transformation was detected, and the increase of Hf leads to a significant increase in transformation temperatures.

The dependence of the electrical NER alloys on their deformation ε, shown in the lower part of the figure, is almost linear. For the alloy with 10% Hf there is always a point of intersection of the lines corresponding to heating and cooling, while for alloy with 20% Hf it disappears at stresses of more than 200 MPa.

Thus, alloys based on titanium nickelide, synthesized by melt quenching and subsequent annealing, are a particularly promising class of nanostructured materials with thermoelastic martensitic transformations and possess favourable functional and structural characteristics. They have the temperature, deformation and strength parameters of narrow-hysteresis SME effects and high strength, plastic, fatigue, corrosion properties essential for practical applications. An important role in the effects of the thermomechanical memory is played by the crystallographic texture and microstructure of these materials. The selection of the process conditions of melt quenching, the temperature and time of subsequent annealing, and the chemical composition of the original alloys determines the state of the structure, the structural and functional characteristics of alloys and their geometric dimensions. The Ti–Ni–Cu, obtained by melt quenching, can be widely used to create new products and processes in engineering and medicine.

References

1.   Glezer A.M., et al., Mechanical behavior of amorphous alloys, Novokuznetsk, Sib-GIU, 2006.
2.   Encyclopedia of Physics, ed. Alexander Prokhorov, Sov. entsikl., 1990, V. 1, 704; V. 2, 705 p.
3.   Frenkel' Ya.I., Introduction to the theory of metals, Leningrad, Nauka, 1972.
4.   Egami T., Structural relaxation in amorphous alloys, Amorphous Metallic Alloys, Ed. F. Luborsky, London, Butterworths, 1983, P. 100–113.
5.   Iveronova V.I., Katsnelson A., Short-range order in solid solutions, Moscow, Nauka, 1977.
6.   Speapen F., Cargill G.S. Defects in metallic glasses, Proc. of the Fifth Intern. Conf. RQM, Elsevier Publ., 1985, V. 1, 581–584.
7.   Hafner J. Short-range oder in metallic glasses, Proc. of the Fifth Intern. Conf. RQM,

Elsevier Publ., 1985, V. 1, 421–425.

8. Belenky A.Ya., Nature. 1987, V. 2, 80–88.

9. Glezer A.M., Molotilov B.V., The structure of amorphous alloys, Fiz. Met. Metalloved., 1990, V. 69, No. 2, 5–28.

10. Wagner C.N.J., Amorphous Metallic Alloys, Ed. F. Luborsky, London, Butterworths, 1983, 58–73.

11. Gaskell P.H., J. Non-Cryst. Solids, 1979, V. 32, No. 1, 207–224.

12. Bakai A.S. Polycluster amorphous structures and their properties, Part 1, Moscow, TsNIIAtominform, 1984.

13. Gaskell P.H., J. Non-Cryst. Solids, 1985, V. 75, No. 2, 329–345.

14. Likhachev V.A., Khairov R.Yu., Introduction to the theory of disclinations, Leningrad, LSU, 1975.

15. Sadoc J.F., J. Phys. Lett., 1983, V. 44, No. 17, 707–715.

16. Dobromyslov A.V., et al., Materialovedenie, 2001, No. 10, 43–46.

17. Thomas G., Gorindge M.G., Transmission electron microscopy of materials, Moscow, Nauka, 1983.

18. Zweck J., Hoffman H., The high-resolution electron microscopy study of atomic structure of Fe–B amorphous alloy, Proc. of the Fifth Intern. Conf. RQM, Elsevier Publ., 1985, V. 1, 509–512.

19. Glezer, A.M., Molotilov B.V., Structure and mechanical properties of amorphous alloys, Moscow, Metallurgiya, 1992.

20. Skakov Yu.A., Glezer A.M., Ordering and intraphase transformations, Itogi nauki i tekhniki, Metalloved. Term. Obrab., Moscow, VINITI, 1975, 5–73.

21. Levashov E.A., Shtansky D.V., et al., Deformatsiya Razrush. Mater., 2009. No. 11, 19–37.

22. Fudjimori H., Magnetic structure of amorphous ferromagnetic alloys, Proc. of the Fifth Intern. Conf. RQM, Elsevier Publ., 1985, V. 1, 300–316.

23. Gubanov A.I., Quantum electronic theory of amorphous and liquid semiconductors, Leningrad, Publishing House of the USSR Academy of Sciences, 1963.

24. Andrievski R.A., Glezer A.M., Scr. Mater., 2001, V. 44, 1621–1624.

25. McHenry M.E., Willard M.A., Laughin D.E., Prog. Mater. Sci., 1999,V. 44, 291–433.

26. Pozdnyakov V.A., Glezer A.M., Fiz. Tverd. Tela, 2002, V. 44, No. 4, 705–710.

27. Scott M.G., Crystallization, in: Amorphous metallic alloys, Moscow, Metallurgiya, 1987, 137–164.

28. Lu K., Luck R., Predel B., J. Non-Cryst. Solids. 1993, V. 156–158, 589–593.

29. Clavaguera-Mora M.T., Clavaguera N., Crespo D., Pradell T., Progress in Materials Science, 2002, V. 47, 559–619.

30. Cusido J.A., Isalque A., Phys. Stat. Sol. (A), 1985, V. 90, No. 1, 127–133.

31. Khonik V.A., Kitagawa K., Morii H., J. Appl. Phys. 2000, V. 87, No. 12, 8440–8443.

32. Guo X., Louzguine D., Inoue A., Mater. Transact. JIM, 2001, V. 42, No. 11, 2406–2409.

33. Christian J., Theory of transformations in metals and alloys, V. 1, Springer-Verlag, 1978.

34. Burke J., The kinetics of phase transformation in metals, London, Pergamon, 1965.

35. Leonova E.A., et al., Izv. RAN, Ser. Fiz., 2001, V. 65, No. 10, 1420–1423.

36. Leonova E.A., et al., Izv. RAN, Ser. Fiz., 2001, V. 65. No. 10, 1424–1427.

37. Kester, W., Harold W., Crystallization of metallic glasses, in: Metallic glasses. Ionic structure, electron transfer and crystallization. No. 1, Academic Press, 1983, 325–371.

38. Abrosimova G.E., et al., Metallofizika, 1980, V. 2, No. 6, 96–101.

39. Abrosimova G.E., et al., Fiz. Met. Metalloved., 1989, V. 68, No. 3, 552–556.
40. Boldyrev V.I., et al., Fiz. Met. Metalloved., 1999, V. 87, No. 5, 83–86.
41. Noskova N.I., et al., Fiz. Met. Metalloved., 1994, V. 77, No. 5, 89–94.
42. Abrosimova G.E., Aronin A., Fiz. Tverd. Tela, 1998, V. 40, No. 10,1769–1772.
43. Gupta P.K., Baranta G., Denry I.L., J. Non–Cryst. Sol., 2003, V. 317, 254–269.
44. Kobelev N.P., et al., Fiz. Tverd. Tela, 2001, V. 43, No. 10, 1735–1738.
45. Proceedings of the 3rd All-Russian Conference on Nanomaterials, Yekaterinburg: Uralsk Publishers, 2009.
46. Lu K., Wei W.D., Wang J.T. Scr. Metall. Mater., 1990, V. 24, 2319–2324.
47. Guo H.Q., et al., Phys. Status. Solidi A, 1991, V. 127, 519–527.
48. Morris D.G., Acta Metall., 1981, V. 29, No. 7, 1213–1220.
49. He I.L., Liu X.N., Acta Electron. Sinica, 1982, V. 4, 70–75.
50. Nicolaus M.M., Sinning H.R., Haessner F., Mater. Sci. Eng. A, 1992, V. 150, 101–108.
51. Greer A.L., Acta Metall., 1982, V. 30, 171–181.
52. Tong H.Y., et al., J. Appl. Phys., 1994, V. 75, 654–657.
53. Koster U., Abel R., Blanke H., Glastech. Ber., 1983, V. K56, 584–596.
54. Koster U., Meinhardt J., Alves H., Conditions of structure refinement at the heating of metallic glasses, Proc. of ISMANAM 94, Grenoble, 1994, P. 85–88.
55. Koster U., Meinhardt J., Mater. Sci. Eng. A, 1994, V. 178, No. 1–2, 271–278.
56. Kulik T., Horubala T., Matyja H., Mater. Sci. Eng. A, 1992, V. 157, No. 1, 107.
57. Gorria P., Orue I., Plazaola F., Barandiaran J.M., J. Appl. Phys., 1993, V. 73, No. 10, 6600–6603.
58. Pustov A., et al., Zashch. Met., 1999, V. 35, No. 6, 565–576.
59. Zhdanova L.I., ibid, 1999, V. 35, No. 6, 577–580.
60. Potapov A.P. Dmitrieva N.V., Glezer A.M., Fiz. Met. Metalloved., 1995, V. 79, No. 2, 51–56.
61. Kekalo I.B., Leffler F., Fiz. Met. Metalloved., 1989, V. 68, No. 2, 280–288.
62. Betekhtin V.I., et al., Pis'ma Zh. Teor. Fiz., 1998, V. 24, No. 23, 58–64.
63. Betekhtin V.I., Kadomtsev A.G., Tolochko O.V., Fiz. Tverd. Tela, 2001, V. 43, No. 10, 1815–1820.
64. Würschum R., et al., Scr. Metall. Mater., 1991, V. 25, 2451–2456.
65. Schaefer H.E., et al., Phys. Rev. B., 1988, V. 38, 9545–9554.
66. Ping D.H., et al., Mater. Sci. Eng. A, 1994, V. 194, 211–217.
67. Sui M.L., Lu K., He Y.Z., Phil. Mag. B, 1991, V. 63, 993–1008.
68. Mutschele T., Kirchheim R., Scr. Metall. Mater., 1987, V. 21, No. 8, 1101–1105.
69. Hahn H., Hofler H.J., Averback R.S., Defect Diffus. Forum. 1989, V. 66–69, 549–553.
70. Ecken J., Holzer J.C., Krill C.E., Johnson W.L., J. Mater. Sci., 1992, V. 7, 1751–1761.
71. Gaffet E., Louison C., Harmelin M., Faudet F., Mater. Sci. Eng. A, 1991, V. 134, 1380–1384.
72. Lu K., Sui M.L., Luck R., Nanostruct. Mater., 1994, V. 4, 465–469.
73. Andrievsky R.A., Uspekhi khimii, 2002, V. 71, No. 10, 967–981.
74. Yamauchi K., Yoshizawa Y., Nanostruct. Mater., 1995, V. 6, No. 1–4, 247–256.
75. Yoshizawa Y., Oguma S., Yamauchi K., J. Appl. Phys., 1988, V. 64, No. 10, 6044–6046.
76. Yoshizawa Y., Yamauchi K., et al., J. Appl. Phys., 1988, V. 64, 6047–6051.
77. Yoshizawa Y., Yamauchi K., Mat. Sci. Eng., 1991, V. A133, 176–182.
78. Yoshizawa Y., Yamauchi K., Mater. Trans., JIM, 1989, V. 31, 3324–3326.

79. Muller M., Mattern M., Kuhn U., JMMM, 1996, V. 157/158, 209–210.
80. Dugaj P., et al., Mater. Sci. Eng., 1991, V. A133, 398–402.
81. Rixecher G., Schaaf P., Gonser U., J. Phys., Condens. Matter., 1992, V. 4, 10295–10310.
82. Muller M., Mattern M., Illgen L., Z. Metallk., 1991, V. 82, No. 12, 895–901.
83. Nemoschkalenko V.V., Vlasenko L.E., Romanova A.V., et al., J. Met. Phys. Adv. Techn., 1998, V. 20, No. 6, 22–34.
84. Illecova E., et al., Mat. Sci. Eng., 1996, V. A205, 166–179.
85. Mat'ko I., et al., Mat. Sci. Eng., 1994, V. A179/A180, 557–562.
86. Maslov V.V., et al., Fiz. Met. Metalloved., 2001, V. 91, No. 5, 47–55.
87. Koster U., Mat. Sci. Eng., 1991, V. A133, 611–615.
88. Hampel G., Pundt A., Hesse J., J. Phys. Condens. Matter.,1992, V. 4, 3195–3214.
89. Fujinami M., et al., Japan J. Appl. Phys., 1990, V. 29, 477–480.
90. Chen L.C., Spaepen F., J. Appl. Phys., 1991, V. 69, No. 2, 679–688.
91. Bakai A.S., Polycluster amorphous structures and their properties. Part 2, Moscow, TsNIIAtominform, 1985.
92. Hiraga K., Kohmoto O., Mater. Trans. JIM, 1991, V. 33, No. 9, 868–871.
93. Vlasenko L.E., et al., Metallofizika i Noveishie Tekhnol., 1998, V. 20, No. 7, 75–82.
94. Kim K.Y., Noh T.H., J. Appl. Phys., 1993, V. 73, 6594–6597.
95. Zhou X.Y., Morrish A.H., et al., J. Appl. Phys., 1993, V. 73, 6597–6601.
96. Hono K., Inoue A., Sakurai T., Appl. Phys. Lett., 1991, V. 58, No. 19, 2180–2184.
97. Ayers J.D., Harris V.G. et al., Appl. Phys. Lett., 1994, V. 64, 974–978.
98. Ayers J.D., Harris V.G., et al., Acta Mater., 1998, V. 46, 1861–1869.
99. Saito Y., Okuda M., et al., J. Phys. Chem., 1994, V. 98, 6696–6705.
100. Hono K., Zhang Y., et al., Metal. Mater., 1992, V. 40, 2137–2145.
101. Kojima A., Horikiri H., et al., Mat. Sci. Eng., 1994, V. A179, 945–952.
102. Iwanabe H., Lu B., et al., J. Appl. Phys. 1999, V. B85, 4424–4429.
103. Willard M.A., Laughlin D.E., et al., J. Appl. Phys., 1998, V. 84, 6773–6780.
104. Willard M.A., Huang M-Q., et al., J. Appl. Phys., 1999, V. 85, P. 4421–4429.
105. Glezer A.M., Manaenkov S.E., et al., Bulletin Tambovsk State University. Ser. Natural and engineering sciences, 2010, V. 15, No. 3, 1169–1176.
106. Chin T.-S., Lin C.Y., et al., Materials Today, 2009, V. 12, No. 1–2, 34–39.
107. Kornilov I.I., et al., Titanium nickelide and other alloys with the memory effect, Moscow, Nauka, 1977..
108. Matveeva N.M., et al., Izv. AN SSSR, Metally, 1989, No. 4, 171.
109. Matveeva N.M., et al., Izv. AN SSSR, Metally, 1991, No. 3. 164.
110. Babanly M.B., et al., Izv. AN SSSR, Metally, 1993, No. 5, 171–178.
111. Glezer A.M., et al., Izv. AN SSSR, Metally, 1998, No. 4, 45–47.
112. Pushin V.G., et al., Fiz. Met. Metalloved., 1997, V. 83, No. 4, 155–166.
113. Pushin V.G., et al., Fiz. Met. Metalloved., 1997, V. 83, No. 6, 157–163.
114. Pushin V.G., et al., Fiz. Met. Metalloved., 1997, V. 84, No. 4, 172–181.
115. Matveeva N.M., et al., Fiz. Met. Metalloved., 1997, V. 83, No. 6, 82–92.
116. Pushin V.G., Queen T.G., Yurchenko L.I., Structure and properties of nanocrystalline materials, Yekaterinburg, Ural Branch of RAS, 1999, 373–376.
117. Rösner H., et al., Scr. Mater., 2000, V. 43, No. 10, 871–876.
118. Rösner H., et al., Mater. Sci. Eng. A, 2001, V. 307, No. 1–2, 188–189.
119. Rösner H., et al., Acta Mater., 2001, V. 49, 1541–1548.
120. Scheil E., Z. Anorg. Algem. Chemie, 1929, V. 183, No. 1–2, 98–120.
121. Maksimova O.P., Nemirovsky V.V., Dokl. AN SSSR, 1967, V. 177, No. 1, 81–84.
122. Voznesenskii V.V., et al., Fiz. Met. Metalloved., 1975, V. 40, No. 1, 92–101.

123. Umemoto M., Owen W.S., Metal. Trans., 1974, V. 5, No. 9, 2041–2046.
124. Maximova O.P., Zambrzhitsky V.N., Fiz. Met. Metalloved., 1986, V. 62, No. 5, 974–984.
125. Leslie W.C., Miller R.L., Trans. ASM, 1964, V. 57, 972–979.
126. Ishida I., Trans. Japan Inst. Metals, 1988, V. 29, No. 5, 365–372.
127. Ishida I., Kiritani M., Acta Met., 1988, V. 36, No. 8, 2129–2139.
128. Easterling K.E., Swann P.R., Acta Met., 1971, V. 19, No. 2, 117–121.
129. Kinsman K.R., Sprys J.W., Asaro R.J., Acta Met., 1975, V. 23, No. 12, 1431–1442.
130. Wusatowska-Sarnek A.M., Miura H., Sakai T., Scripta. Mat., 1998, V. 39, No. 10, 1457–1461.
131. Zhou Y.-H., Harmelin M., Bigot J., Mater. Sci. Eng., 1990, V. A124, 241–245.
132. Kajiwara S., Ohno S., Honma K., Phil. Mag., 1991, V. 63, No. 4, 625–644.
133. Zhao X., Liang Y., Hu Z., Liu B., Japan J. Appl. Phys., 1996, V. 35, 4468–4473.
134. Dong X.L., Zhang Z.D., Zhao X.G., Chuang Y.C., J. Mater. Res., 1999, V. 14, No. 2, 398–406.
135. Inokuti Y., Cantor B., Acta Met., 1982, V. 30, No. 2, 343–356.
136. Samuel F.H., Pract. Met., 1987, V. 24, 58–67.
137. Hayzelden C., Rayment J.J., Cantor B., Acta Met., 1983, V. 31, No. 3, 379–386.
138. Samuel F.H., J. Mater. Sci., 1987, V. 22, 3885–3892.
139. Inoue A., Kojima Y., Minemura T., Masumoto T., Trans. ISIJ, 1981, V. 21, No. 9, 656–663.
140. Blinov E.N., et al., Fiz. Met. Metalloved., 1999, V. 87, No. 4, 49–54.
141. Glezer A.M, et al., Izv. RAN, Ser. fiz., 2002, V. 66, No. 9, 1263–1275.
142. Glezer A.M.,et al., J. Nanopart. Research, 2003, V. 5, 551–560.
143. Glezer A.M., Blinova E.N., Dokl. RAN, 2004, V. 396, No. 1, 41–43.
144. Pozdnyakov V.A., Izv. RAN, Ser. fiz., 2005, V. 69, No. 9, 1282–1291.
145. Lyubov B.Ya., Roitburd A.L., Dokl. AN SSSR, 1958, V. 120, No. 5, 1011–1014.
146. Kaufman L., Cohen M., in Advances in Physics of Metals, V. IV, Moscow, GN-TILChTsM, 1961, 192–289.
147. Muskhelishvili N.I. Some basic problems of the mathematical theory of elasticity, Moscow, Nauka, 1966.
148. Mura T., Micromechanics of defects in solids, Dordrecht, Boston, Lancaster, Martinus Nijhoff Publ., 1987.
149. Fisher J.C., Turnbull D., Acta Met., 1953, V. 1, No. 3, 310–314.
150. Roitburd A.L., Imperfections of the crystal structure and martensitic transformations, Moscow, Nauka, 1972.
151. Pozdnyakov V.A., Izv. RAN, Ser. fiz., 2005, V. 69, No. 9, 1282–1291.
152. Glezer A.M., et al., Materialovedenie, 2007, No. 12, 3–9.
153. Pechkovsky E.P., Trefilov V.I., Influence of the structure of austenite on martensite transformation in iron-based alloys. Preprint IPM, No. 71.4, Kiev, 1971.
154. Pankova M.N., et al., Fiz. Met. Metalloved., 1983, V. 55, No. 3, 576–582.
155. Glezer A.M., Aleshin D.N., Deformatsiya Razrush. Mater., 2005, No. 9, 43–45.
156. Meyers M.A., Mishra A., Benson D. J., Prog. Mater. Sci., 2006, V. 51, 427–556.
157. Zang H.Y., Hu Z.Q., Lu K., J. Appl. Phys., 1995, V. 77, 2811–2813.
158. Noskova N.I., et al., Fiz. Met. Metalloved., 1997, V. 81, No. 1, 116–121.
159. Greer A.L., Changes in structure and properties associated with the transition from the amorphous to the nanocrystalline state, Nanostructured Materials. Science and Techonology, Eds. Chow G.M., Noskova N.I, Dordrecht, Kluwer Acad. Publ., 1998, 143–162.
160. Fougere G.E., Weertman J.R., Siegel R.W., Scr. Met., 1992, V. 6, 1879–1881.

161. Chang H., Altstetter C. J., Averback R.S., J. Mater. Res., 1992, V. 7, 2962–2970.
162. Palumbo G., Erb U., Aust K.T., Scr. Met. Mater., 1990, V. 24, 2347–2350.
163. Koch C.C., Morris D.G., Lu K., Inoue A., MRS Bulletin, 1999, V. 24, No. 2, 52–54.
164. Hillenbrand H.G., et al., Influence of soft crystalline particles on the mechanical properties of (Fe, Co, Ni)–B metallic glasses, Proc. of the Fourth Intern. Conf. RQM, Sendai, Japan, 1981, V. 2, 1369–1372.
165. Glezer A.M., et al., Dokl. RAN, 2008, V. 418, No. 2, 181–184.
166. Glezer A.M., Permyakova I.E., Modern ideas on how to study the mechanical properties of metallic glasses, Deformatsiya Razrush. Mater., 2006. No. 3, 2–11.
167. Fedorov V.A., et al., Metally, 2004. No. 3, 108–113.
168. Finkel' V.M., Physical basis of failure inhibition, Moscow, Metallurgiya, 1977.
169. Merk N., Morris D. G., Morris M.A., J. Mater. Sci., 1988, V. 23, No. 11, 4132–4140.
170. Kovneristyi J.K., Metally, 2001, No. 5, 19–23.
171. Kekalo I.B., et al., Fiz. Met. Metalloved., 1987, V. 64, No. 5, 983–990.
172. Skakov Yu.A., Kraposhin V.S., Itogi Nauki i Tekhniki, Metalloved. Term. Obrab. Met., Moscow, VINITI, 1980, V. 13, 13–78.
173. Freed R.I., Vander J.B., Met. Trans., 1979, V. 10A, No. 11, 1621–1626.
174. Lyakishev N.P., Alymov M.I., Rossiiskie nanotekhnologii, 2006, V. 1, No. 1–2, 72–81.
175. Kalin B.A., Fedotov V.T., Molokanov V.V., in: Problems of research of the structure of amorphous alloys, Moscow, MISiS, 1988, 324–325.
176. Donovan P.E., Stobbs W.M., Acta Metall., 1983, V. 31, No. 1, 1–8.
177. Noskova N.I., et al., Strength and ductility of the alloy Pd–Cu–Si amorphous and nanocrystalline states, Fiz. Met. Metalloved., 1996, V. 81, No. 1, 163–170.
178. Surinach S., Otero A., Baro M.D., et al., Nanostruct. Mater., 1995, V. 6, No. 1–4, 461–464.
179. Noskova N.I., Fiz. Met. Metalloved., 1998, V. 86, No. 2, 101–116.
180. Golovin Yu.I., Introduction to nanotechnology, Moscow, Mashinostroenie, 2003.
181. Gryaznov V.G., Trusov L.I., Progr. Mater. Sci., 1993, V. 37, No. 4, 289–401.
182. Yamasaki T., Schlossmacher P., et al., Mater. Sci. Forum, 1998, V. 269–272, 975–980.
183. Rice R.W., J. Mater. Sci., 1997, V. 32, 1673–1692.
184. Zhang H.Y., Hu Z.Q., Lu K., J. Appl. Phys., 1995, V. 77, No. 6, 2811–2813.
185. Sui M. L., Patu S., He Y.Z.,Scr. Metall. Mater., 1991, V. 25, No. 7, 1537–1542.
186. Bresson L., Chevalier J.P., Fararel M., Scr. Met., 1982, V. 16, 499–505.
187. Glazer A.A., et al., Fiz. Met. Metalloved., 1992. No. 8, 96–100.
188. Noskova N.I., et al., Fiz. Met. Metalloved., 1992, No. 2, 83–88.
189. Noskova N.I., et al., Nanostruct. Materials, 1995, V. 6, No. 5–8, 969–972.
190. Liu X.D., et al., J. Appl. Phys, V. 74, No. 7, 4501–4505.
191. Christman T., Scr. Met. Mater., 1993, V. 28, 1495–1500.
192. Suryanarayana C., Int. Mater. Rev., 1995, V. 40, 41–64.
193. Siegel R.W., Fouger G.E., Nanostruct. Mater., 1995, V. 6, 205–216.
194. Wang D.I., Kong Q.P., Shui J., Scr. Metall. Mater., 1994, V. 31, 47.
195. Inoue A., Kim Y.H., Masumoto T., Mater. Trans. JIM, 1992, V. 33, No. 5, 487–490.
196. Andrievski R.A., J. Mater. Sci., 1994, V. 29, 614–631.
197. Mayo M.J., Siegel R.W., Narayanasamy A., Nix W.D., J. Mater. Res., 1990, V. 5, 1073–1082.
198. Hahn H., Averback R.S., J. Amer. Cer. Soc., 1991, V. 74, 2918–2921.
199. Mayo M.J., Superplasticity of nanostructured ceramics, Mechanical properties and deformation behavior of materials having ultrafine microstructure, Eds. Nastasi M.,

Parkin D.M., Gleiter H., Dordrecht, Kluwer Acad. Publ., 1993, 361–380.

200. Prabhu G.B., Bourell D.L., Scr. Metall. Mater., 1995, V. 33, 761–766.
201. Huang Z., Gu L.Y., Weertman J.R., Scr. Mater., 1997, V. 37, 1071–1075.
202. Kaiser A., Vassen R., Stover D., Buchkremer H., Nanostruct. Mater., 1997, V. 8, 489–497.
203. Kolobov Yu.R., et al., Izv. VUZ, Fizika, 1998. No. 3, 77–82.
204. Mishra R.S., Valiev R.Z., McFadden S.X., Mukhejee A.K., Mater. Sci. Eng., 1998, V. A252, 174–178.
205. Betz U., Hahn H., Nanostruct. Mater., 1999, V. 11, 376–388.
206. Guermazi M., Hofler H.J., Hahn H., Averback R.S., J. Amer. Cer. Soc., 1991, V. 74, 2672–2674.
207. Kumplmann A., Gunhter V., Kunze H.-D., Mater. Sci. Eng., 1993, V. A168, 165–168.
208. Ping D.M., Li D.X., Ye H.G., J. Mater. Sci. Lett., 1995, V. 14, 1536–1540.
209. Chen Z., Ding J., Nanostruct. Mater., 1998, V. 10, No. 2, 205–215.
210. Alexandrov I.V., et al., Features of the structure of nanocrystalline materials produced by severe plastic deformation, in: Structure, phase transformation and properties of nanocrystalline alloys, Yekaterinburg, Russian Academy of Sciences, 1997, 57–69.
211. Lojkowski W., Acta Met. Mater., 1991, V. 39, No. 8, 1891–1899.
212. Chokshi A.H., Rosen A., et al., Scr. Met., 1989, V. 23, 1679–1683.
213. Nazarov A.A., Scr. Met., 1996, V. 34, No. 5, 697–701.
214. Lu K., Sui M.L., Scr. Met. Mater., 1993, V. 28, 1465–1470.
215. Lu K., Zhang H., Zhong Y., Fecht H.J., J. Mater. Res., 1997, V. 12, 923–930.
216. Ma E., Scr. Mater., 2003, V. 49, 941–946.
217. Koch C.C., Scr. Mater., 2003, V. 49, 657–662.
218. Valiev R.Z., Alexandrov I.V., Bulk nanostructured metallic materials: preparation, structure and properties, Moscow, Akademkniga, 2007.
219. Koch C.C., J. Mat. Sci., 2007, V. 42, 1403–1414.
220. Wang Y., Chen M., Zhou F., Ma E., Nature, 2002, V. 419, 912–915.
221. Shen Y. F., Lu L., Lu Q. H., Jin Z.H., Lu K., Scr. Mater., 2005, V. 52, 989–994.
222. Valiev R.Z., et al., J. Mater. Res., 2002, V. 17, 5–8.
223. Pozdnyakov V.A., Izv. RAN, Ser. fiz., 2007, V. 71, 1751–1763.
224. Gil Sevillano J., Aldazabal J., Scr. Mater., 2004, V. 51, No. 8, 795–800.
225. Pozdnyakov V.A., Pis'ma Zh. Teor. Fiz., 2007, V. 33, No. 23, 36–42.
226. Malygin G.A., Fiz. Tverd. Tela, 2008, V. 50, No. 6, 990–996.
227. Gutkin M.Yu, Ovid'ko I.A., Physical mechanics of deformed nanostructures, V. 1. Nanocrystalline materials, St. Petersburg, Janus, 2003.
228. Estrin Y., Kim H.S., Encyclopedia of Nanoscience & Nanotechnology, V. 8, Ed. H. Nalwa, USA, Amer. Science Publication, 2004, 489.
229. Zhao Y.H., et al., Apl. Phys. Lett., 2006, V. 89, Paper 121906.
230. Glezer A.M., Pozdnyakov V. A., Nanostruct. Mater., 1995, V. 6, 767–769.
231. Glezer A.M., Izv. RAN, Ser. fiz., 2003, V. 67, No. 6, 810–817.
232. Glezer A.M., Deformatsiya Razrush. Mater., 2006, No. 2, 10–16.
233. Kolesnikov A., Ovid'ko I.A., Romanov A.E., Pis'ma Zh. Teor. Fiz., 2007, V. 33, No. 15, 26–33.
234. Malygin G.A., Fiz. Tverd. Tela, 2007, V. 49, No. 6, 961–982.
235. Malygin G.A., Fiz. Tverd. Tela, 2008, V. 50, No. 0, 1013–1017.
236. Malygin G.A., Fiz. Tverd. Tela, 2007, V. 49, No. 8, 1392–1397.
237. Conrad H., Nanotechnology. 2007, V. 18, No. 32. Paper 325701.

238. Gutkin M.Yu, Ovid'ko I.A., Defects and mechanisms of plasticity in nanostructured and non-crystalline materials, St. Petersburg, Logos 2001.
239. Gutkin M.Yu, Ovid'ko I.A., Physical mechanics of deformable nanostructures, V. II., Nanocrystalline films and coatings, St. Petersburg, Janus, 2005.
240. Gutkin M.Yu, Ovid'ko I.A., Appl. Phys. Lett., 2006, V. 88, No. 21. Paper 211901.
241. Gutkin M.Yu, Ovid'ko I.A., Fiz. Tverd. Tela, 2008, V. 50, No. 4, 630–638.
242. Bobylev S.V., Ovid'ko I.A., Fiz. Tverd. Tela, 2008, V. 50, No. 4, 617–623.
243. Gertsman V.Y., et al., Scr. Met., 1994, V. 30, 229–234.
244. El-Sherik A.M., Erb U., Palumbo G., Aust K.T., Scr. Met. Mater., 1992, V. 27, 1185–1188.
245. Zaichenko S.G., Glezer A.M., Fiz. Tverd. Tela, 1997, V. 39, 2023–2028.
246. Zaichenko S.G., Glezer A.M., Mater. Sci. Forum, 1998, V. 269–272, 687–692.
247. Gryaznov V.G., Gutkin M.Y., Romanov A.E., J. Mater. Sci., 1993, V. 28, 4359–4365.
248. Carsley J.E., Ning J., Milligan W.W., et al., Nanostruct. Mater. 1995, V. 5, 441–448.
249. Konstantinidis D.A., Aifantis E.C., Nanostruct. Mater., 1998, V. 10, 1111–1118.
250. Masumura R.A., Hazzledine P.M., Pande C.S. Acta Mater., 1998, V. 46, 4527–4534.
251. Kumar K.S., Van Swygenhoven H., Suresh S., Acta Mater., 2003, V. 51, No. 19, 5743–5774.
252. Wolf D., Yamakov V., Phillpot S.R., Mukherjee A., Gleiter H.. Acta Mater., 2005, V. 53, 1–40.
253. Van Swygenhoven H., Weertman J.R., Materials Today, 2006, V. 9, 24–31.
254. Schiotz J., Di Tolla F.D., Jacobsen K.W., Nature, 1998, V. 391/5, No. 2, 561–563.
255. Szlufarska I., Nakano A., Vashista P., Science. 2005, V. 309, 911–914.
256. Van Swygenhoven H., Caw A., MD computer simulation of elastic and plastic behavior of n-Ni, Nanophase and Nanocomposite Materials II, Eds. Komarneni S., Parker J.C., Wollenberger H. J. MRS, Pittsburgh, 1997, V. 457, 193–198.
257. Van Swygenhoven H., Spaczer M., Farkas D., Caro A., Nanostruct. Mater., 1999, V. 11, 417–429.
258. Ke M., Hackney S.A., Milligan W.W., Aifantis E.C., Nanostruct. Mater., 1995, V. 5, 689–697.
259. Pozdnyakov V.A., Glezer A.M., Dokl. RAN 2002, V. 384, No. 2, 177–180.
260. Orlov A.N., et al., Grain boundaries in metals, Moscow, Metallurgiya, 1980.
261. Jang J.S., Koch C.C., Scr. Metall. Mater., 1990, V. 24, No. 8, 1599–1604.
262. Argon A.S., Acta Metall., 1979.V. 27, No. 1, 47–58.
263. Pozdnyakov V.A., Materialovedenie, 2002, No. 11, 39–47.
264. Pozdnyakov V.A., Fiz. Met. Metalloved., 2003, V. 96, No. 1, 114–128.
265. Malygin G., Fiz. Tverd. Tela, 1995, V. 37, 2281–2292.
266. Gryaznov V.G., et al., Phys. Rev. B, 1991, V. 44, 42–46.
267. Glezer A.M., Deformatsiya Razrush. Mater., 2010. No. 2, 1–8.
268. Cai B., et al., Mater. Sci. Eng. A, 2000, V. 286, 188–192.
269. Suryanarayana C., J. Mater. Res., 1992, V. 7, No. 8, 2114–2118.
270. Noskova N.I., Mulyukov R.R., Submicrocrystalline and nanocrystalline metals and alloys, Yekaterinburg, Izd. IMP UB RAS 2003.
271. Chen H., He Y., Shiflet G.Y., Poon S.J., Scr. Metall. Mater., 1991, V. 25, No. 6, 1421–1424.
272. Abrosimova G.E.,et al., Fiz. Tverd. Tela, 1998, V. 40, No. 1, 10–16.
273. Glezer, A.M., et al., Deformatsiya Razrush. Mater., 2010. No. 8, 1–10.
274. Cherniavsky K.S., Stereology in metals science, Moscow, Metallurgiya, 1977, 95–100.

275. Diagrams of binary metallic systems, a handbook, Ed. N.P. Lyakishev, V. 2, Mashinostroenie, 2000.
276. Samsonov G.V., Vinnitskii I.M., Refractory compounds, Moscow, Metallurgiya, 1976.
277. Alekhin V.P., Khonik V.A., The structure and physical relationships of deformation of amorphous alloys, Moscow, Metallurgiya, 1992.
278. Glezer, A.M., Manaenkov S.E., Permyakova I.E., Izv. RAN, Ser. fiz., 2007, V. 71, No. 12, 1745–1748.
279. Pozdnyakov V.A., Fiz. Met. Metalloved., 2004, V. 97, No. 1, 9–17.
280. Goldshtein M.I., et al., Metal physics of high-strength alloys, Moscow, Metallurgiya, 1986.
281. Glezer A.M., et al., Dokl. AN SSSR, 1985, V. 283, No. 1, 106–109.
282. Takayama S., J. Mater. Sci., 1981, V. 16, No. 9, 2411–2418.
283. Argon A.S., Inelastic deformation mechanisms in glassy and microcrystalline alloys, Proc. of the Fifth Intern. Conf. RQM, Elsevier Sci. Publ., 1985, V. 2, 1325–1335.
284. Skakov Yu.A., Finkel' M.V., Izv. VUZ, Chern. Met., 1986. No. 9, 84–88.
285. Davis L.A., Mechanics of metallic glasses, Prep. Second Intern. Conf. RQM, Cambridge, Cambridge Univ., 1975.
286. Neuhäuser H., Scr. Met., 1978, V. 12, No. 5, 471–474.
287. Alekhin V.P., Pompe W., Vettsig K., Metalloved. Term. Obrab. Met., 1982. No. 5, 33–36.
288. Bian Z., He G., Chen G.L., Scr. Mater., 2002, V. 46, 407–412.
289. Glezer A.M., et al., Izv. RAN, 2008, V. 72, No. 9, 1335–1337.
290. Glezer A.M., et al., in: Materials XLVI Intern. conference Actual problems of strength, Vitebsk, VGTU, 2007, Part 1, 11–14.
291. Chuvil'deev V.N., Fiz. Met. Metalloved., 1996, V. 81, No. 5, 5–14.
292. Li J.M., Quan M.X., Hu Z.Q., Appl. Phys. Lett., 1996, V. 69, 1559–1561.
293. Golovin Yu.I., Fiz. Tverd. Tela, 2008, V. 50, No. 12, 2113–2142.
294. Pozdnyakov V.A., Pis'ma Zh. Teor. Fiz., 2003, V. 29, No. 4, 46–51.
295. Pozdnyakov V.A., Izv. RAN, Ser. fiz., 2003, V. 67, No. 6, 868–876.
296. Li Z., Ramasamy S., Hahn H., Siegel R.W., Mater. Lett., 1998, V. 6, 195–201.
297. Gan Y., Zhou B., Scr. Mater., 2001, V. 45, No. 6, 625–630.
298. Novikov I.I., Ermishkin V.A., Micromechanisms of failure of metals, Moscow, Nauka, 1991.
299. Cottrell B., Rice J.R., Intern. J. Fract., 1980, V. 16, 155–169.
300. Ashby M.F., Acta Metal. Mater., 1993, V. 41, No. 5, 1313–1335.
301. Vladimirov V.I., Romanov A.E., Disclinations in crystals, Leningrad, Nauka, 1986.
302. Rybin V.V., et al., Fiz. Met. Metalloved., 1990, V. 69, No. 1, 5–26.
303. Gutkin M.Yu., Ovid'ko I.A., Phil. Mag. A, 1994, V. 70, 561–575.
304. Richter A., Romanov A.E., Pompe W., Vladimirov V. I., Phys. stat. sol. (B), 1987, V. 143, 43–53.
305. Nazarov A.A., Romanov A.E., Valiev R.Z., Nanostruct. Mater., 1995, V. 6, 775–778.
306. Pande C.S., Masumura R.A., Armstrong R.W., Nanostruct. Mater. 1993, V. 2, 323–331.
307. Rice J.R., J. Appl. Mech., 1968, V. 35, 379–386.
308. Krasovskii A.Y., Brittleness of metals at low temperatures, Kiev, Naukova Dumka, 1980.
309. Oljjl T., Jeong Y.K., Choa Y.II., Niihara K., J. Amer. Ceram. Soc., 1998, V. 81, 1453–1460.
310. Pozdnyakov V.A., Glezer A.M., Fiz. Tverd. Tela, 2005, V. 47, No. 5, 793–800.

311. Hanlon T., Kwon Y. N., Suresh S., Scr. Mater., 2003, V. 49, No. 7, 675–680.
312. Argon A.S., Acta Metall., 1979, V. 27, No. 1, 47–55.
313. Gusev A.I., Usp. Fiz. Nauk, 1998, V. 168, 29–58.
314. Kronmuller H., Nanostruct. Mater. 1995, V. 6, 157–168.
315. Ustinov V,V., Kravtsov E.A., in: Nanostructured Materials, Science and Technology, Eds. Chow G.M., Noskova N.I, Dordrecht, Kluwer Acad. Publ., 1998, 441–456.
316. Becker J.I., J. Appl. Phys., 1984, V. 55, No. 6, 2067–2072.
317. Glazer A.M., Fiz. Met. Metalloved., 1979, V. 48, No. 6, 1165–1172.
318. Drozdov M.A., et al., Fiz. Met. Metalloved., 1989, V. 67. No. 5, 896–901.
319. Yagodkin Yu.D., Izv. VUZ, Chern. Met., 2007. No. 1, 37–46.
320. Herzer G., J. Magn. Magn. Mat., 1986, No. 2–3, V. 62, 143–151.
321. Herzer G., Hilzinger H.R., Physica Scripta, 1989, V. 39, 639–642.
322. Herzer G., IEEE. Trans. Mag., 1989, V. 25, 3327–3329.
323. Herzer G., IEEE. Trans. Mag., 1990, V. 26, 1397–1402.
324. Herzer G., J. Magn. Magn. Mat., 1992, V. 112, No. 1–3, 258–262.
325. Herzer G., Warlimont H., Nanostruct. Mat., 1992, V. 1, No. 3, 263–268.
326. Coey J.M.D., Rare-earth iron permanent magnets, Oxford: Oxford Science Publ., Clarendon Press, 1996.
327. Kneller E.F., Hawig R., IEEE. Trans. Mag., 1991, V. 27, 3588–3560.
328. Skomski R., Coey J.M.D., Phys. Rev. B, 1993, V. 48, 15812–15816.
329. McHenry M.E., et al., in: PV 96-10, ECS Symposium Proceedings, Eds. Kadish K.M., Ruoff R.S., Penmngton, NJ, 1996, 703.
330. Host J.J., et al., J. Mat. Res., 1997, V. 12, No. 5, 1268–1273.
331. Bertotti G., Ferrara E., Fiorillo F., Tiberto P., Mat. Sci. Eng. A, 1997, V. 226–228, 603–613.
332. Herzer G., Mat. Sci. Eng. A, 1991, V. 133, 1.
333. Herzer G., Scr. Metall. Mater., 1995, V. 33, No. 10–11, 1741–1756.
334. Hoffmann H., Fujii T., J., Magn. Magn. Mat., 1999, V. 128, 395–400.
335. Fujii Y., Fujita H., Seki A., Tomida T., J. Appl. Phys., 1991, V. 70, 6241–6243.
336. Tsuei C.C., Duwez P., J. Appl. Phys., 1966, V. 37, 435.
337. Turgut Z., et al., J. Appl. Phys. 1997, V. 81, 4039–4041.
338. Inoue A., Zhang T., J. Appl. Phys., 1998, V. 83, 6326–6332.
339. Gallagher K.A., et al., J. Appl. Phys., 1999, V. 85, 5130–5132.
340. Turgut Z., et al., J. Appl. Phys., 1998, V. 83, 6468–6470.
341. Turgut Z., et al., J. Appl. Phys., 1999, V. 85, No. 8, 4406–4408.
342. Varga L.K., Bakos E., Kiss L.F., Bakonyi I., Mat. Sci. Eng. A, 1994, V. 179–180, 567–571.
343. Stolloff N.S., Mater. Res. Soc. Proc., 1985, V. 39, 3–12.
344. Suzuki K., Kataoka N., Inoue A., Makino A., Masumoto T., Mat. Trans. JIM, 1990, V. 31, 743–746.
345. Smith C.H., IEEE. Trans. Mag., 1982, V. 18, 1376–1381.
346. Hernando A., Kulik T., Phys. Rev. B, 1994, V. 49, 7064–7067.
347. Hernando A., Navarro I., Nanophase materials, Eds.: G.C. Hadjipanyis, R.W. Siegel, The Netherlands, Kluwer, 1994, 703.
348. Hernando A., Navarro I., Gorria P., Phys. Rev. B, 1995, V. 51, 3281–3284.
349. Gomez-Polo C., et al., Phys. Rev. B, 1996, V. 53, 3392–3397.
350. Malkinski L., Slawska-Waniewska A., J. Magn. Magn. Mat., 1996, V. 160, 273–279.
351. Livshits B.G., et al., Physical properties metals and alloys, Moscow, Metallurgiya, 1980, 320.
352. Hono K., et al., Acta Met. Mater., 1992, V. 40, 2137–12144.

353. Glezer A.M., Kiriyenko V.I., Metally, 1998, No. 2, 44–48.

354. Ayers J.D., Konnert J.H., et al., J. Mater. Sci., 1995, V. 30, 4492–4498.

355. Zusman, A.I, Artsishevsky M.A., Thermomagnetic processing of iron–nickel alloys, Moscow, Metallurgiya, 1984..

356. Grognet S., Atmani H., Teilett J., Nanocrystallization by nitriding treatment of FeSiB-based amorphous ribbon, UMR 6634, CNRS.

357. Gonzalez J., et al., J. Appl. Phys., 1994, V. 76, 1131–1134.

358. Sadchikov V.V., et al., Stal', 1997, No. 11, 58–61.

359. Makarov V.A., et al., Fiz. Met. Metalloved., 1991, No. 9, 139–149.

360. Kekalo I.B., Samarin B.A., Physical metallurgy of precision alloys. Alloys with special magnetic properties, Moscow, Metallurgiya, 1989.

361. Kojima A., et al., Mat Sci. Eng., 1994, V. A179–A180, 511–515.

362. Gorria P., et al., IEEE. Trans. Mag., 1993, V. 29, 2682–2684.

363. Suzuki K., et al., J. Appl. Phys., 1991, V. 70, 6232–6237.

364. Garcia-Tello P., et al., IEEE. Trans. Mag., 1997, V. 33, 3919-3921.

365. Garcia-Tello P., et al., J. Appl. Phys., 1998, V. 83, 6338–6340.

366. Iwanabe H., et al., J. Appl. Phys., 1999, V. 85, 4424–4426.

367. Kim K.S., Yu S.C., J. Appl. Phys., 1997, V. 81, 4649–4652.

368. Yoshizawa Y., Yamauchi K., Mat. Trans. JIM, 1990, V. 31, 307–310.

369. Buehler W.J., Gilfrich L.W., Wiley R.C., J. of Appl. Phys., 1963, V. 34, No. 5, 1475–1477.

370. Materials with shape memory, Ed. V.A. Likhachev, V. 1, St. Petersburg, NIIKh St. Petersburg State University, 1997.

371. Kurdyumov G.V., Khandros L.G., Dokl. AN SSSR, 1949, V. 66, No. 2211–215.

372. Estrin E.I., Izv. RAN, 2002, V. 66, No. 9, 1243–1249.

373. Tikhonov A.S., et al., Application of the shape memory effect in modern engineering, Moscow, Mashinostroenie, 1981.

374. Wasilewski R.J., Scr. Metall.. 1975, V. 9, No. 4, 417–422.

375. Khachin B.H., et al., Fiz. Met. Metalloved., 1976, V. 42, No. 3, 658–661.

376. Wang Z.G., Mater. Lett., 2002, V. 56, 284–288.

377. Pushin V., Stolyarov V.V., et al., Physics of Metals and Metallography, 2002, V. 94. Suppl. 1, 54–S68.

378. Pushin V.V., Kourov N.I., Physics of Metals and Metallography, 2002, V. 94. Suppl. 1, 107–S118.

379. Shelyakov A.V., Matveeva N.M., Larin S.G. in: Shape Memory Alloys: Fundamentals, Modeling and Industrial Applications, Eds. F. Trochu and V. Brailovski, Canadian Inst. of Mining, Metallurgy and Petroleum, 1999, 295–303.

380. Antonov V.A., et al., Zh. Teor. Fiz., 1991, V. 61, No. 9, 87–93.

381. Shelyakov A.V., in: Proc. of Intern. Conf. on Shape Memory and Superelastic Technologies (SMST-94), Pacific Grove, CA, USA, 7–10 March, 1994, Eds. A.R. Pelton, D. Hodgson, T. Duerig, MIAS, Monterey, CA, 1995, 335–340.

382. Shelyakov A.V., et al., J. Phys. IV, France, 2003, V. 112, 1169–1172.

383. Kareev S.A., Glezer, A.M., Shelyakov A.V., Materialovedenie, 2006, No. 12, 25–30.

384. Glezer A.M., et al., Izv. RAN, Ser. fiz., 2009, V. 73, No. 9, 1364–1367.

385. Krupp GmbH, Fried Essen, Patentschrift DE 4006076, CI 1990.

386. Abujudom D.N., et al., U. S. Patent No. 5,114,514, Johnson Service Company. Milwaukee, WI, 1992.

387. Shelyakov A., et al., in: Proc. of Intern. Conf. on Shape Memory and Superelastic Technologies (SMST-97), Pacific Grove, CA, USA, 2–6 March, 1997, Eds. A.R. Pelton, D. Hodgson, S.M. Russel, T. Duerig, SMST, Santa Clara, CA, 1997, 89–94.

388.   Potapov P., et al., Mater. Lett., 1997, V. 32, 247–250.
389.   Kareev S.I., Shelyakov A.V., Glezer A.M., Deformatsiya Razrush. Mater., 2007, No. 7, 22–27.
390.   Han X.D., et al., Acta Mater., 1996, V. 44, 3711–3721.

4

Nanocrystals produced by megaplastic deformation of the amorphous state (Type IV nanocrystals)

Extreme effects have a significant impact on the structure and properties of solids [1]. These should certainly include the impact of very large plastic deformations. Over the last years the interest in this method of processing materials increased significantly, as it gives the opportunity to significantly improve the physical and mechanical properties of metallic materials. [2] This is largely due to the formation of nanostructured states of different types and, in particular, the processes for nanocrystallisation in processing of melt-quenched amorphous alloys.

4.1. The nature of large (megaplastic) strains

4.1.1. Terminology

With a light hand the pioneers in the field of study of superhigh plastic strain B. Segal and R.Z.Valiev [3, 4], plastic deformation, in which the true plasticity value e is greater than 1 and can reach 7–8, is referred to in the Russian literature as high-intensity plastic deformation (SPD). However, it seems that this term is not entirely successful. In fact, under intensive processes in nature we understand, as a rule, the processes taking place at high speed [5]. In case of very large deformation, as shown by the assessments, the strain rate is in the range 10^{-1}–10^1 s^{-1}, i.e. in the transition region between the static and dynamic strain rates corresponding to implemented, for example, in conventional rolling. In this context, this type of deformation can not be called severe. The term *severe plastic deformation* [6], used in foreign scientific literature, is apparently more suitable, because it can be translated into Russian as 'strict', 'tough', 'deep', but rather 'strong' plastic deformation [7].

In [8] another, more physically rigorous Russian term was proposed instead of 'high-intensity plastic deformation'. The new term goes back to the general philosophical concept of our ideas about the matter. As is well known [5], science examines three major levels of the material world (Fig. 4.1): MICROworld, which is realized on the scale of individual atoms and molecules; MEGAworld – the scale of the human perception of the world: meter, kilogram, second, and MEGAworld – astronomical scale. There is a direct analogy between these scales of organization of matter and the levels of plastic deformation. In fact, the well-known process of MICROplastic deformation is observed before the value of the macroscopic yield point is reached, and the process of MACROplastic deformation is realised at stresses higher than the yield stress [9]. Thus, continuing this analogy, very high plastic deformation should be termed MEGAplastic deformation (MPD), which corresponds to the general logic of any material phenomenon (Fig. 4.2).

If the border between microplastic deformation and macroplastic deformation is clearly defined – the degree of deformation corresponding to the macroscopic yield point (relative strain $\varepsilon = 0.05$ or 0.2%), the boundary between macroplastic and megaplastic

Microworld   Macroworld   Megaworld

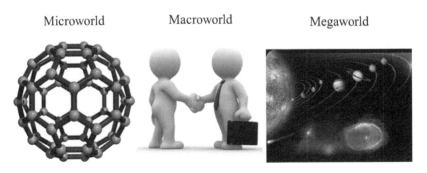

Fig. 4.1. Scale levels of the material world.

DEFORMATION

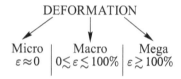

Micro	Macro	Mega
$\varepsilon \approx 0$	$0 \lesssim \varepsilon \lesssim 100\%$	$\varepsilon \gtrsim 100\%$

Fig. 4.2. Scale levels of plastic deformation.

deformation remains uncertain. Conventionally, we assume the boundary region is the relative strain $\varepsilon \approx 100\%$ or true strain $e \approx 1$. Later, we will present be given a more rigorous, physically reasonable definition of plastic deformation, corresponding to the transition to the region of the MPD.

4.1.2. What is known about megaplastic deformation

Figure 4.3 schematically shows the three most common methods of producing huge degrees of deformation: pressure torsion in the Bridgman chamber (PTBC) (a), equal channel angular pressing (ECAP) (b) and accumulated rolling (c). In the first case, the sample is placed between two anvils one of which is rotated slowly while creating very high hydrostatic pressure (a few GPa). In second case, the sample is forced through two equal-sized channels, located at an angle to each other (up to 90°). The plastic strains formed in this case are so high that the conventional values of the relative degrees of deformation loose any meaning and we need to make the transition to true logarithmic strains e. Their values are determined as follows.

For PTBS [4]

$$e = \ln\left(1 + \left(\frac{\varphi \cdot r}{h}\right)^2\right)^{0.5} + \ln\left(\frac{h_0}{h}\right),$$

(4.1)

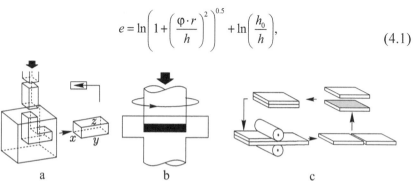

a   b   c

Fig. 4.3. Scheme of the most common methods of creating ultrahigh strains, and a – equal channel angular pressing, b – pressure torsion in a Bridgman chamber, c – accumulated rolling.

where r and h are the radius, the height of the sample in the form of a disk, processed in the Bridgman chamber, and φ is the angle of rotation of the movable anvil. The number of complete rotations of the mobile anvil N corresponds to the deformation at which $\varphi = 2\pi N$.

For ECAP [3]

$$e = \text{arsh}\left(n \cdot \text{ctg } \varphi\right), \qquad (4.2)$$

where n is the number of passes and φ is the angle of rotation of the channels.

In recent years, a new effective method for creating megaplastic deformation – screw extrusion [10] – has been proposed.

Formed at such enormous deformations these structural conditions are unusual and hard to predict. Unfortunately, the vast majority of authors, exploring the impact of ultrahigh plastic strains, confined themselves to study of finite structures and related properties of materials, without analysing the physical processes that occur directly at the gigantic degrees of plastic flow. Classic dislocation and even disclination approaches to understanding the structural processes at very high plastic deformations are not effective and needs rethinking.

Summing up the large number of experimental studies studying the structure of materials subjected to MPD, we can say that there is a complex combination of defect structures, containing low-angle and high-angle grain boundaries in different proportions, as well as defective structures within the grains with varying degrees of perfection. The three-dimensional statistical evaluation of this grain structure, popular in the literature, gives in the best case the ratio between the high-angle and low-angle boundaries in the structure of the material and little information about the nature of the physical processes that occur directly during MPD. In addition, this information is, unfortunately, quite contradictory and ambiguous, since is different experiments the authors observed as a rule different structural states in the same materials under similar deformation conditions. In this case, the true nanostructured state ($d < 0.1$ µm) does not form in quite a number of instances. In steels and alloys MPD is often accompanied by phase transitions (separation and dissolution of phases, martensitic transformation, amorphisation) [11, 12]. The latter usually occurs in intermetallic compounds or multicomponent systems.

The most coherent concept of large plastic deformations was proposed by V.V. Rybin [13]. On the basis of the representations of the dominant role of the dislication mode in the implementation of

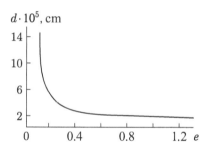

Fig. 4.4. Dependence of the grain (fragment) size on the degree of plastic deformation (V.V. Rybin).

the large plastic deformations and related processes of fragmentation he was able to correctly describe the phenomena that occur at significant degrees of deformation close to $\varepsilon = 100\%$. In accordance with the concept of disclinations the size of the fragments – the main structural elements – gradually increases with increasing deformation, reaching a constant minimum value of 0.2 μm (Fig. 4.4). This means that the transition to the region of the nanostructured state and the formation of the fragments (grains) with the size smaller than 100 nm (0.1 μm) are hardly possible under the effect of the disclination mode. Analysis of the large number of experimental data, especially for BCC crystals, made V.V. Rybin to suggest the limiting (critical) fragmented structure, with further evolution of this structure within the framework of the disclination mode regarded as not possible [13]. Failure regions form at the boundaries of the fragments which separate regions usually free from dislocations. According to the author, the critical fragmented structure is the final product of plastic deformation, is not capable to resist the increasingly strong effect of external and internal stresses, and should result in failure. It should be mentioned that the previously described examination was related in fact to the early stages of megaplastic deformation ($e \leqslant 2$) and for the case of uniaxial tensile loading of rolling at a relatively small contribution of compressive stresses.

In studies by S.A. Firstov et al [14] it is reported that the transition to the highly deformed state is accompanied by jump-like changes in the structure of the material which take place at some critical value e_c. For commercial purity iron $e_c \approx 1$. In addition to this, there are changes in the mechanical behaviour of the material: strain hardening instead of the parabolic law at high strains is governed by the linear law.

A detailed systematisation of the defective structures, formed in different materials with increase of the degree of plastic deformation, was carried out by E.V. Kozlov and N.A. Koneva [15]. They showed that on approaching the range of megaplastic deformation there is, depending on the nature of the material, a gradual change from one structural state to other (cellular, banded, fragmented structures, etc), as in the case of structural phase transitions. This is accompanied by changes in the internal stress and the conditions for the manifestation of anomalies of the mechanical behaviour of crystals. Here it is also important to mention the study [16] in which it is shown that the increase of the degree of megaplastic deformation results in the formation in the structure of a very large number of excess point defects (mainly vacancies), capable of stimulating the occurrence of diffusion phase transformations during deformation.

Special attention has been given to the hypothesis according to which megaplastic deformation results in the formation of 'peculiar' non-equilibrium grain boundaries [17]. According to many authors, these boundaries are responsible for the anomalous phenomena of sliding, diffusion, interaction with lattice the effects and, consequently, may be responsible for the high values of strength and plasticity.

It is important to mention the results of the somewhat forgotten and only rarely cited study by V.A. Likhachev et al [18] in which megaplastic deformation was applied to copper wire (e = 1.6 and 3.7). The authors detected in the cyclic changes of the structure with increase of strain: the fragmented structure (d = 0.2 μm) $\Rightarrow$ the recrystallised structure $\Rightarrow$ fragmented structure (d = 0.1 μm), where c is the mean size of the fragments. It is important to mention that the fragmented structure of the 'second generation' is twice as dispersed as the structure of the first generation (Fig. 4.5). In principle, this assumes the establishment of the nanostructured stayed in the third and further cycles.

So far, there have been only a small number of serious attempts to describe the phenomena taking place at high strains which are situated only at the 'threshold' of megaplastic deformation. The author of the previously mentioned studies analysed the relationships governing plastic deformation at $e \leq 1.5$–2.0, and the investigators, who carried out experiments in the range of megaplastic deformation ($e \geq 1.0$–1.5) restricted themselves to the descriptive analysis of the final structures without discussing the mechanism of gigantic plastic deformation. The only exception is the study [18].

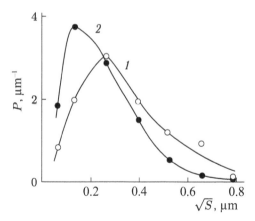

Fig. 4.5. The density of probability P of appearance of the fragments with the size of $S^{1/2}$ for smooth (1) and notched specimens directly in the mouth of the crack (2): $1 - e = 1.6$; $2 - e = 3.7$ (V.A. Likhachev, et al).

4.1.3. Energy principles of the mechanical effect on solids

The efficient theory of megaplastic deformation should, in our view, be capable of providing unambiguous answers to the following questions.

- what are the structural and phase transformations taking place in the processes of megaplastic deformation;
- what are the prerequisites for the realisation of megaplastic deformation in accordance with a specific scenario;
- what are the conditions of formation during megaplastic deformation of the true nanostructured state with the crystal size smaller than 100 nm, separated by high-angle boundaries or by another phase [19]?
- what are the special structural features of the processes of megaplastic deformation, what is the difference between megaplastic deformation and 'normal' plastic deformation?
- what determines the boundary value of the strain starting at which we can assume that we are in the range of megaplastic deformation?

Initially, we consider the energy aspects of the behaviour of solids under loading (Fig. 4.6) [20]. In the mechanical effect on the solid of finite dimensions, specific elastic energy is 'pumped' into the material. The evident 'dissipation channel' of this energy is plastic deformation. If this channel is exhausted, another channel – mechanical failure – can be realised. However, at high values of elastic energy, plastic deformation can in principle initiate

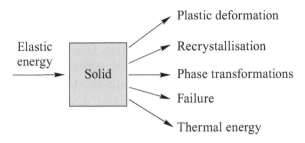

Fig. 4.6. Possible channels of dissipation of elastic energy in the mechanical effect on the solid.

additional 'dissipation channels: dynamically crystallisation, phase transformations and the generation of thermal energy. In the case of megaplastic deformation when the component of the uniform compression stresses is high, the formation and growth of splitting cracks is partially or completely suppressed and, consequently, the realisation of the failure processes is made more difficult. In other words, using ECAP, PTBC or similar loading methods, we stop deformation of the solid so that it does not fracture. If the concept proposed by V.V. Rybin [13] is correct, the plastic deformation is effective up to a specific limit, corresponding to the formation of the critical defective structure and, subsequently, the main channels of dissipation be represented by other physical processes: dynamic recrystallisation, phase transformations and/or generation of heat. In [21] the authors propose three possible scenarios of the development of further events (Fig. 4.7). In cases in which the processes of

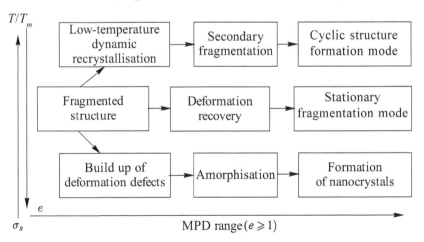

Fig. 4.7. Main scenarios of the development of structural processes in megaplastic deformation.

dislocation (disclination) rearrangement in the material (for example, in pure metals) are made easier, plastic deformation is followed by low-temperature dynamic recrystallisation (the upper branch in Fig. 4.7). Local areas of the structure are 'cleaned' from defects, and the process of plastic flow starts again in the new recrystallised grains with the help of dislocation and disclination modes. In this case, dynamical crystallisation plays the role of an additional powerful channel of dissipation of elastic energy. In cases in which the mobility of the carriers of plastic deformation is relatively low (for example, in solid solutions or intermetallics), the phase transition (the lower branch in Fig. 4.7) can be a powerful additional channel of dissipation of elastic energy. In most cases, it is the crystal $\Rightarrow$ amorphous state transition. Consequently, plastic flow is localised in the amorphous matrix without the effects of strain hardening and buildup of high internal stresses. Evidently, there is an intermediate case (the middle branch in Fig. 4.7) in which the additional dissipation channel may be represented by disclination rearrangement leading to the stabilisation of the fragmented structure observed in a number of experiments with the development of megaplastic deformation.

Obviously, the transition from one scenario of structural rearrangement to another also depends on the parameter (T_{MPD}/T_m), where T_{MPD} is the temperature of megaplastic deformation taking into account the possible effect of generation of heat, and T_m is the melting point.

4.1.4. Low-temperature dynamic recrystallisation

Proposing the first scenario of structural changes in megaplastic deformation, we have *a priori* assumed that the processes of crystallisation in severe plastic deformation can be realised even at room temperature. Since the recrystallisation process (including dynamic) is thoroughly diffusion [22], it has also been assumed that the processes of diffusion and self-diffusion of the substitutional atoms, required for the formation of the recrystallisation nuclei and their subsequent growth, can be efficiently realised in iron, nickel, aluminium, titanium and other metals, and also in alloys based on these metals for which the MPD experiments were carried out. At first sight, the claim of this type seems to be incorrect. However, we will present a number of considerations confirming the accuracy of this hypothesis.

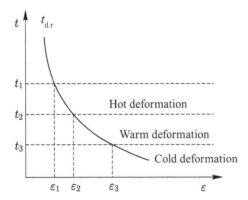

Fig. 4.8. Temperature dependence of the degree of deformation for the onset of dynamical crystallisation; $t_{d.r}$ – the temperature for the start of dynamic recrystallisation.

1. It is well-known [22] that the temperature for the onset of dynamic recrystallisation depends on the degree of deformation at the given temperature. As the degree of deformation increases, the temperature of deformation at which recrystallisation starts to take place decreases (Fig. 4.8). In the cold deformation region, this dependence is also valid and there are no physical restrictions on applying this dependence also to the range of room and similar temperatures. In this case, dynamical crystallisation should be accompanied by very high degrees of deformation and these strains (megaplastic deformation) are also the subject of our considerations.

2. Many investigators, studying the possibilities of the occurrence of specific diffusion processes in the megaplastic deformation conditions, have not taken into account the strong effect of internal stresses on the diffusion flows. At the same time, the term 'diffusion under stress' has been known for a long time [23]. Since the processes of plastic deformation are characterised by heterogeneities, a significant role is played by the gradients of elastic stresses, which are especially strong in megaplastic deformation. The resultant gradient of the chemical potential in accordance with the well-known second Onsager postulate [24] should result in the formation of diffusion flows. It is fully possible that the diffusion processes are accelerated even more as a result of the colossal supersaturation of the material, subjected to megaplastic deformation, by point defects [16]. The additional contribution to the acceleration of diffusion may also be associated with the dynamic capture of

the atoms by the ensembles of the individually and collectively moving dislocations and disclinations.

3. There is a large number of examples in which the acting stresses displace the physical processes to low temperatures. In section 3.3 of the previous chapter we examined in detail the mechanism of plastic deformation of the nanocrystals as a result of low-temperature grain boundary microsliding as a method of plastic flow of the nanocrystals at room temperatures. It is well-known that the grain boundary sliding is a process controlled by diffusion, and in the normal conditions takes place at high temperatures. Nevertheless, the experiments with computer modelling showed that sliding may also take place at the grain boundaries at room temperature in the conditions of high deformation stresses [25, 26].

As shown in the review in [27], after megaplastic deformation of pure copper the activation energy of a number of diffusion processes in this material is considerably lower than in the conventional material. For grain boundary diffusion it is 0.64–0.69 eV/atom, for Coble creep 0.72 eV/atom, for the grain growth process 0.7 eV/at. The general tendency for the large reduction of temperature at which the diffusion processes may take place in the conditions of very high plastic strains is clearly evident, as shown in the following, also in megaplastic deformation of amorphous alloys.

4. Figure 4.9 shows the electron microscopic images of the structure of pure iron subjected to treatment at the room temperature by the PTBC method (four complete rotations, $e = 5.6$). At the background of the matrix with a high defect density there are small (100–200 nm) regions completely free from the dislocations and representing in all likelihood recrystallisation nuclei. Evidently, the formation of these images is a very rare event because the process of megaplastic deformation was arrested when the recrystallisation nuclei only appeared and did not manage to grow and/or to get the dislocations as a result of the continuing processes of megaplastic deformation. The experiments show [28] that the transition from the pure metal to the solid solution based on this metal complicates dynamical crystallisation. The martensitic transformation, induced by deformation, shows the same behaviour. The martensitic transformation however leads to extensive dispersion of the structure and transfers the structure to the nanocrystalline state (Fig. 4.9c).

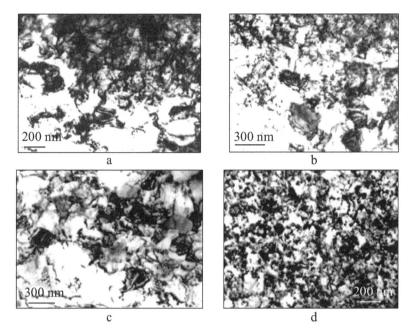

Fig. 4.9. Electron microscopic images of the early stages of dynamical recrystallisation in pure iron (a–c) and nanoparticles of strain martensite in the Fe–32% Ni alloy (d) after four complete rotations at room temperature by the PTBC method. The bright field.

The mechanisms of formation of the recrystallisation nucleus and of its subsequent growth in the processes of megaplastic deformation may coincide with the mechanisms already known for the 'conventional' dynamical recrystallisation at high temperatures, but may prove to be completely different and possible only at low temperatures in the megaplastic deformation conditions.

5. In [18] the authors convincingly demonstrated the phenomenon of dynamical recrystallisation in the processes of megaplastic deformation of pure copper at room temperature. Although the authors created the conditions for megaplastic deformation by a very original method (by local deformation in the zone of the growing crack), this nevertheless does not reduce the importance of this result, also taking into account the fact that they also discovered the phenomenon of secondary fragmentation.

Thus, the results which show that the process of megaplastic deformation of pure metals (Fe, Al, Cu, etc.) and of solid solutions based on these metals is accompanied by the process of dynamical recrystallisation can be regarded as fully confirmed by the experiments and theoretical considerations. This process is the powerful additional channel of the suppression of elastic energy introduced into the solid in the processes of megaplastic deformation.

4.1.5. The principle of cyclicity in megaplastic deformation

The classic assumptions of plastic deformation are based on the fact that the increase of the degree of deformation is accompanied by the buildup of dislocation defects. As the degree of plastic deformation increases, the number of defects which should be found in the deformed crystals also increases. The first exception from this rule was found in applying high plastic strains with the active participation of the disclination modes: fragments had thin boundaries and were almost completely free from the dislocations. However, in transition to the region of megaplastic deformation, it was shown that cardinal structural rearrangement takes place as a result of the additional channels of the dissipation of elastic energy. The jump-like variation of the structure and properties in transition to the range of megaplastic deformation has also been reported by the authors of [14]. If we examine the specific microvolume of the deformed specimens, it appears that after dynamic recrystallisation or amorphisation the process of plastic deformation appears to start from the 'clean sheet' in the newly formed recrystallised grain or in the region of the amorphous phase. Subsequently, the investigated microvolume shows the buildup of defects under the effect of the deformation stresses and the process is repeated. A similar cyclic behaviour in megaplastic deformation was directly observed by the authors of [18].

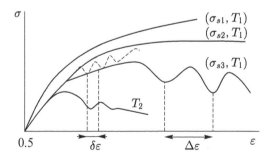

Fig. 4.10. Possible forms of the strain curves in the region of megaplastic deformation for materials with different Peierls barriers $\sigma_{s1} > \sigma_{s2} > \sigma_{s3}$ and at temperatures of $T_2 \gg T_1 = 300$ K.

Figure 4.10 shows the calculated curves of plastic flow taking into account the existence of the dislocation channel and obtained for the materials with different mobilities of the dislocations and at different temperatures [29]. It may be seen that at low values of the Peierls barrier σ_s, and in the presence of the effective dislocation channel, the plastic flow curve is cyclic with the wavelength $\Delta\varepsilon$. The fact that such curves are almost never recorded in the experiments does not contradict this conclusion. The process of plastic flow in all stages of its development is, as is well-known, highly heterogeneous and different areas of the deformed crystals are situated at different stages of their evolution.

The cyclic behaviour of the crystal–amorphous state transition was most distinctive in the mechanoactivation processes very similar to megaplastic deformation [30] (Fig. 4.11). In treatment of the powder of the intermetallic compound $Co_{75}Ti_{25}$ in a ball mill with the treatment time of up to 720 ks, x-ray diffraction examination revealed the cyclic phase transitions BCC-$Co_{75}Ti_{25}$ → amorphous

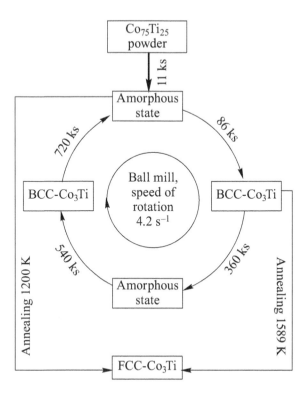

Fig. 4.11. Cyclic nature of the phase transitions amorphous state → BCC crystals in mechanoactivation of Co_3Ti alloy.

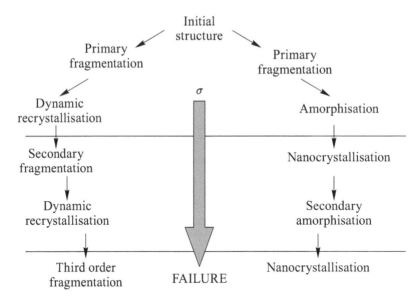

Fig. 4.12. General scheme of possible structural transformations in the process of megaplastic deformation.

state $\rightarrow$ BCC-$Co_{75}Ti_{25}$ $\rightarrow$ the amorphous state $\rightarrow$ BCC-$Co_{75}Ti_{25}$ $\rightarrow$ the amorphous state, and so on.

Figure 4.12 shows the principal scheme of the structural processes, generalising all the above assumptions, and demonstrating the cyclic principle in megaplastic deformation [8]. The failure process compensated by the uniformed compression stresses and is not considered. Two branches of the structural transformations in megaplastic deformation correspond to either dynamical recrystallisation or amorphisation of the alloys. Evidently, the scheme is simplified and does not take into account a number of additional conditions which may make the overall pattern more complicated. However, it is believed that the cyclicity principle is of fundamental importance in the examination of megaplastic deformation.

It is now possible to provide two exhaustive answers to all the questions formulated at the beginning of section 4.1 [31].

- Additional (in addition to plastic deformation) channels of elastic energy dissipation should operate efficiently in the process of megaplastic deformation. The structural changes in megaplastic deformation are characterised are cyclic.

- The specific route of structural rearrangement in megaplastic deformation is determined by a number of factors: temperature, the value of the Peierls barrier and their capacity for diffusion rearrangement, the difference of the free energies of the crystalline and amorphous state.

- The occurrence of megaplastic deformation does not guarantee at all the formation of the nanocrystalline state with the crystal size smaller than 100 nm, separated by the high angle or interphase boundaries. For example, this is almost completely impossible in pure metals with the high dislocation mobility. An important factor of the formation of the nanostructures in megaplastic deformation is the occurrence of the phase transformations of the martensitic and diffusion types, and also transition to the amorphous state. Stimulating the phase transformations by varying the temperature and chemical composition of the materials, it is possible to produce nanostructures of different types.

- The distinguishing feature of megaplastic deformation is the existence of additional effective channels of elastic energy dissipation. It is believed that there are four such channels (if the processes of mechanical failure are disregarded): dynamical recrystallisation, disclination rearrangement, phase transformations (including the transition to the amorphous state) in the generation of the latent heat of deformation origin. Conventional (macroplastic) deformation results in the buildup of elastic energy, and powerful dissipation processes start to operate only in the stage of megaplastic deformation.

- It is possible to determine quite accurately the boundary deformation region in which macroplastic deformation changes to megaplastic. Figure 4.13 shows the slightly modernised dependence of the temperature of the start of dynamic recrystallisation on the strain at the given temperature, shown in Fig. 4.8 [8]. It is assumed that deformation is carried out at

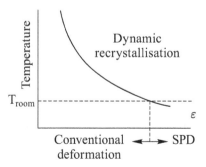

Fig. 4.13. The graph explaining the location of the boundary region of deformation separating macroscopic and megaplastic deformation.

room temperature (in principle, the temperature can be of any value, satisfying the ratio $T_d/T_m < 0.4$). At strains lower than the boundary value ε_b, dynamical recrystallisation does not take place and we are in the region of macrodeformation. At $\varepsilon > \varepsilon_b$, the process of plastic deformation starts to include dynamic recrystallisation and we transfer to the range of megaplastic deformation. Thus, the boundary of realisation of megaplastic deformation in the case of operation of one of the powerful dissipation channels is strictly determined. In the case of the second channel (amorphisation), an indication of the transition to the range of megaplastic deformation can be the appearance of microregions of the amorphous phase in the structure. If both previously mentioned dissipation channels operate at the same time (this is a very rare case), the boundary value of deformation corresponds to the channel with the lower efficiency.

We will now attempt the exact definition of megaplastic deformation.

Megaplastic (severe) deformation is a processes of plastic flow at temperature $T_d < 0.4\ T_m$, satisfying two conditions:

1. The stress state of the deformed solid contains a significant component of uniform compression stresses, preventing mechanical failure.
2. The value of the plastic deformation so high that plastic flow is accompanied by cyclic processes of dynamical recrystallisation and/or amorphisation of the structure which take place at the same temperatures taking into account the effects of generation of latent heat.

To conclude this section, it is useful to make a number of comments [31].

1. In the framework of the considered model of megaplastic deformation it is not necessary to use the considerations of 'special' highly non-equilibrium grain boundaries. Although undoubtedly the boundaries, formed in dynamic recrystallisation are far from perfect, they should have the same properties as any other grain boundaries.
2. The deformation behaviour of the material in the megaplastic deformation conditions is in its nature very similar to the behaviour of the material in the superplastic state. A similar analogy may prove to be useful when explaining the nature of superplasticity.

3. Megaplastic deformation is a phenomenon which take place only in late stages of deformation; it can be realised by any scheme of the stress state (for example, conventional rolling) under the condition of formation of high hydrostatic stresses.

4.2. The phenomenon of nanocrystallisation in amorphous alloys, subjected to megaplastic deformation

The amorphisation process may be an additional channel of dissipation in intermetallics and other materials with low mobility of dislocations. The most typical example is the titanium nickelide in which the transition to the amorphous state was recorded after PTBC [11] and after cold rolling [32]. The transition to the amorphous state in megaplastic deformation is most marked in the alloys susceptible to amorphisation in superfast quenching from the melt. Evidently, the crystal, containing a very high concentration of the linear and point defects, is thermodynamically unstable with respect to the transition to the amorphous state, especially if the difference between the free energies of the crystalline and amorphous states is not large.

What will take place if the amorphous state produced, for example, by melt quenching or by some other method, is deformed in the megaplastic deformation conditions? Taking the above considerations into account, the amorphous state should remain amorphous. However, as shown in [33–35], in deformation by PTBC nanocrystallisation takes place: nanocrystals with the size of approximately 10–20 nm homogeneously or heterogeneously distributed in the amorphous matrix are observed.

4.2.1. Amorphous alloys of the metal–metalloid type

Structure and mechanical properties
The formation at room temperature of nanocrystals with the size of up to 20 nm, uniformly distributed throughout the entire volume of the amorphous matrix, is difficult to explain on the basis of the classic considerations regarding the thermally activated nature of the crystallisation processes. In [36] it was attempted to analyse in detail the special features of the structure and properties under the effect of megaplastic deformation on a number of amorphous alloys of the metal–metalloid type. In particular, $Ni_{44}Fe_{29}Co_{15}Si_2B_{10}$ alloy produced by melt quenching was investigated. However, in addition to this, in [36] the authors investigated the alloys $Fe_{74}Si_{13}B_9Nb_3Cu_1$ (Finemet),

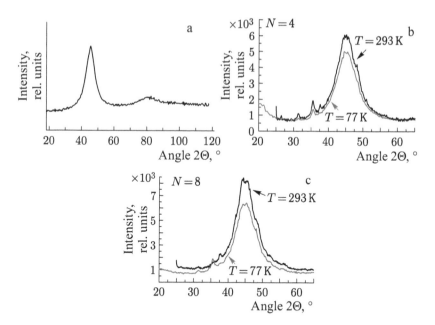

Fig. 4.14. X-ray diffraction diagrams of the $Ni_{44}Fe_{29}Co_{15}Si_2B_{10}$ amorphous alloy in the initial condition after melt quenching (a), after $N = 4$ at 293 K and 77 K (b) and after $N = 8$ at the same deformation temperatures (c).

$Fe_{57.5}Ni_{25}B_{17.5}$, $Fe_{49.5}Ni_{33}B_{17.5}$ and $Fe_{70}Cr_{15}B_{15}$. Figure 4.14 shows the x-ray diffraction diagrams of the Ni–Fe–Co–Si–B amorphous alloy in the initial condition (a), after $N = 4$ at 293 K and 77 K (b) and after $N = 8$ at the same deformation temperature (c) (N is the number of total revolutions in the Bridgman chamber). It may be seen that after megaplastic deformation, the processes of crystallisation started to take place in the alloy and were far more intensive after deformation at room temperature. Using a special computer program [6] it was possible to determine the volume fraction and the size of the crystal phase in the case of x-ray diffraction diagrams shown in Fig. 4.14. For example, for $N = 4$ and $T = 293$ K, the fraction of the crystal phase and the mean size of the crystals d was 8% and 3 nm, respectively. It is interesting to note that the values of the same parameters correspond almost completely to those obtained after $N = 8$, but at $T = 77$ K ($\alpha = 6\%$ and $r = 2$ nm). Electron microscopic studies (Fig. 4.15) confirmed these results both qualitatively and quantitatively.

Figure 4.16 shows the variation of microhardness HV of the amorphous and partially crystallised alloy (preliminary annealing of the amorphous state at temperatures higher than T_{cr} in relation to the deformation N in the Bridgman chamber at different temperatures).

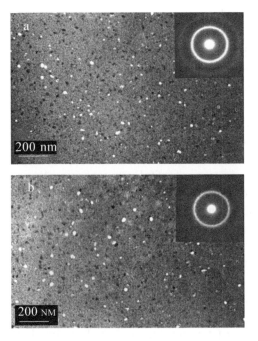

Fig. 4.15. Dark field images of the $Ni_{44}Fe_{29}Co_{15}Si_2B_{10}$ amorphous alloy after $N = 4$ at 293 K (a) and after $N = 8$ at 77 K (b).

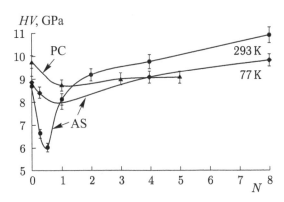

Fig. 4.16. Variation of the microhardness HV of the amorphous (AS) and partially crystalline (PC) $Ni_{44}Fe_{29}Co_{15}Si_2B_{10}$ alloy in relation to the value of N after megaplastic deformation at a temperature of 293 and 77 K.

Attention should be given to the large difference in the form of the curves in the initially amorphous and initially partially crystalline states. In the former case, the HV values greatly decrease in the initial stages of megaplastic deformation ($N = 0.5$) and subsequently monotonically increase, and both the decrease effect and the effect of subsequent increase are far more pronounced at $T = 293$ K (-2.8 and $+2.1$ GPa at 293 K and -0.8 GPa and $+1.1$ GPa at 77 K). As regards the initial partially crystalline alloy, its *HV* values were 0.75 GPa higher than in the initial amorphous alloy, but after megaplastic deformation they initially rapidly and then slowly decreased without any extreme manifestations to the value which was almost identical with the *HV* for the amorphous non-deformed state.

In the region of the rapid decrease of *HV* inhomogeneous plastic deformation with the formation of coarse shear bands took place; this is typical of all amorphous alloys at temperatures considerably lower than the point of transition to the crystalline state [37]. The local shear bands were observed by the method of transmission electron microscopy in the deformed specimens as a result of the fact that they showed the crystallisation effects (Fig. 4.17). Otherwise, the contrast on the electron microscopic image could be only of the absorption nature; initially, it is necessary to prepare the specimen for electron microscopic studies and then deform it [37] (section 3.4.6). The theoretical estimates show that the local increase of temperature in the shear bands may reach 500°C [9]. In this case, the local temperature in the zone of plastic shearing zone can exceed the crystallisation temperature of the amorphous alloy (in the present case 420°C) and lead to the formation of the primary FCC phase in the shear bands.

In later stages of megaplastic deformation ($N \geqslant 1.0$), the deformation pattern drastically changes. The shear bands are no longer observed. Instead of this, there are nanoparticles of the crystalline phase with the size of up to 10 nm homogeneously

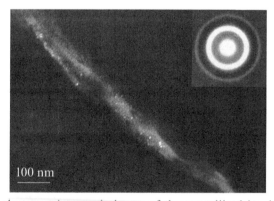

100 nm

Fig. 4.17. The electron microscopic image of the crystallised local shear band in the $Ni_{44}Fe_{29}Co_{15}Si_2B_{10}$ amorphous alloy after megaplastic deformation ($N = 0.5$; $T = 293$ K).

distributed throughout the entire volume of the specimen (Fig. 4.15). It may be concluded that the process of plastic deformation of the amorphous alloy ceased to be strongly localised, inhomogeneous and, in all likelihood, transforms into a 'quasi-homogeneous' process. This nature of the plastic flow is typical of the amorphous alloys at very high temperatures, close to the glass transition point in the conditions of the large reduction of the dynamic viscosity of the metallic glass [37]. In this case, it is almost impossible to obtain the similar 'softened' state at room temperature and even more so at 77 K. Obviously, here we are concerned with the manifestation of the completely new structural mechanism of plastic deformation of the amorphous alloys which operates only in the megaplastic deformation conditions.

One of the possible explanations of this course of the events is as follows (Fig. 4.18). With the propagation of the shear band in the amorphous matrix in the processes of megaplastic deformation, the temperature of the band constantly increases and the temperature at the front of the shear band is always maximum. In this phase of propagation of the band the local temperature of the front reaches the crystallisation temperature (II in Fig. 4.18), and a nanocrystal forms at the front of the growing band and greatly inhibits the movement of the zone of plastic flow because the resultant crystal is nanosized and not capable of dislocation plastic flow. Two variants of the further course of the events are possible. Firstly, under the effect of the shear band the nanocrystal will accumulate a high level of elastic stresses and, consequently, a new shear band will form by the elastic accommodation mechanism in the amorphous matrix (III in Fig. 4.18). In this case, the process of plastic flow will take place by the relay mechanism generating the formation of nanocrystals in the shear band equidistantly distributed along the

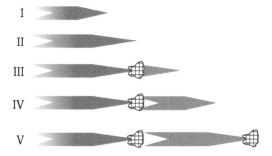

Fig. 4.18. Mechanism of 'self-blocking' of a shear band propagating in the amorphous matrix.

trajectory of movement of the shear band in the amorphous matrix. This is confirmed by the electron microscopic image in Fig. 4.19 which actually shows the chains of the equidistantly distributed nanocrystals formed as a result of megaplastic deformation. Secondly, the process of branching of the shear bands, inhibited as a result of the frontal formation of the nanocrystals, may also operate. This process appears to resemble the multiplication of dislocations in the non-intersected particles shown schematically in Fig. 4.20. As a result of such a 'self-retardation' of the shear bands on the frontally located nanocrystals delocalisation of the inhomogeneous plastic flow takes place in later stages of megaplastic deformation. The observed effect of transition to homogeneous nanocrystallisation on the shear bands indicates that the plastic flow is characterised by the high volume density of the shear bands and, consequently, by the homogeneous precipitation of nanocrystals in the 'thinner' shear bands.

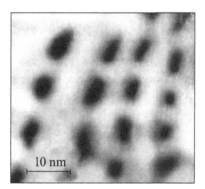

Fig. 4.19. Chains of equidistant nanocrystals formed during megaplastic deformation of the Fe–Ni–B alloy alloy. Transmission electron microscopy.

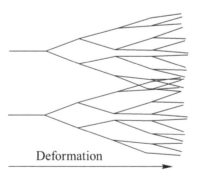

Fig. 4.20. The mechanism of multiplication of the shear bands, interacting with the frontal nanocrystals.

Thus, nanocrystallisation is the consequence of the local generation of heat as a result of the plastic deformation processes. The thermally activated nature of nanocrystallisation and, consequently, the local increase of temperature, resulting from this process, is also indicated by the fact that the structural state and the value of microhardness after $N = 4$ at room temperature accurately correspond to the structural state and the microhardness values after $N = 8$ in deformation at 77 K [36]. In other words, the higher values of deformation compensate the deficit of temperature in the diffusion processes of nanocrystallisation.

Investigation of the megaplastic deformation of the partially crystallised alloy resulted in unexpected results (Fig. 4.21). The x-ray diffraction patterns show clearly that the partially crystalline state, formed after annealing of the amorphous alloy, again becomes amorphous when the deformation is increased to $N = 5$. The electron microscopic experiments unambiguously confirm this tendency: the size of the nanocrystal rapidly decreases while the volume density of the nanocrystals remains unchanged. Such unusual evolution of the amorphous–nanocrystalline structure with the increase of N is clearly indicated by the dependence of the mean size of the nanoparticles and of their volume density with the increase of N in the partially crystallised alloy (Fig. 4.22). At the almost constant number of the nanoparticles in the unit volume, the disappearance of the crystalline phase takes place as a result of a large decrease of the size of the nanoparticles. In other words, this takes place as a result of the 'dissolution' of these nanoparticles in the amorphous

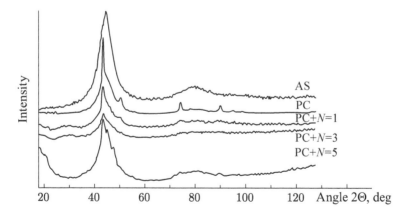

Fig. 4.21. Evolution of the x-diffraction patterns of the partially crystallised (PC) state of the $Ni_{44}Fe_{29}Co_{15}Si_2B_{10}$ alloy after megaplastic deformation with different values of N ($T = 293$ K).

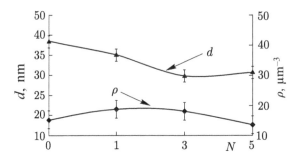

Fig. 4.22. Dependence of the mean size (d) and volume density (ρ) of the nanoparticles of the crystalline phase in the initial partially crystallised alloy $Ni_{44}Fe_{29}Co_{15}Si_2B_{10}$ on the value of N in megaplastic deformation ($T = 293$ K).

matrix. The effect of 'dissolution' of the nanoparticles in megaplastic deformation can be visualised even more accurately by comparing the histograms of the distribution of the nanoparticles of the crystalline phase produced after megaplastic deformation in different conditions (Fig. 4.23). It can be seen that with increase of N each subsequent histogram loses its 'tail' corresponding to the largest nanoparticles (crosshatched in the histograms shown in Fig. 4.23).

Thus, these are, at first sight, clearly contradicting results. On the one hand, the megaplastic deformation of the $Ni_{44}Fe_{29}Co_{15}Si_2B_{10}$ amorphous alloy results in partial transition of the alloy to the crystalline (more accurately, nanocrystalline) state. On the other hand, megaplastic deformation of the same partially crystallised alloy results in the dissolution of the crystalline phase, i.e. there is a tendency to return to the initial amorphous state. This contradiction is apparent and can be logically explained taking into account the specific features of the structural processes taking place during megaplastic deformation.

A large amount of elastic energy is supplied to the solid in the process of megaplastic deformation. The possible channels of dissipation in this case include plastic deformation, phase transformations and generation of heat. Crystallisation may be caused by both the local increase of temperature and the presence of high local stresses in the amorphous matrix. The stresses stimulate the processes which depends on temperature, and as the stress increases, the temperature at which the thermally activated crystallisation process takes place decreases. In addition to this, it is also important to take into account the fact that the activation energy of the crystallisation process Q^* is lower than the conventional temperature as a result of the considerably higher concentration in

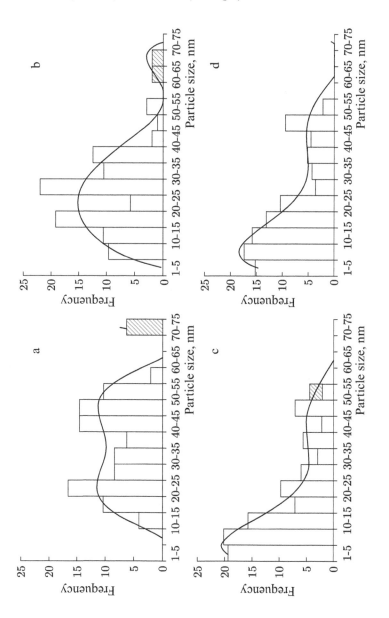

Fig. 4.23. Histograms of the size distribution of the nanocrystals, produced in different stages of megaplastic deformation of the partially crystallised $Ni_{44}Fe_{29}Co_{15}Si_2B_{10}$ alloy; $N = 0$ (a), 1 (b), 3 (c), 5 (d); the crosshatched dimensional fractions, which disappear with increasing deformation.

the shear bands of the regions of the excess free volume [38]. Finally, another important factor should be considered: the local atomic structure of the amorphous matrix in the shear bands can differ from the' classic' structure for the amorphous state. It is fully possible that the amorphous matrix contains (even prior to crystallisation) the deformation-induced regions with a higher correlation in the distribution of the atoms – the nuclei of the crystalline phase with a completely different degree of composition and topological short-range order. This is indirectly indicated by the results published in [39] in which it is shown that the chemical composition of the crystalline phase in the amorphous alloy based on aluminium after conventional annealing and after megaplastic deformation greatly differs.

Thus,

$$Q^* = Q_c - G\tau - \Delta Q_{fv} - \Delta Q_{so} \tag{4.3}$$

where Q^* is the effective activation energy of crystallisation in the shear band, Q_c is the activation energy of crystallisation as a result of thermal fluctuations, G is the shear modulus, τ is the shear stress in the region of the zone of plastic shearing, ΔQ_{so} is the contribution to the reduction of the activation energy of crystallisation associated with the presence of the short-range order (topological and/or composition) in the zone of the shear bands, and ΔQ_{fv} is the contribution to the reduction of the activation energy of crystallisation as a result of significant enrichment of the shear band with the excess free volume.

The formation of the crystals in the shear bands takes place during megaplastic deformation and not after completion of this deformation. Consequently, the shear bands, newly formed in the amorphous matrix, start to interact with the crystals formed previously in the earlier stages of megaplastic deformation. As discussed previously, a similar interaction may take place by several mechanisms (inhibition of the shear bands at the particles of the crystalline phase, intersection or bending of the shear bands around these particles, and also the primary and secondary accommodation effects). In any case, a similar interaction may cause the formation of dislocations in the crystalline particles. The highest dislocation density will form evidently in the vicinity of the interphase boundary where the strength of the effect of the shear bands is the greatest. Finally, a moment arises in which the dislocation density in the boundary zone is extremely high and the zone (or the entire crystalline particle) spontaneously transforms to the amorphous state because the free energy of the section of the highly defective crystal appears to be higher than the free energy of the amorphous state. In reality, this will be regarded as the 'dissolution' of the crystals in the amorphous matrix under the effect of shear bands actively acting in the amorphous matrix during megaplastic deformation. In particular, this process of 'dissolution' is detected in megaplastic deformation of the partially crystallised amorphous alloy (Fig. 4.22 and 4.23). It should be remembered that deformation 'dissolution' of the crystals can hardly continue to the end. It is well-known [40] that the very small crystalline particles (less than 10 nm) are not capable of accumulating the dislocation-type defects as a result of the presence of very high imaging forces. In the present case, this means that the nanocrystalline particles smaller than 10 nm, distributed in the amorphous matrix, will not 'dissolve' in the process of megaplastic deformation simply because of the fact that they will effectively 'push' the dislocations to the interphase boundary and will be without defects. In other words, the nanocrystals smaller than 10–20 nm, formed in the amorphous matrix, will be structurally stable and will remain in the material throughout the long stages of megaplastic deformation. This clearly explains the fact that in all investigations without exception in which the attention was given to the processes of formation of the crystals in megaplastic deformation, the size of the crystals was always smaller than 20 nm.

At the same time, as shown by the experiments, cases are possible in which the balance between the precipitation of the crystals at the shear bands and subsequent deformation 'dissolution' may be disrupted. In this case, the cyclicity principle starts to operate in megaplastic deformation (section 4.1.5) and the structural states with the large or small volume fraction of the nanocrystals in the amorphous matrix periodically replace each other with increasing deformation.

Figure 4.16 shows that in the stage preceding nanocrystallisation ($N = 0.5$) the value of hardness HV greatly decreases, especially in megaplastic deformation at 293 K. This means that the amorphous state in the initial stages of megaplastic deformation is structurally rearranged in such a manner that the processes of plastic shearing become easier. Preliminary annealing of the amorphous alloy and its partial crystallisation fully remove this effect (Fig. 4.16). The following structural model of the investigated effects can be proposed. In the process of application of hydrostatic pressure the regions of the free volume are redistributed (or coalesce) in such a manner that the application of a moderate shear stress improves the conditions for the processes of formation of shear bands inside which the concentration of the regions of the free volume should be considerably (by several orders of magnitude) higher than in the surrounding matrix. This is supported by the local atomic rearrangement (the variation of the topological and composition short-range order) as a result of the combined effect of hydrostatic and shear components of the stresses. This leads to the formation of nanoclusters (associates) with the dominance of the metallic and covalent nature of the interatomic interaction. Evidently, the process of plastic shearing in the former should be easier. Similar transformations in the amorphous matrix are confirmed by the fact that in particular at $N = 0.5$ there are large changes in the width and position of the maximum of the halo on the x-ray diffraction diagram. In addition to this, as shown later, the magnetic characteristics greatly change in this deformation range. Since similar rearrangements are to a certain extent of the thermally activated nature, the effect of this rearrangement in low-temperature deformation (77 K) is considerably weaker (Fig. 4.16).

Magnetic properties
The amorphous and nanocrystalline alloys based on iron and cobalt belong in the group of magnetically soft materials whose magnetic characteristics are considerably superior to those of the crystalline analogues [41]. The best example is the amorphous nanocrystalline Fe–Si–B–Nb–Cu (Finemet) alloy whose magnetic permeability at the normal and high magnetisation frequencies is considerably higher than that of the permalloy, Sendast, ferrites and other industrial crystalline magnetically soft materials [42] (section 3.4). At the same time, it is well-known that the 'weak link' of the magnetically soft amorphous nanocrystalline alloys is their low saturation magnetisation determined by the need for the presence in the composition of the high concentration (up to 20 at.%) of the magnetic atoms–metalloids (boron, phosphorus, silicon, etc).

Plastic deformation, especially megaplastic deformation, is a complicated process which leads not only to changes in the shape of the deformed solid but also to large changes in the structure and properties of the material [43]. In particular, deformation stimulates mass transfer and changes of the chemical composition both on the macroscale and microscale levels [44]. In turn, the redistribution of the components of the solid solution in the process of plastic deformation may change a number of physical properties of the materials and, in particular, their magnetic properties [45].

In [46] the effect of megaplastic deformation in the Bridgman chamber and of the structural changes taking place during this process on the magnetic properties of a number of amorphous alloys of the metal–metalloid type which are of considerable practical importance was studied in detail. Investigations were carried out on five amorphous alloys $Ni_{44}Fe_{29}Co_{15}Si_2B_{10}$ (alloy 1), $Fe_{74}Si_{13}B_9Nb_3Cu_1$ (Finemet alloy), $Fe_{57.5}Ni_{25}B_{17.5}$ (alloy 2A), $Fe_{49.5}Ni_{33}B_{17.5}$ (alloy 2B) an $Fe_{70}Cr_{15}B_{15}$ (alloys 3). The magnetic properties were determined with a vibration magnetometre in a direct magnetic field with the strength of up to 720 kA/m.

Figure 4.24 shows as an example of the typical magnetic hysteresis curves obtain for alloy 1 in the initial condition (after melt quenching) (a) and after treatment in the Bridgman chamber (b). The magnetic properties of alloy 1 were analysed, as in [36], after megaplastic deformation at 273 K and 77 K, and also in the partially crystalline state (annealing at 380°C, 1 h) after megaplastic deformation at 273 K. The dependence of I_s on the number of revolutions in the Bridgman chamber N for alloy 1 is shown in Fig. 4.25. The value of I_s rapidly increases (approximately 3 times) followed by saturation with the increase of N after megaplastic deformation at 77 K and then there is a sharp maximum of the values of I_s at $N = 1$ followed by a reduction to the initial value after megaplastic deformation at 293 K. It is characteristic that the maximum value of I_s after megaplastic deformation at 293 K and the limiting value of I_s with a small increase after megaplastic deformation at 77 K are approximately identical. The dependence I_s

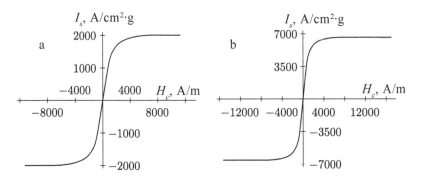

Fig. 4.24. Magnetic hysteresis loops, obtained for the alloy 1 in the initial condition (after melt quenching) (a) and treated in the Bridgman chamber ($N = 4$, $T = 293$ K) (b).

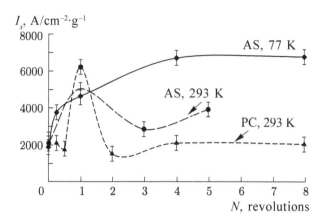

Fig. 4.25. Dependence of I_s on the number of revolutions in the Bridgman chamber N for alloy 1: AS – the initial amorphous state with subsequent megaplastic deformation at 77 K and 293 K; PC – the initial partially crystallised state followed by megaplastic deformation at 293 K.

(N) for the partially crystallised state of alloys 1 after megaplastic deformation at 293 K is in qualitative agreement with the one which corresponds to the initial amorphous state at the same deformation temperature but is most marked in both the growth stage and in the reduction stage. The maximum value of I_s in this case also corresponds to $N = 1$, but as regards the absolute value it is smaller by the extent by which the volume fraction of the amorphous phase in the state is smaller (the volume fraction of the crystalline and amorphous phases of the partial crystallisation were determined in [36]).

The dependence $H_c(N)$ for the alloy 1 is shown in Fig. 4.26. It may be seen that the form of the dependence for the initial state (amorphous or amorphous–crystalline) and for different temperatures of megaplastic deformation is qualitatively identical: initially the value H_c rapidly increases at $N = 0.5$ and this is followed by a gradual reduction to the values slightly higher than the initial values. As in the case of I_s, the value of the maximum for the partially crystalline state is considerably lower and also corresponds to $N = 1$.

The dependence $I_s(N)$ for the Finemet amorphous alloy after megaplastic deformation at 77 K and 273 K is shown in Fig. 4.27. Here, as in the alloy 1, magnetisation rapidly increases with increase of N but, firstly, the increase is no more than 40% and 30% and, secondly, the dependences show a sharp maximum at $N = 3$ and $N = 2$ for megaplastic deformation at 273 K and 77 K, respectively.

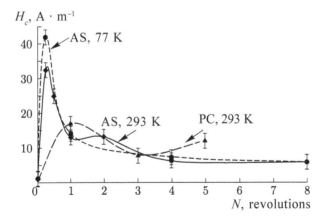

Fig. 4.26. Dependence of H_c on the number of revolutions in the Bridgman chamber N for the alloy 1; the symbols at the same as in Fig. 4.25.

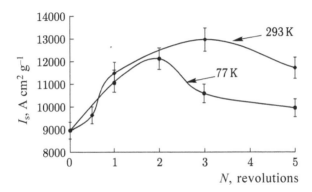

Fig. 4.27. Dependence of I_s on the number of revolutions N in the Bridgman chamber for the alloy 2; megaplastic deformation at 77 and 293 K.

Figure 4.28 shows the dependences $I_s(N)$ for the alloys 2A, 2B and 3 after megaplastic deformation at 273 K. The form of these dependences for different compositions greatly differs. If in alloy 2A the megaplastic deformation has almost no effect on magnetisation at all values of N, then in the alloy 2B the values rapidly increase in a 'jump' (by 120%) at $N = 1$. In alloy 3, on the other hand, at $N = 1$ the value I_s rapidly decreases (by 250%). The nature of variation of H_c with the increase of N in all investigated alloys was approximately the same and corresponded to the dependence as shown in Fig. 4.26 for the alloy 1.

The structural state of the alloys, corresponding to the maximum value of I_s after megaplastic deformation is characterised by the presence on the x-ray diffraction diagrams of the distinctive

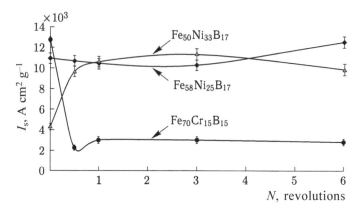

Fig. 4.28. Dependence of I_s on the number of revolutions N in the Bridgman chamber for the alloys 2A, 2B and 3; megaplastic deformation at 293 K.

asymmetry of the profile of the intensity of the main halo, corresponding to the amorphous state. Both the x-ray and electron microscopic studies do not show (with a small number of exceptions) the existence of the nanocrystalline phases in these states. It was shown (Fig. 4.29) that in the structural states, corresponding to the large increase of the saturation magnetisation, the asymmetry of the profile of the x-ray halos is determined in all likelihood by the superposition of two maxima, corresponding to different amorphous structures. In other words, the initially homogeneous and amorphous phase separates under the effect of megaplastic deformation into two amorphous phases with different chemical compositions and, possibly,

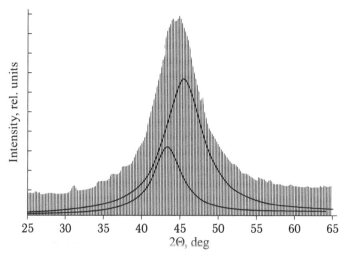

Fig. 4.29. Calculation of the profile of the x-ray line for the amorphous alloy 1, subjected to megaplastic deformation ($N = 4$, $T = 77$ K).

different structures. Since the phase separation of the matrix into two amorphous phases results in a large change (increase or decrease) of saturation magnetisation, it is rational to assume that the processes of megaplastic deformation is accompanied by the formation in the structure of regions enriched or, correspondingly, depleted in the ferromagnetic components.

An interesting relationship is found: in the alloy 1 which contains three ferromagnetic components (Fe, Ni and Co), the maximum effect of increase of I_s in megaplastic deformation treatment is 300%; in the alloy 2B with two ferromagnetic components (Fe and Ni) it is 120%, in the Finemet alloy containing one ferromagnetic component it is 40%, and finally, in the alloy 3 with one ferromagnetic (Fe) and one and antiferromagnetic component (Cr) the increase is replaced by the reduction of I_s reaching 250%. Figure 4.30 shows the dependence of the observed effect (ΔI_s) on the number of the ferromagnetic components in the alloy n. It was assumed that for the alloys $3n = l_F + l_A = 0$, where l_F and l_A are the numbers of respectively the ferromagnetic (Fe) and antiferromagnetic (Cr) components, they have the opposite sign. The dependence is almost linear although its explicit physical meaning should be explained by further investigations.

Comparison of the values of (ΔI_s) in the alloys 2A and 2B shows that one can propose the assumption on the local chemical composition of the amorphous phase, enriched with the ferromagnetic components in the Fe–Ni–B system. It can be seen (Fig. 4.30) that the positive 'jump' (ΔI_s) in the alloy 2B takes place after megaplastic deformation ($N \geqslant 1$) at the same value of saturation magnetisation as in the alloy 2A in the initial state. At the same, alloy 2A was completely indifferent to megaplastic deformation. Consequently, the alloys of this system containing two amorphous phases are advantageous from the energy viewpoint. One of these phases, rich in the metallic components, has the ratio of the ferromagnetic atoms Fe:Ni the same as in the alloy 1, namely 57.5:25 = 2.3.

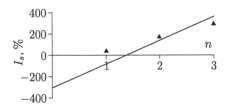

Fig. 4.30. Dependence of the effect of variation of saturation magnetisation ΔI_s on the number of ferromagnetic components n in the investigated amorphous alloys.

The prospects for practical application of these results will be discussed briefly. By megaplastic deformation in the optimum conditions it is possible to increase greatly the saturation magnetisation – 'the weak link' of the amorphous and nanocrystalline magnetically soft alloys – whilst retaining the very low values of coercive force. For example, for the Finemet alloy which is produced on the industrial scale (in Russia several tens of tonnes per annum) and used widely in electronics and instrument making, the resultant value of (ΔI_s) was up to 40% which consequently makes it possible to improve greatly the magnetic properties of this alloy after the heat treatment in the optimum conditions, including the treatment in the magnetic field. The innovation value of the magnetic alloys, subjected to megaplastic deformation, is very high. Being inferior in the individual magnetic parameters to the Fe–Co alloys (as regards the parameter I_s) and the cobalt-based amorphous alloys (parameter μ), the resultant alloys are superior to these alloys (including the standard Finemet alloy) by the set of these parameters. It is therefore justified to assume that the results of the study [46] can form the basis of formation of a new class of advanced magnetically soft materials.

4.2.2. Ti–Ni–Cu amorphous alloys

It has already been mentioned in section 3.2.9 that recently special attention has been paid to the alloys based on titanium nickelide characterised by the shape memory effect [47]. It has been shown that the TiNi intermetallic subjected to megaplastic deformation by shear under pressure in the Bridgman chamber or in cold rolling, can transfer partially or completely to the amorphous state [48]. Subsequently, this effect, typical of titanium nickelide, has been confirmed several times by other investigators [49, 50]. Figure 4.31 shows the linear dependence of the volume fraction of the amorphous phase, formed in different alloys in the vicinity of the TiNi composition on the degree of deformation in cold rolling [51].

At the same time, studies have been published in which the alloy based on titanium nickelide $Ti_{50}Ni_{25}Cu_{25}$ was produced by melt quenching in the amorphous state and subsequently subjected to megaplastic deformation in the Bridgman chamber. The transition from the amorphous to nanocrystalline structure was detected in a specific stage of deformation [34, 52]. Thus, on the one side, megaplastic deformation leads to the crystal $\rightarrow$ amorphous state

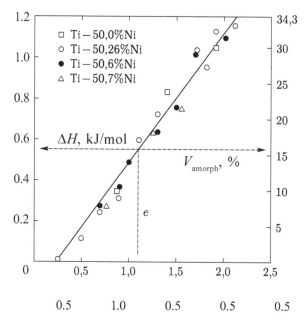

Fig. 4.31. Variation of the amount of the amorphous phase with the increase of the degree of deformation in cold rolling in Ti–Ni alloys (S.D. Prokoshkin, et al).

phase transition and, on the other hand, the amorphous state → crystal (nanocrystal) phase transition. This is evident contradiction which we discussed previously already for the amorphous alloys of the metal–metalloid type, was the reason why the authors of [53] carried out detailed and systematic studies of the structural–phase transformations in the alloys based on titanium nickelide under the effect of megaplastic deformation in the Bridgman chamber with the variation of the chemical composition, initial structure and also the temperature and degree of megaplastic deformation. The $Ti_{50}Ni_{25}Cu_{25}$ alloy was selected, and this alloy could be both in the crystalline and amorphous state prior to deformation in the Bridgman chamber. The amorphous state can be produced by vacuum melt quenching by the spinning method with a speed of 10^6–10^7 deg/s [54].

The results published in [53] show unambiguously that as a result of the megaplastic deformation of the $Ti_{50}Ni_{25}Cu_{25}$ alloy, produced by melt quenching, phase transformations of different type take place in this alloy (Fig. 4.32).

The application of hydrostatic pressure only without shear deformation already results in the formation of a small amount of the crystalline phase of the type B19 in the amorphous matrix. In

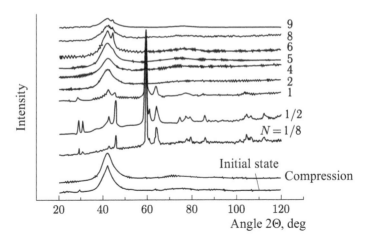

Fig. 4.32. Profiles of the x-ray spectrum of the initial amorphous alloy $Ti_{50}Ni_{25}Cu_{25}$ after hydrostatic compression ($P = 4$) without shear and after shear under pressure with different number of revolutions ($N = 1/8$, 1/2, 1, 2, 4, 5, 6, 8, 9).

the initial stage of megaplastic deformation ($N = 1/8$), the volume fraction of the crystalline phase (type B19 and B2) greatly increases (up to ~70%). The highest value of the volume fraction of the crystals (~80%) is observed after $N = 1/2$, and subsequently (after $N = 1$) there is a large reduction of the amount of the crystalline phase (~30%) (Fig. 4.33). This tendency to the reduction of the fraction of the crystalline phase is retained with increasing deformation, and after $N = 2$, the crystalline phase disappears almost completely from the structure (Fig. 4.34a). The structure corresponds to the x-ray amorphous spectrum shown in Fig. 4.35. The spectrum is typical of the amorphous state of the solid, and only the unusual nature of a number of electron diffraction micropatterns and dark field high-resolution electron microscopy indicates the presence of a small fraction of the nanocrystalline phase. Deformation at $N = 4$ results in complete 'dissolution' of the nanocrystals and in the formation of the amorphous state which differs only in certain details from the initial amorphous state, produced by melt quenching. A further increase of megaplastic deformation ($N = 5$) again leads to the formation of a small amount of the crystalline phase in the structure which can be detected only on the electron microscopic level (Fig. 4.34b). At an even larger increase of deformation ($N = 6$) the presence of the crystalline phase is confirmed by both x-ray diffraction examination and electron microscopy (Fig. 4.36). The theoretical analysis of the x-ray spectra with the split halo together

Fig. 4.33. Electron diffraction patterns and dark field images of the structure of the $Ti_{50}Ni_{25}Cu_{25}$ alloy at $N = 5$ in $(110)_{B2}$ reflection; zone axis $[001]_{B2}$ (a) and in $(100)_{B19}$ reflection; zone axis $[010]_{B19}$ (b).

Fig. 4.34. Electron diffraction patterns and dark field images of the structure of the $Ti_{50}Ni_{25}Cu_{25}$ alloy under the effect of the section of the main (first) diffraction halo, corresponding to megaplastic deformation at $N = 2$ (a) and $N = 5$ (b).

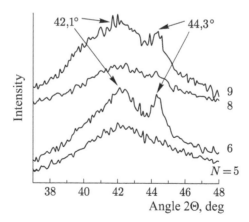

Fig. 4.35. Evolution of the main (first) halo of the x-ray amorphous spectra of the $Ti_{50}Ni_{25}Cu_{25}$ alloy for different degrees of megaplastic deformation. The number of revolutions in the Bridgman chamber is shown in the figure.

with the electron microscopic data indicate that here we are concerned
with the two-phase amorphous–crystalline structure. A further
increase of megaplastic deformation results in new 'dissolution' of
the crystalline phase ($N = 8$) and to its subsequent appearance ($N = 9$). All these results are clearly indicated by the graph in Fig. 4.37
which shows the dependence of the volume fraction of the crystalline
phase in the initially amorphous $Ti_{50}Ni_{25}Cu_{25}$ alloy on the number
of revolutions N in the Bridgman anvil. The results, shown in the
graph, were obtained by the methods of x-ray diffraction analysis
(a high volume fraction of the crystalline phase) and by transmission
electron microscopy (a small volume fraction).

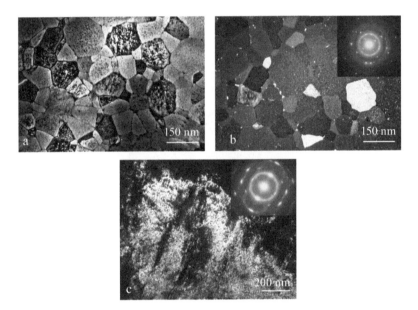

Fig. 4.36. Electron diffraction micropatterns and electron microscopic images of
the structure of the $Ti_{50}Ni_{25}Cu_{25}$ alloy, corresponding to megaplastic deformation at
$N = 6$; a, c) brightfield, b) dark field.

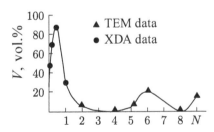

Fig. 4.37. Dependence of the volume fraction of the crystalline phase on the degree
of deformation of the $Ti_{50}Ni_{25}Cu_{25}$ amorphous alloy.

In the investigations in [56] it was evidently possible for the first time to detect the three cycles of mutual amorphous–crystalline phase transitions. Previously, the identical effect had been detected in experiments with mechanoactivation of the powder of the Co_3Ti intermetallic compound [12] (Fig. 4.11). A similar cyclic dependence, i.e. tendency for the phase transition to the crystalline state of the initially amorphous structure and subsequently the existence of the completely opposite tendency in the specific stages of megaplastic deformation explains in our view the apparent contradiction in the experimental results of the studies [48–50] on the one hand, and the studies [34–52] on the other hand.

It will be attempted to explain what determines the similar cyclic dependence. The 'initial amorphous state $\rightarrow$ crystal (nanocrystal)' phase transition is evidently caused by two reasons. Firstly, as mentioned previously, the applied hydrostatic pressure is capable of stimulating the phase transition in which the equilibrium phase has a smaller specific volume. The fact that the $Ti_{50}Ni_{25}Cu_{25}$ amorphous alloy is partially transferred to the crystalline state without shear deformation and also as a result of the application of hydrostatic pressure ($P = 4$ GPa) confirms this hypothesis. Secondly, the appearance of additional channels of dissipation of elastic energy, characteristic of the processes of megaplastic deformation [8], leads in the conditions of the highly localised plastic flow in the shear bands to the local generation of thermal energy and the corresponding local increase of the temperature of the amorphous matrix. In these conditions, the shear bands, as already mentioned, are capable of undergoing crystallisation.

After formation of the crystalline phase in the processes of megaplastic deformation, this phase is subjected to very high plastic strains. One of the actual energy dissipation channels in the conditions of low mobility of the dislocations may be amorphisation [20]. A similar phase transition (solid phase melting) becomes possible especially when the free energy of the crystal, containing the extremely high defect density (vacancies, dislocations, disclinations, fragment boundaries, etc.), becomes higher than the free energy of the disordered state of the system. The negative volume effect is evidently secondary and is not controlling. These transitions were detected in experiments especially in the intermetallics and, in particular, in titanium nickelide [48–50].

Further, with the development of megaplastic deformation, the situation is evidently repeated but with some special features associated with the continuously increasing elastic energy and high (gigantic) internal stresses. Evidently, the described tendency for cyclic transformations is common for the behaviour of metallic materials, susceptible to amorphisation, in the course of high-energy effects, in particular in megaplastic deformation realised by different methods.

It is interesting to mention that the amorphisation of the crystalline phase is not the only additional channel of dissipation of elastic energy in the processes of megaplastic deformation. As shown in [21] all these functions can be fulfilled by the processes of dynamical recrystallisation, if the mobility of the dislocations is sufficiently high. In the present case, the amorphisation process was dominant but in other cases evidently, in the conditions of the local increase of temperature, the consequences of dynamic recrystallisation were also found in the structure (Fig. 4.34a).

An important special feature of the processes taking place during megaplastic deformation is their nonuniformity typical of any type of plastic deformation [10]. For this reason, the cyclic amorphous –crystalline phase transitions are 'mismatched' in the volume of the deformed material and take place in different microvolumes at different deformation parameters. This can be used to explain the results which shows that it is almost impossible to record the states with the limiting content of either the amorphous or crystalline phase. In any stage of megaplastic deformation there are local areas of the matrix which either 'outstrip or 'lag behind' the adjacent regions. As the deformation increases, the strength of this mismatch effect becomes greater. Finally, we can reach the specific dynamic equilibrium in which it will not be possible to record any changes in the phase composition within the framework of the given averaging scale.

It is of considerable interest to analyse the special features of deformation amorphisation and crystallisation of the same material with the variation of its initial state. This would make it possible to produce a structural model of the cyclic phase transformations in the process of megaplastic deformation.

The structural-phase transformations with the variation of the intensity of megaplastic deformation using the Bridgman chamber were carried out on $Ti_{50}Ni_{25}Cu_{25}$ alloy which in contrast to the investigations published in [55] was in the crystalline (not amorphous) state prior to the start of deformation experiments.

The complete x-ray spectra, corresponding to all the investigated state, are shown in Fig. 4.38. The main results of these investigations can be summarised as follows:

1. The investigated $Ti_{50}Ni_{25}Cu_{25}$ alloy has the initial crystalline structure represented mainly by the lath martensite B19. In the process of megaplastic deformation, the laths rotate, break up, are refined and, finally, completely disappear (Fig. 4.39).

2. The alloy starts to amorphise already after $N = 0.25$ (Fig. 4.38). After deformation $N = 1$ ($e = 2.15$), the mass degradation of the plate structure in the transition to the amorphous state becomes visible.

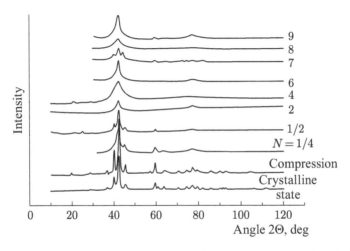

Fig. 4.38. Complete profiles of the x-ray diffraction patterns of the $Ti_{50}Ni_{25}Cu_{25}$ alloy in different stages of the experiment.

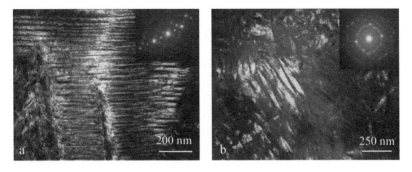

Fig. 4.39. Electron microscopic images of the structure (dark field) of the martensitic structure of the B19 phase and the appropriate electron diffraction micropatterns prior to (a) and after megaplastic deformation ($N = 1$) (b).

3. Degradation of the martensite laths is accompanied by the plastic deformation of the resultant amorphous phase as a result of which, starting at $N = 0.5$ ($e = 1.80$), electron microscopy shows the formation of the nanocrystalline B2-phase with the size of the individual particles up to 10 nm (Fig. 4.40). In addition to this, sometimes there are globular regions of the B2 phase with the size of approximately 300 nm containing a large amount of deformation origin defects.

4. Further deformation $n = 2$–4 ($e = 2.51$–2.90) is characterised structurally by the superposition of the amorphous phase in the nanocrystals of the B2 phase frequently formed in the shear bands of the amorphous matrix (Fig. 4.41).

5. In later stages of deformation after $N = 6$ ($e = 3.5$) electron microscopic studies show the state of local instability of the B2 phase which is an intermediate state of the B2 $\rightarrow$ B19 martensitic transformation (Fig. 4.41).

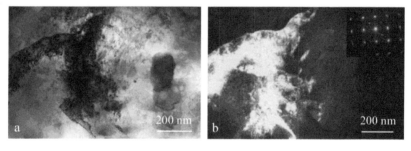

Fig. 4.40. Electron microscopic images (a, b) showing the presence of the B2 phase after megaplastic deformation ($N = 0.5$); a and b are respectively the bright and dark field images in the reflection of the B2 phase.

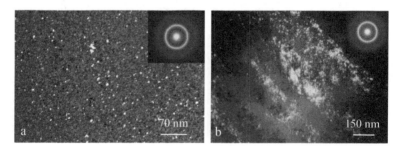

Fig. 4.41. Electron microscopic images (dark field) of the nanocrystals of the B2 phase, separated uniformly in the volume (a) and in the shear bands (b) after megaplastic deformation ($N = 2$).

6. The x-ray spectra of the later stages of deformation after $N = 7$ ($e = 4.0$) are already characterised by the presence of the wide maxima of the B19 phase on the background of the x-ray amorphous state.

7. The x-ray spectra in the final stages of deformation ($N = 9$, $e = 5.3$) are again completely x-ray amorphous, and electron microscopic examination shows the presence of the crystals of the B2 phase (Fig. 4.38 and 4.42).

Thus, as in the case of the initial amorphous alloy $Ti_{50}Ni_{25}Cu_{25}$ [53], the crystalline $Ti_{50}Ni_{25}Cu_{25}$ alloy shows the following cyclic sequence of the phase transition with the increase of the intensity of megaplastic deformation in the Bridgman chamber:

$$B19 \rightarrow AS \rightarrow B2 \rightarrow B19 \rightarrow AS \rightarrow B2,$$

where AS is the amorphous state of the solid.

In the literature, it is assumed that the periodicity of the structural changes in megaplastic deformation in a general case is determined by the activation of different dissipation channels (relaxation) of elastic energy, stored by the material during deformation [8]. Evidently, the special features of the variation of the structure in the $Ti_{50}Ni_{25}Cu_{25}$ crystalline alloy, observed in [55] in the processes

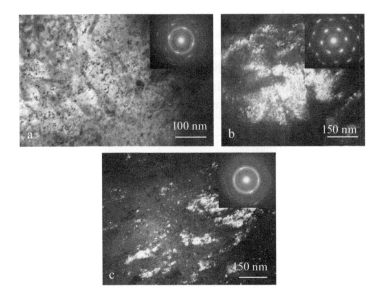

Fig. 4.42. Electron microscopic images (bright field (a) and dark field (b, c) under the effect of reflections from the crystalline phases) and the appropriate microdiffraction patterns in different stages of megaplastic deformation; $N = 4$ (a), 6 (b) and 9 (c).

of megaplastic deformation are associated with the special features of the course of the direct and reversed phase transformations of both the diffusion and martensitic types. Figure 4.43 shows the diagram providing information on the nature of cyclic transitions in megaplastic deformation from the crystalline to amorphous state and further from the amorphous to nanocrystalline state, with subsequent periodic repetition of the processes, but already on the nanoscale level.

We will discuss the question as to why the process of secondary crystallisation of the amorphous phase during megaplastic deformation is accompanied by the formation of initially the B2 phase instead of the B19 phase equilibrium of the room temperature. The process of crystallisation of the amorphous state in heating in shear bands takes place evidently by the diffusion mechanism at temperatures of 500–510°C where the crystalline phase of the B2 type is the equilibrium phase [47]. In subsequent cooling to room temperature in the range 50°C the thermoelastic martensitic transformation with the formation of the B19 phase takes place [47]. In the present case, the martensitic transformation is suppressed in certain stages of megaplastic deformation, and the B2 phase is stable at room temperature. In [56] a dimensional effect was detected in the B2 $\rightarrow$ B19 thermoelastic transformation in the $Ti_{50}Ni_{25}Cu_{25}$ alloy (section 3.4). The nanoparticles smaller than 20 nm did not undergo transformation in cooling to room temperature and had the structure of the high-temperature B2 phase. In the present case, the nanocrystals, formed in the shear bands or by a different mechanism and having the size smaller than 10 nm were stable in specific stages of megaplastic deformation because of the small (nanosize) size of

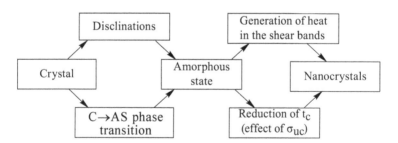

Fig. 4.43. Diagram of the processes leading to the transition from the crystalline to amorphous state and further from the amorphous state to crystalline during megaplastic deformation; C — crystal, AS — amorphous state, t_c — the temperature of transition from the amorphous to crystalline state, σ_{uc} — the stress of uniform (hydrostatic) compression.

the particles of the initial B2 phase. In subsequent stages, the B2 → B19 shear transformation takes place in later stages of megaplastic deformation, evidently as a result of the high acting stresses or growth of the particles of the B2 phase. The latter fact is confirmed by studies of the microdiffraction patterns (Fig. 4.42c), corresponding to the pre-martensitic state of the B2 phase after $N = 6$ ($e = 3.5$) and by the appearance of wide lines of B19 on the x-ray spectrum with a further increase of deformation to $N = 7$ ($e = 4.0$).

We believe that the most interesting aspect is the explanation of the structural mechanism of the phase transition in the process of megaplastic deformation from the crystalline to amorphous state. In [55] the rotations, distortion and fragmentation of the initially regularly distributed particles of the B19 phase in the process of megaplastic deformation were detected in the experiments (Fig. 4.39). However, the final act of amorphisation – 'the dissolution' of the nanosized 'fragments' of the martensitic laths – remains speculative. Evidently, one of the methods of verifying this assumption is the experimental examination of the dissolution of the nanosized crystals in deformation or computer modelling of the process of solid phase solution in the course of shear deformation in the conditions of uniform compression.

Thus, we can conclude that both the deformation of intermetallics and complex phases, susceptible to amorphisation during megaplastic deformation, and the deformation of the amorphous alloys in the processes of megaplastic deformation are accompanied by successive transition from the amorphous to crystalline state and, vice versa, from the crystalline to amorphous state. This leads to the formation of a stable amorphous–nanocrystalline structure undergoing quantitative changes with a further increase of deformation.

References

1.   Fortov V.E., Extreme states of matter, Moscow, Fizmatliz, 2009.
2.   Dobatkin S.V., Lyakishev N.P., Prospects and use of nanostructured steels , Proc. of the Second All-Russian Conference on Nanomaterials, Novosibirsk, IKhTTM SD RAS, 2007, 35–36.
3.   Segal V.M., et al., Processes plastic structuring of metals, Minsk: Science and Technology, 1994.
4.   Valiev R.Z., Alexandrov I.V., Nanostructured materials produced by severe plastic deformation, Moscow, Logos, 2000.
5.   Golovin Y.I., Universal principles of science, Tambov, Publ TGU, 2002.
6.   Valiev R.Z., et al., V. 58, No. 4, 33–39.
7.   The Large English–Russian dictionary, Moscow, Russkii Yazyk, 1988, V. 2, 427.

8.   Glezer A.M., Izv. RAN, Ser. fiz., 2007, V. 71, No. 12. 1764–1772.
9.   Shtremel' M.A., Strength of alloys, Part 2, Moscow, MISiS, 1997.
10.   Beloshenko V.A., Varyukhin V.N., Spuskanyuck V.Z., Theory and practice of hydro-extrusion, Kiev, Naukova Dumka, 2007.
11.   Tatyanin E.V., Kurdyumov V.G., Fedorov V.B., Fiz. Met. Metalloved., 1986, V. 62, No. 1, 133–137.
12.   Smirnova N.A., et al., Fiz. Met. Metalloved., 1986, V. 61, No. 4. 1170–1178.
13.   Rybin V.V., Severe plastic deformation and fracture of metals, Moscow, Metallurgiya, 1986.
14.   Firstov S.A., et al., Izv. VUZ, Fizika, 2003, No. 3, 41–48.
15.   Konev N.A., Kozlov E.V., Physical nature of the stages of plastic deformation, in: Structural levels of plastic deformation and failure, Ed. V.E. Panin, Novosibirsk, Nauka, 1990, 123–186.
16.   Gapontsev V.L., Kondrat'ev V.V., Dokl. RAN, 2002, V. 385, No. 5, 684–687.
17.   Valiev R.Z., Ross. nanotekhnologii, 2006, V. 1, No. 1–2, 208–216.
18.   Bykov V.M., et al., Fiz. Met. Metalloved., 1978, V. 45, No. 1, 163–169.
19.   Andrievsky R.A., Glezer A.M., Fiz. Met. Metalloved., 2000, V. 89, No. 1, 91–112.
20.   Glezer A.M., Izv. VUZ Fizika, 2008 V. 51, No. 5, 36–46.
21.   Pozdnyakov V.A., Glezer A.M., Izv. RAN, Ser. fiz., 2004, V. 68, No. 10, 1449–1455.
22.   Gorelik S.S., Dobatkin S.V., Kaputkina L.M., Recrystallization of metals and alloys, Moscow, MISiS, 2005, 431 p.
23.   Umansky Y.S., Finkelstein B.N., Blanter M.E., Physical basis of metals science, Moscow, Metallurgiya, 1949.
24.   Bockshtein B.S., Diffusion in Metals, Moscow, Metallurgiya, 1978.
25.   Pozdnyakov V.A., Glezer A.M., Fiz. Tverd. Tela, 2002, No. 4, 705–710.
26.   Glezer A.M., Deform. Razrush. Mater., 2005, No. 2, 10–15.
27.   Kozlov E.V., Konev N.A., Zhdanov A.N., Fiz. Mezomekh., 2004. V. 7, No. 4, 93–113.
28.   Blinov E.N., Glezer A.M., Materialovedenie, 2005, No. 5, 32–39.
29.   Glezer A.M., Pozdnyakov V.A., Deform. Razrush. Mater., 2005, No. 4, 9–15.
30.   Sherif El-Eskandarany M., Aoki K., Sumiyama K., Suzuki K., Acta Met., 2002, V. 50, 1113–1123.
31.   Glezer A.M., Metlov L.S., Physics of Solid State, 2010, V. 52, No. 6, 1162–1169.
32.   Brailovski V., et al., Materials Transaction, 2006, V. 47, No. 3, 795–804.
33.   Chen H., He Y., Shiflet G.J., Poon S.J., Lett. Nature, 1994. V. 367, No. 2, 541–543.
34.   Gunderov D.V., et al., Deform. Razrush. Mater., 2006, No. 4, 22–25.
35.   Glezer A.M., et al., Mater. Sci. Forum, 2008, V. 584–586, 227–230.
36.   Glezer, A.M., et al., Izv. RAN, Ser. fiz., 2009, V. 73, No. 9, 1302–1309.
37.   Glezer, A.M., et al., Mechanical behavior of amorphous alloys, Novokuznetsk SGIU, 2006.
38.   Kovneristy Y.K., et al., Deform. Razrush. Mater., 2008, No. 1, 35–41.
39.   Donovan E., Stobbs W.M., Acta Met., 1981, V. 29, No. 6, 1419–1424.
40.   Gryaznov V.G., et al., Pis'ma Zh. Teor. Fiz., 1989., V. 15, No. 2, 1256–1261.
41.   McHenry M. E., Willard M.A., Laughlin D.E., Prog. in Mater. Sci., 1999, V. 44, 291–433.
42.   Yoshizawa Y., Oguma S., Yamauchi K., J. Appl. Phys., 1989, V. 64, 6044–6051.
43.   Zhilyaev A.P, Langdon T.G., Prog. Mater. Sci., 2008, V. 53, 893–979.
44.   Derjagin A.I., et al., Fiz. Met. Metalloved., 2008, V. 106, No. 3, 301–311.
45.   Inoue A., Acta Mater., 2000, V. 48, 279–286.

46. Glezer, A.M., et al., Izv. RAN, Ser. Fiz., 2009, V. 73, No. 9, 1310–1314.
47. Pushin V.G., et al., Shape memory titanium nickelide alloys, Part 1. The structure, phase transformations and properties, Yekaterinburg, IMP UB RAS, 2006.
48. Tat'yanin E.V., et al., Fiz. Tverd. Tela, 1997, V. 39, 1237–1243.
49. Prokoshkin S.D., et al., Acta Mater., 2005, V. 53, 2703–2714.
50. Zel'dovich V.I., et al., Fiz. Met. Metalloved., 2005, V. 99, 90–98.
51. Prokoshkin S.D., et al., Fiz. Met. Metalloved., 2010, V. 110, No. 3, 305–320.
52. Gunderev D.V., Electronic scientific Journal 'Issledovano v Rossii', 2006. 151.pdf.
53. Nosova G.I., et al.. Crystallography, 2009, V. 54, No. 6, 1111–1119.
54. Pushin V.G., et al., Fiz. Met. Metalloved., 1997, V. 83, No. 6, 149–156.
55. Glezer A.M., et al., Izv. RAN, Ser. fiz., 2010, V. 74, No. 11, 1576–1582.
56. Glezer A.M., et al., J. Nanoparticle Research, 2003, V. 5, 551–560.

5

Application of melt-quenched nanomaterials

Any scientific developments will sooner or later be the subject of practical application. This is why in the final section of this book we discuss briefly a number of distinctive examples of the application of nanocrystals, produced by different melt quenching methods. We will discuss both the existing and future developments.

Recently, the methods of producing the rapidly quenched ribbons have been regarded as the most efficient and fastest method of producing products in the form of sheets, strips and wires, bypassing the tranditional stages of metallurgical processing. The advances in the investigation of the structure and physical–mechanical properties of the nanocrystalline metallic alloys determine the areas of application in technology and medicine. It is now evident that as a result of the unique physical properties of these materials, they have obvious advantages in practical application in comparison with the conventional crystalline materials. The production of these materials is no longer in the stage of laboratory investigations and small-series production. The nanocrystalline materials of a new generation are used on an increasing scale in various areas of technology.

The main areas of application of the melt-quenched metallic crystals at present are associated with the application of these materials as functional materials. Undoubtedly, the widest range of applications is in electrical engineering, electronics and instrument making. As a result of the unique electrical, magnetic and mechanical properties, these materials have become irreplaceable in advanced radio electronics. Figure 5.1 shows the main areas of application of nanocrystalline magnetically soft materials in electrical engineering. The largest effect comes from the application of the ribbons of

- Transformers for power sources
- Miniature cores for noise suppressors
- Saturation cores
- Filtering choke coils
- Choke coils suppressing sinphase interferences
- Choke coils for correcting the power factor
- Control transformers
- Current transformers

Fig. 5.1. Main areas of application of the nanocrystalline magnetically soft materials in electrical engineering.

nanocrystalline alloys as the cores of transformers and secondary power sources [1]. At the present time, several tens of thousands toones of Finemet alloy and its analogues are produced in the world per annum (including Russia). These materials are characterised by the unique parameters of magnetically soft materials. The Russian analogue, characterised by the higher magnetic characteristics, as already mentioned in chapter 3, has been named 5BDSR. The Finemet-type alloys are evidently the most widely used type of nanocrystalline alloy in the world (as regards the volume of production). It has been shown that the application of these alloys as the core of choke coils has been highly efficient and has resulted in a reduction of the pulsed stresses in comparison with the amorphous alloys based on Fe or Mn–Zn ferrites [2]. These choke coils can be used in a wide frequency range and also as protection against the noise caused by discharges. The same alloy can be used in linear indicators of the circuits of switching regulators, requiring that the permeability is independent of the strength of the polarising magnetic field [3]. Reports have also been published on the very low losses of 0.066 W/kg at 1 T and 50 Hz for the $Fe_{86}Zr_7B_6Cu_1$ nanocrystalline alloys of the Nanoperm type. These values are superior to the properties (permeability, induction) of the $Fe_{78}Si_9B_{13}$ amorphous alloy and crystalline electrical engineering steel Fe–3%Si, used as the core of transformers [2]. It has been proposed to use these materials as the cores of the transformers, including high-frequency transformers, in which the core at the moment is produced from the Fe–Si–B amorphous alloys. In addition to other factors, these alloys are characterised by higher thermal stability and higher saturation induction. The films of nanocrystalline alloys are used as the material of the cores for the thin-film magnetic heads and also high-frequency transformers and inductors [1]. The following possible applications of the Nanoperm and Finemet alloys have been proposed [2]: power transformers, the choke coils of the conventional type, pulsed transformers, magnetic metres with the saturation core.

The Nanoperm alloys combine the permeability of the order of $1.6 \cdot 10^5$, typical of the Finemet alloy, with zero magnetostriction, characteristic of the cobalt-based amorphous alloys. The saturation induction of the Nanoperm alloys is greater than 1.57 T which is superior to the properties of the iron-based amorphous alloys. The application of the Nanoperm alloy in the choke coils results in the values of B_s comparable with the Fe–Si–B amorphous alloy and in lower magnetic losses so that it is possible to reduce the mass of the core and, correspondingly, the dimensions (the volume) of the choke coils.

The nanocrystalline alloys based on Fe–Co are efficient mostly as the high-temperature magnetic materials for space engines [4]. Permendur and Supermendur alloys are used at present for these applications. However, extensive investigations are being carried out into the development of high-temperature magnetically soft materials for rocket engines. The aim of these developments is the reduction of the rate and the replacement of liquid cooling by air cooling. The appropriate electronic systems have the components in which the role of the magnetic materials is dominant. As a result of the magnetically soft properties, it is assumed that the nanoparticles are characterised by the volume interaction and, consequently, the technologies ensuring reliable compacting should be used. The application of nanocrystallisation of the amorphous alloys for the manufacture of components of the required shape (composite) also includes the compacting operations because after nanocrystallisation the ribbons often become brittle. The magnetically soft properties, required for publication in the rotors have some parameters lower than those obtained for the Finemet and Nanoperm alloys. The permeability at a frequency of 1 kHz should be 10^2–10^3. On the other hand, it is important to retain these high induction values at 500 and 600°C. Undoubtedly, the requirement for the high mechanical properties must be taken into account. The combination of high magnetic permeability with high elasticity modulus and the absence of the degradation of the properties after mechanical effect enabled the effective application of the nanocrystalline alloys as magnetic screens [5].

Although the main method of melt quenching is undoubtedly the spinning method or the method of quenching with a flat jet, several other methods are also used for the production of amorphous and nanocrystalline filaments. We examine the examples of application of the nanomaterials as the absorbers of electromagnetic waves (AEW) in greatly differing spheres. The main spheres are [6]:

– masking of military technology to protect against radio location detection;

– protection of information, i.e. protection against unsanctioned downloading in the electromagnetic channel;

– solution of the problem of electromagnetic compatibility of radio electronic equipment;

– solution of the problem of medical–biological electromagnetic safety (protection against the harmful effect of secondary radiation of electronic devices).

Here, it should be mentioned that the most efficient absorbers of electromagnetic waves should have the required combination of the dielectric and magnetic losses in a wide frequency range. The application of nanomaterials in the AEW devices makes it possible to produce shielding and masking coatings on the basis of the technologies used in flexible displays. For example, the mobile particles of a pigment, coating the surface of the hidden objects, can change their position or orientation, develop a new colour similar to the movement of the wings of insects when the colour depends on the observation direction [7]. This 'active' camouflage may be used not only in the clothing of any personnel but also for masking special types of equipment. Phototonics methods can already be used to develop filaments and fabrics absorbing radiation in the visible and infrared ranges, and the reflection coefficient for such a coating can be regulated on the real timescale. These coatings can also produce some 'reflecting patterns' in other frequency ranges. These patterns or images can be visualised using special devices with the standard 'friend or foe' principle. It is assumed that these coatings will already be used in the next five years [7].

The Central Design Bureau of Radio-absorbing Materials has developed an ultrabroadband radio-absorbing materials based on the nanostructured ferromagnetic microconductor (NFMC) and glass insulation [6]. The main radio-absorbing elements in this material is the NFMC in the form of a thin metallic core in the glass insulation. The technology of production of NFMC (Fig. 5.2) ensures the simultaneous melting of the metal, softening of the glass pipe, surrounding the metal charge, and quenching of the resultant composite at a speed of 10^6 °/s. This results in the formation of a three-layer composite consisting of the metallic conductor

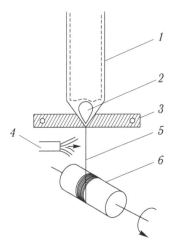

Fig. 5.2. Production of the micro-conductor in glass insulation: 1) glass capillary; 2) metal charge; 3) HF induction coil; 4) crystalliser; 5) micro conductor; 6) reception device.

Fig. 5.3. The electron micrograph of the structure of the amorphous ferromagnetic micro conductor in glass insulation. The length of the scale on the photograph 20 µm.

d = 1–30 µm, the nanostructured transition layer with a thickness of approximately 5 nm, and glass insulation, with a thickness of 2–30 µm (Fig. 5.3). As a result of the difference in the coefficient of thermal expansion (CTE) of the metal and the glass, and also the presence of the nanostructured transition layer, the material of the metallic core is subjected to the effect of very high stresses (1000 MPa) and is characterised by the unique electrophysical characteristics in the microwave range [6].

The magnetic properties of the cast microconductor in the glass insulation depend strongly on the magnetic structure which is determined mostly by two factors: magnetoelastic anisotropy and shape anisotropy [5]. The shape anisotropy includes the anisotropy of the shape of the domains and the anisotropy caused by the cylindrical shape of the strand. The magnetoelastic anisotropy depends mainly on the difference in the CTE and the sign and magnitude of magnetostriction of the strand material. Maximum permeability is obtained in the compensation of the shape anisotropy by magnetoelastic anisotropy. Experiments were carried out to determine the dependence of electrical resistivity of the nanostructured ferromagnetic microconductor on the frequency and the microwave range. The anomalous increase in relation to the linear

Fig. 5.4. Specimens of radio-absorbing materials.

electrical resistance, detected for the NFMC based on Fe and Co, was used to interpret this phenomenon as the natural ferromagnetic resonance (NFMR) with the frequency in the range 5–7 GHz and determined by the composition of the alloy, the glass insulation and the geometrical factors [8].

The estimation of the maximum magnetic permeability of the microconductor in the microwave range shows that the magnetic permeability of the Fe-based alloys is greater than 300 [6]. This level of the magnetic properties in this frequency range was obtained for the first time. The microconductor with the amorphous strand in the glass insulation is a unique material combining the required properties. Consequently, the microconductor is a very attractive object for the development on its basis of woven absorbers of electromagnetic waves.

Figure 5.4 shows the radio engineering design of the absorber which efficiently absorbs or scatters incident electromagnetic waves [6]. In addition to this, a technology was developed for the manufacture of ecologically clean screening fabrics for medical and biological protection of personal and population working and living in the conditions of the harmful effect of electromagnetic fields of different frequency and intensity, and also for solving the problem of information protection [9].

The uniquely high magnetic permeability of the nanocrystalline alloys, produced by melt quenching, enables them to be used in fire safety systems, installed in shops, libraries, banks and other similar organisations [10]. In addition to this, these materials are efficiently used in testers and other devices capable of rapid and accurate identification of different products (Fig. 5.5). They can also be used for protecting bank notes and other valuable papers.

The operating principle of the system is based on the unique magnetic properties of the superfine magnetic fibres with a thickness equal to that of the human hair. These fibres are used for the production of marking material which can be deposited, for example, on the internal surface of thermal shrinkage sleeves, can be added to the adhesive layer of excise stamps or positioned below a hologram. The information, coded in the making material can be analysed by the detector of nanocrystalline fibres. The electronic circuit of the device analyses the magnetic response of the marking material and generates sonic and light signals of the authenticity of the product and its working parameters. The extensive application of these devices is a way of solving one of the most serious economic and social problems of Russia associated with the falsification of company products (for example, medical preparations) which results not only in considerable economic losses but also creates often danger for the health and life of the population. Consequently, this would ensure improved economic safety of Russia.

The rapidly quenched nanocrystalline alloys based on Ni and Al can be used to produce light, very strong structural materials [11]. The amorphous and microcrystalline ribbons are used widely as high-temperature solders for different important joints. It is very promising to use rapidly quenched material as catalysts in chemical industry [12].

The wide range of application is also typical of nanocrystalline ribbon alloys with the shape memory effect (SME). They can be used efficiently in a large number of thermally sensitive devices, especially with high operating speed parameters. We mention several examples of the devices and systems in which the ribbon alloys with the shape memory effect can be used [13]: temperature indicators, thermal relays, signalising devices in the systems of fire prevention in companies, households, railway and automobile transport (Fig. 5.6); thermal regulators in the systems for thermostatic control of

Fig. 5.5. Fire prevention systems and testers using nanomaterials.

greenhouses, elevators, food storage and warehouses; sensors of superheating of liquids and vapours in water cooling radiators, in oil jackets for cooling of industrial transformers, in heat exchangers, in steam boilers; thermal switches for carburettors, ventilators and the systems for heating automobile seats; heat sensors for the control of technological processes. There is also a large range of applications in medicine which can be solved or have already been solved using thin materials with the shape memory effect [14]. It is in particular securing, prosthesis of walls of blood vessels, including in varicose expansion of veins, narrowing of arterial vessels; the filters for trapping blood clots; removal of tumours in brain arteries, arteriosclerotic plaque; prosthesis of heart valves, production of artificial heart muscles, production of miniature medical instruments, in particular for endoscopic surgery (clamps, holders, scalpels, etc.). In addition to this, the ribbons of the alloys with the shape memory effect can be used as working elements of light modulators in the optical system for information processing; drives in thermal machines, in refrigerators; damping materials and components of composite materials; actuating mechanisms in a number of products for extensive application (thermometers, thermostats, toys, jewellery,

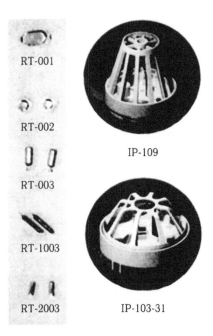

Fig. 5.6. Thermal relay and fire alarms with the nanomaterial with the shape memory effect.

etc) [15]. As an example, Fig. 5.6 shows heat shielding relays and detectors in which the active element is the nanocrystalline ribbon material with the shape memory effect [16].

The requirement for the further miniaturisation and improvement of the efficiency of operation of electronic devices imposes new tasks for the developers of nanomaterials;

– reduction of the thickness of the rapidly quenched strips to 15–20 μm which is an additional reserve for reducing the energy capacity of the magnetic circuit;

– development of new compositions of the alloys, increasing the magnetic induction 1.5–1.8 times, with a further reduction of the energy losses in magnetic reversal (in comparison with the 5BDSR alloy);

– further development of fundamental studies in order to develop completely new structures which would make it possible in turn to find new reserves for drastic improvement of the properties of the nanomaterials for successful solution of important applied tasks.

It is evident that the effective application of the materials using the method of melt quenching would result in a considerable economic effect and in a huge jump in the quality of the constructional and functional materials.

References

1. Greer A.L., Changes in structure and properties associated with the transition from the amorphous to the nanocrystalline state, Nanostructured Materials, Science & Technologies, NATO ASI Series, Dordrecht, Boston, London, Kluwer Acad. Publish., 1998, V. 3/50, 143–162.
2. Tsakalakos T., Lehrman R.L., et al., Applications of functional nanomaterials, Nanostructures: Synthesis, Functional Properties and Application. NATO Science Series, 2002, V. II/128, 675–690.
3. Frolov G., Zhigalov V.S., Physical properties and applications magnitoplenochnyh nanocomposites. - Novosibirsk: Publishing House of the Russian Academy of Sciences, 2006, 187.
4. Glezer A.M., et al., Structure and mechanical properties of doped alloys FeCo. Novokuznetsk: NPK, 2009. 142 p.
5. Gorynin I.V., Ross. mamotekhnologii, 2007, V. 2, No. 34, 36–57.
6. Vladimirov D.N., Handogina E.N., Mir tekh. tekhnol., 2007, No 5, 46–48.
7. Altman Yu., Military nanotechnology. Possible applications and preventive arms control, Moscow, Tekhnosfera, 2006.
8. Khandogina E.N., Petelin A.P., J. Magn. Magn. Mater., 2002, V. 249, Issue 12, 55–59.
9. Ustimenko L.G., Handogina E., Vladimirov D.N., Probl. chern. metall. materialoved., 2009, No. 2, 81–85.
10. Antonenko A., Manov V., et al., Mater. Sci. Engin., A, 2001, V. 304–306. 975–978.

11.   Vassiliev V.A., et al., High-speed solidification of melts, Moscow, SP Intermet Engineering, 1998.
12.   Shmyreva T.P., Birch E.Yu., Rapidly cooled eutectic alloys, Kiev, Tehnika, 1990.
13.   Shelyakov A.V., et al., Intern. J. Smart and Nanomater., 2011, V. 2, 68–77.

Index